T0188853

# Proceedings
# of the Ninth North American
# Blueberry Research
# and Extension Workers
# Conference

*Proceedings of the Ninth North American Blueberry Research and Extension Workers Conference* has been co-published simultaneously as *Small Fruits Review*, Volume 3, Numbers 1/2 and 3/4 2004.

# Proceedings of the Ninth North American Blueberry Research and Extension Workers Conference

Charles F. Forney
Leonard J. Eaton
Editors

*Proceedings of the Ninth North American Blueberry Research and Extension Workers Conference* has been co-published simultaneously as *Small Fruits Review*, Volume 3, Numbers 1/2 and 3/4 2004.

CRC Press
Taylor & Francis Group
Boca Raton London New York

CRC Press is an imprint of the
Taylor & Francis Group, an **informa** business

*Proceedings of the Ninth North American Blueberry Research and Extension Workers Conference* has been co-published simultaneously as *Small Fruits Review*, Volume 3, Numbers 1/2 and 3/4 2004.

CRC Press
Taylor & Francis Group
6000 Broken Sound Parkway NW, Suite 300
Boca Raton, FL 33487-2742

First issued in paperback 2020

© 2004 by Taylor & Francis Group, LLC
CRC Press is an imprint of Taylor & Francis Group, an Informa business

No claim to original U.S. Government works

ISBN-13: 978-1-56022-114-2 (hbk)
ISBN-13: 978-1-56022-115-9 (hbk)

Visit the Taylor & Francis Web site at
http://www.taylorandfrancis.com

and the CRC Press Web site at
http://www.crcpress.com

Cover design by Lora Wiggins

**Library of Congress Cataloging-in-Publication Data**

North American Blueberry Research and Extension Workers Conference (9th : 2002 Halifax, Nova Scotia) Proceedings of the Ninth North American Blueberry Research and Extension Workers Conference / Charles F. Forney, Leonard J. Eaton, editors.
     p. cm.
     Proceedings of the Ninth North American Blueberry Research and Extension Workers Conference has been simultaneously published as Small Fruits Review, vol. 3, nos. 1/2 & 3/4, 2004.
     Includes bibliographical references (p. ).
     ISBN 1-56022-114-3 (hard cover : alk. paper) – ISBN 1-56022-115-1 (soft cover : alk. paper)
     1. Blueberries–Congresses. 2. Blueberries–North America–Congresses. I. Forney, Charles F. II. Eaton, Leonard J. III. Title.
SB386.B7N67 2002
634′.737–dc22

                                       2004008995

# Proceedings of the Ninth North American Blueberry Research and Extension Workers Conference

## CONTENTS

# ABOUT THE EDITORS

**Charles F. Forney, PhD,** has been a Postharvest Physiologist at the Atlantic Food and Horticulture Research Centre in Kentville, Nova Scotia, for the past 12 years. He also served as Study Leader of their Postharvest Program. He is a former research scientist in postharvest physiology for the United States Department of Agriculture in Fresno, California. Dr. Forney's research involves the development of technologies to maintain the quality of fresh fruits and vegetables throughout storage and marketing. He has conducted extensive research on the postharvest handling and storage of fresh highbush and lowbush blueberries, including the development of controlled atmosphere storage technologies.

**Leonard J. Eaton, PhD,** was Professor of Biology at the Nova Scotia Agricultural College, 1968-1996, and Department Head, 1989-1996. He was Director of the Nova Scotia Wild Blueberry Institute, 1982-1996. Dr. Eaton has done extensive research on fertility and management of the wild blueberry since 1974 with papers and reports on short- and long-term fertilizer effects, nitrogen cycling, second cropping, pollination, effects of frost, deicing, and marine salt spray on developing fruit buds.

# Introduction
# to the Proceedings
# of the Ninth North American Blueberry
# Research and Extension Workers Conference

Charles F. Forney
Leonard J. Eaton

The North American Blueberry Research and Extension Workers Conference (NABREW) has convened nine times over the past 42 years (Table 1). The proceedings of the ninth meeting, which occurred in Halifax, Nova Scotia during the summer of 2002 will be presented in two parts of the *Small Fruits Review*, Volume 3. In Part I, papers on the management of lowbush, highbush and rabbiteye blueberries are presented.

The conference was attend by about 100 people representing experts in taxonomy, breeding, physiology, culture, management, storage, processing, and marketing of blueberries from throughout North America, Japan, China, and Korea. The conference included 27 oral and 28 poster presentations addressing ongoing research on highbush, southern highbush, rabbiteye, lowbush, and other native species of blueberries. Tours

---

Charles F. Forney is affiliated with the Agriculture & Agri-Food Canada, Atlantic Food and Horticulture Research Centre, 32 Main Street, Kentville, NS B4N IJ5, Canada. Dr. Leonard J. Eaton is affiliated with the Department of Environmental Sciences, Nova Scotia Agricultural College, Truro, NS B2N 5E3, Canada.

[Haworth co-indexing entry note]: "Introduction to the Proceedings of the Ninth North American Blueberry Research and Extension Workers Conference." Forney, Charles F., and Leonard J. Eaton. Co-published simultaneously in *Small Fruits Review* (Food Products Press, an imprint of The Haworth Press, Inc.) Vol. 3, No. 1/2, 2004, pp. 1-2; and: *Proceedings of the Ninth North American Blueberry Research and Extension Workers Conference* (ed: Charles F. Forney, and Leonard J. Eaton) Food Products Press, an imprint of The Haworth Press, Inc., 2004, pp. 1-2. Single or multiple copies of this article are available for a fee from The Haworth Document Delivery Service [1-800-HAWORTH, 9:00 a.m. - 5:00 p.m. (EST). E-mail address: docdelivery@haworthpress.com].

of the Nova Scotia blueberry industry also were conducted before and after the conference.

The next NABREW will be sponsored by the University of Georgia and will be held in Georgia in the spring of 2006.

TABLE 1. Previous meetings of the North American Blueberry Research and Extension Workers Conference.

| Conference No. | Date | Location | Sponsor |
|---|---|---|---|
| 1 | 6-7 July 1960 | New Brunswick, NJ | Rutgers University |
| 2 | 6-7 April 1966 | Orono, ME | University of Maine |
| 3 | 6-7 Nov. 1974 | East Lansing, MI | Michigan State University |
| 4 | 16-81 Nov. 1979 | Fayetteville, AR | University of Arkansas |
| 5 | 1-3 Feb. 1984 | Gainsville, FL | University of Florida |
| 6 | 10-12 July 1990 | Porland, OR | Washington State University |
| 7 | 5-8 July 1994 | Beltsville, MD | USDA Fruit Laboratory, Beltsville, MD and Blueberry Cranberry Research Station, Chatsworth, NJ |
| 8 | 27-29 May 1998 | Wilmington, NC | University of North Carolina |
| 9 | 18-21 Aug. 2002 | Halifax, Nova Scotia | Nova Scotia Agricultural College, Truro, NS and Agriculture and Agri- Food Canada, Kentville, Nova Scotia |

# Application of Paper Mill Biosolids, Wood Ash and Ground Bark on Wild Lowbush Blueberry Production

### J. Lafond

**SUMMARY.** Soils in wild lowbush blueberry production are prone to wind erosion and have very low nutrient and water storage capacities. An experiment was initiated to assess paper mill biosolids (PB) mixed with wood ash and ground bark as a soil amendment/fertilizer for wild lowbush blueberry (*Vaccium angustifolium* Ait.) in the Lac St-Jean area, Quebec, Canada. A mixture of PB was applied during spring (mid-May) of the sprout year (1998) on 120 m$^2$ plots at a rate of 15 t ha$^{-1}$ (wet basis) with wood ash (1 and 2 t ha$^{-1}$) and ground bark (0, 3, 6, 9 and 15 t ha$^{-1}$, wet basis). Blueberry leaves were sampled in the first year and wet digestion and dry ashing were performed to determine foliar nutrient concentration. In 1999 and 2000, fruit yields tended to increase with PB with wood ash and ground bark application (31% in 1999 and 29% in 2000). Foliar N, P and K concentrations were increased whereas Ca and Mg were unaffected compared to control. Other nutrients were also determined and only Fe tended to increase with PB application whereas Ni tended to decrease. This study indicated that PB mixed with wood ash and ground bark is a potential nutrient source for blueberry on these poor sandy soils without short-term loss in crop yield. *[Article copies available for a fee from The Haworth Document Delivery Service: 1-800-HAWORTH. E-mail address: <docdelivery@haworthpress.com> Website: <http://www. HaworthPress.com> © 2004 by The Haworth Press, Inc. All rights reserved.]*

J. Lafond is Researcher, Agriculture and Agri-Food Canada, Soils and Crops Research and Development Center, Normandin, Quebec, G8M 4K3, Canada.

[Haworth co-indexing entry note]: "Application of Paper Mill Biosolids, Wood Ash and Ground Bark on Wild Lowbush Blueberry Production." Lafond, J. Co-published simultaneously in *Small Fruits Review* (Food Products Press, an imprint of The Haworth Press, Inc.) Vol. 3, No. 1/2, 2004, pp. 3-10; and: *Proceedings of the Ninth North American Blueberry Research and Extension Workers Conference* (ed: Charles F. Forney, and Leonard J. Eaton) Food Products Press, an imprint of The Haworth Press, Inc., 2004, pp. 3-10. Single or multiple copies of this article are available for a fee from The Haworth Document Delivery Service [1-800-HAWORTH, 9:00 a.m. - 5:00 p.m. (EST). E-mail address: docdelivery@haworthpress.com].

Digital Object Identifer: 10.1300/J301v03n01_02

**KEYWORDS.** *Vaccinium angustifolium*, paper mill biosolids, wood ash, ground bark, foliar nutrient concentration

## INTRODUCTION

In the Saguenay-Lac-St-Jean area (Québec, Canada), soil degradation severely limits wild blueberry (*Vaccinium angustifolium* Ait.) fruit production. Excessive burning decreased soil organic matter (SOM), which has a direct impact on soil fertility and water retention capacity. The loss of SOM also increased wind erosion risk. Residues of the paper mill industry are abundant in the area and can be used to restore the SOM under the wild blueberry stand. The mixture of primary and secondary sludge was found to be sources of plant nutrients (N and P) (Cabral et Vasconcelos, 1993; Simard et al., 1997) and organic matter for crop production (Philips et al., 1997). Wood ash is primarily used as a liming amendment but it also adds plant nutrients to soil (Huang et al., 1992; Naylor and Schmidt, 1989). Ligneous materials, such as ground bark, are used as organic amendment in agricultural crops but N immobilization remained the main factor on crop production (Beauchemin et al., 1992). The objective of this study was to assess paper mill biosolids (PB) mixed with wood ash and ground bark as a soil amendment/fertilizer for wild lowbush blueberry.

## MATERIALS AND METHODS

The experiment was carried out in Saguenay-Lac-Saint-Jean area, province of Quebec on wild lowbush blueberry (*Vaccinium angustifolium* Ait.). A mixture of primary and secondary paper mill biosolids (PB) was applied during spring (mid-May) of the sprout year (1998) on 120 $m^2$ plots at a rate of 15 t ha$^{-1}$ (wet basis) with wood ash (1 and 2 t ha$^{-1}$) and ground bark (0, 3, 6, 9, and 15 t ha$^{-1}$, wet basis). The mixture of PB, wood ash and ground bark composition is shown in Table 1. Paper mill biosolids and ground bark *respected* criteria of MEF (Ministère de l'Environnement et de la Faune du Québec, 1997). Wood ash also *respected* criteria of MEF (1997) and BNQ (Bureau de normalisation du Québec, 1997). Blueberry leaves were sampled in the first year (1998) and wet digestion and dry ashing were performed to determine foliar nutrient concentration. The treatments were arranged in a randomized complete block design in four replicates. Data were treated separately

TABLE 1. Paper mill biosolids (PB), wood ash, and ground bark composition.

| Chemical properties | | PB | Wood ash | Ground bark | Criteria[z] | NP/TE[y] |
|---|---|---|---|---|---|---|
| Solids | % | 74 | 10 | 51 | _x | - |
| Neutralizing power | % | 88 | - | - | - | - |
| pH (water) | | 5.59 | 13.07 | 3.16 | - | - |
| Total N | % | 2.29 | 0.05 | 0.22 | - | - |
| Total P | % | 0.50 | 0.40 | 0.04 | - | - |
| Organic C | % | 51.64 | 1.36 | 54.3 | - | - |
| C/N | | 23 | 27 | 247 | - | - |
| K | mg kg$^{-1}$ | 2700 | 33400 | 1100 | - | - |
| Ca | mg kg$^{-1}$ | 8800 | 147800 | 10400 | - | - |
| Mg | mg kg$^{-1}$ | 800 | 34600 | 500 | - | - |
| Fe | mg kg$^{-1}$ | 2059 | 4.29 | 249 | - | - |
| Mn | mg kg$^{-1}$ | 1255 | 1663 | 2449 | - | - |
| Cd | mg kg$^{-1}$ | 1.63 | 6.80 | 0.60 | 3 | 2.50 |
| Co | mg kg$^{-1}$ | 2.47 | 14.84 | 1.99 | 34 | 0.333 |
| Cr | mg kg$^{-1}$ | 14.74 | 245 | 4.69 | 210 | 0.047 |
| Cu | mg kg$^{-1}$ | 2.10 | 0.21 | 1.05 | 100 | 0.066 |
| Ni | mg kg$^{-1}$ | 8.66 | 255 | 2.62 | 62 | 0.278 |
| Pb | mg kg$^{-1}$ | 6.92 | 46.50 | 9.32 | 150 | 0.100 |
| Zn | mg kg$^{-1}$ | 9.05 | 11.27 | 55.50 | 500 | 0.027 |

[z] Criteria for land application of residuals (MEF, 1997).
[y] NP/TE: Ratio of neutralizing power/trace element (BNQ, 1997).
[x] Not applicable.

for each year using analysis of variance with a priori comparisons using the GLM procedure (SAS Institute Inc, 1990).

## RESULTS

*Yield.* In 2000, a severe frost occurred at the beginning of June and fruit yield was three times lower than in 1999. The effect of ground bark application on fruit yield was not significant in both years. However,

FIGURE 1. Effect of paper mill biosolids (PB) with wood ash (1 t ha$^{-1}$ of wood ash: —●—; 2 t ha$^{-1}$ of wood ash: --■--) and ground bark on blueberry yield in 1999. Regression curves are not significant at P ≤ 0.05.

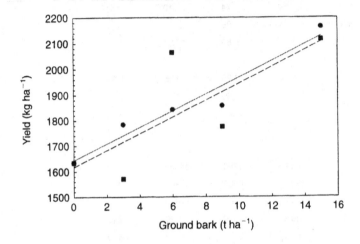

FIGURE 2. Effect of paper mill biosolids (PB) with wood ash (1 t ha$^{-1}$ of wood ash: —●—; 2 t ha$^{-1}$ of wood ash: --■--) and ground bark on blueberry yield in 2000. Regression curves are not significant at P ≤ 0.05.

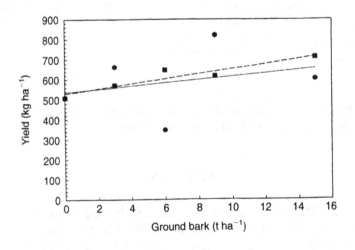

yields tended to increase with ground bark application (Figures 1 and 2). Compared to control, fruit yields were 31% and 29% higher in 1999 and 2000, respectively, with the highest rate of ground bark. The effect of wood ash on crop yield was also not significant.

*Foliar nutrient concentrations.* The treatments had no significant effect on foliar N concentration, but a slight increase was associated with ground bark application (Tables 2 and 3). Phosphorus and potassium increased significantly with the application of 2 t ha$^{-1}$ of wood ash. Calcium and magnesium concentrations were unaffected even with 2 t ha$^{-1}$ of wood ash. Iron concentration was higher with 2 t ha$^{-1}$ of wood ash compared to 1 t ha$^{-1}$ of wood ash but only increased significantly with ground bark application mixed with 2 t ha$^{-1}$ of wood ash.

Heavy metal concentrations were not affected by treatments. Only Ni was significantly affected by treatments and concentration decreased with ground bark mixed with 2 t ha$^{-1}$ of wood ash application.

Foliar nutrient concentrations were in the satisfactory range as described by Trevett (1962) and Lockhart and Langille (1962) except for Mn where the concentration was two times lower than the lowest limit of satisfactory range.


## CONCLUSIONS AND GROWER BENEFITS

Use of paper mill biosolids mixed with wood ash and ground bark as a nutrient source and soil amendment allowed to maintain fruit yield. However, there was no significant effect of wood ash rates on fruit yield. Nitrogen, phosphorus, and potassium concentration in leaves increased with PB mixed with wood ash and ground bark. Heavy metal status in leaves was unaffected by treatments and foliar nutrient concentrations were in the satisfactory range. These results indicated that PB mixed with wood ash and ground bark can be used as fertilizer/amendment on poor sandy soils without short-term loss in crop yield.

TABLE 2. P values of treatments and degree of significance of selected contrasts of the effect of paper mill biosolids (PB) with wood ash and ground bark on foliar nutrient concentration in the sprout year (1998).

| Source of variations | N | P | K | Ca | Mg | Fe | Mn | Cd | Co | Cr | Cu | Ni | Pd | Zn |
|---|---|---|---|---|---|---|---|---|---|---|---|---|---|---|
| Treatments | 0.62 | 0.01 | 0.23 | 0.87 | 0.44 | 0.05 | 0.86 | 0.73 | 0.27 | 0.29 | 0.60 | 0.10 | 0.49 | 0.66 |
| Contrasts | | | | | | | | | | | | | | |
| 1 vs. 2 t ha$^{-1}$ of wood ash | 0.49 | 0.24 | 0.73 | 0.75 | 0.74 | 0.02 | 0.28 | 0.68 | 0.99 | 0.12 | 0.14 | 0.34 | 0.70 | 0.30 |
| 1 t ha$^{-1}$ of wood ash | | | | | | | | | | | | | | |
| Bark linear | 0.06 | 0.01 | 0.24 | 0.81 | 0.14 | 0.21 | 0.75 | 0.87 | 0.11 | 0.66 | 0.57 | 0.51 | 0.20 | 0.12 |
| Bark quadratic | 0.88 | 0.27 | 0.40 | 0.99 | 0.53 | 0.95 | 0.37 | 0.69 | 0.45 | 0.56 | 0.68 | 0.15 | 0.60 | 0.35 |
| 2 t ha$^{-1}$ of wood ash | | | | | | | | | | | | | | |
| Bark linear | 0.82 | 0.01 | 0.04 | 0.38 | 0.37 | 0.01 | 0.70 | 0.55 | 0.12 | 0.12 | 0.60 | 0.02 | 0.12 | 0.47 |
| Bark quadratic | 0.69 | 0.87 | 0.36 | 0.18 | 0.12 | 0.11 | 0.17 | 0.20 | 0.99 | 0.99 | 0.28 | 0.16 | 0.74 | 0.52 |

TABLE 3. Effect of paper mill biosolids (PB) with wood ash and ground bark on foliar nutrient concentration in the sprout year (1998).

| Wood ash | | | 1 t ha⁻¹ | | | | 2 t ha⁻¹ | | | |
|---|---|---|---|---|---|---|---|---|---|---|
| Ground bark (t ha⁻¹) | | 0 | 3 | 6 | 9 | 15 | 3 | 6 | 9 | 15 |
| Element | | | | | | | | | | |
| N | % | 1.98 | 2.07 | 2.02 | 2.12 | 2.20 | 2.14 | 2.00 | 2.04 | 2.05 |
| P | % | 0.17 | 0.18 | 0.18 | 0.19 | 0.20 | 0.18 | 0.18 | 0.18 | 0.19 |
| K | % | 0.69 | 0.75 | 0.73 | 0.73 | 0.74 | 0.72 | 0.77 | 0.71 | 0.76 |
| Ca | % | 0.34 | 0.34 | 0.34 | 0.34 | 0.34 | 0.34 | 0.35 | 0.36 | 0.32 |
| Mg | % | 0.16 | 0.14 | 0.15 | 0.15 | 0.16 | 0.15 | 0.16 | 0.16 | 0.14 |
| Fe | mg kg⁻¹ | 32.6 | 30.0 | 32.3 | 27.6 | 32.1 | 31.4 | 33.5 | 34.1 | 34.3 |
| Mn | mg kg⁻¹ | 253 | 244 | 245 | 247 | 242 | 256 | 253 | 253 | 247 |
| Cd | mg kg⁻¹ | 0.16 | 0.12 | 0.12 | 0.12 | 0.09 | 0.09 | 0.13 | 0.04 | 0.16 |
| Co | mg kg⁻¹ | 0.23 | 0.42 | 0.47 | 0.50 | 0.53 | 0.42 | 0.36 | 0.48 | 0.51 |
| Cr | mg kg⁻¹ | 0.65 | 0.67 | 0.52 | 0.40 | 0.57 | 0.60 | 0.55 | 0.70 | 0.77 |
| Cu | mg kg⁻¹ | 6.26 | 5.83 | 5.74 | 6.52 | 5.59 | 5.18 | 5.40 | 5.63 | 5.59 |
| Ni | mg kg⁻¹ | 1.23 | 0.97 | 1.14 | 1.10 | 1.27 | 1.19 | 1.04 | 1.26 | 1.06 |
| Pb | mg kg⁻¹ | 0.36 | 0.61 | 0.32 | 0.51 | 0.37 | 0.46 | 0.59 | 0.29 | 0.32 |
| Zn | mg kg⁻¹ | 13.9 | 14.2 | 14.0 | 15.0 | 14.9 | 14.1 | 13.4 | 14.2 | 14.0 |

## LITERATURE CITED

Beauchemin, S., A. N'dayegamiye, and M.R. Laverdière. 1992. Effets d'amendements ligneux sur la disponibilité d'azote dans un sol sableux cultivé en pomme de terre. Can. J. Soil Sci. 72: 89-95.

Bureau de normalisation du Québec (BNQ). 1997. Amendements calciques ou magnésiens provenant de procédés industriels. NQ 0419-090.

Cabral, F. and E. Vasconcelos. 1993. The use as fertilizer of combined primary/secondary pulp-mill sludge. Pp. 77-81, In: M.A.C. Fragoso and M.L. van Beusicheim. (eds.) Optimization of plant nutrition. Kluwer Academic Publishers, Dordrecht, The Netherlands.

Huang, H., A.G. Campbell, R. Folk, and R.L. Mahler. 1992. Wood ash as a soil additive and liming agent for wheat: Field studies. Commun. Soil Sci. Plant Anal. 23:25-33.

Ministère de l'Environnement et de la Faune du Québec (MEF). 1997. Critères provisoires pour la valorisation des matières résiduelles fertilisantes. Gouvernement du Québec. Avril 1997.

Naylor, L.M. and E. Schmidt. 1989. Paper mill wood ash as a fertilizer and liming material: Field trial. Tappi J. 62:199-206.

Lockhart, C.L. and W.M. Langille. 1962. The mineral content of lowbush blueberry Can. Plant Dis. Survey 42: 124-128.

Phillips, V.R., N. Kirkpatrick, I.M. Scotford, R.P. White, and R.G.O. Burton. 1997. The use of paper-mill sludges on agricultural land. Bioresource Technol. 60:73-80.

SAS Institute Inc. 1990. SAS/Stat user's guide. Vol. 2 GLM-VARCOMP. Version 6. 4th ed. SAS Institute, Cary, N.C.

Simard, R.R., J. Coulombe, R. Lalande, B. Gagnon, and S. Yelle. 1997. Use of fresh and composted de-inking sludge in cabbage production. Pp. 349-362, In: S. Brown, J.S. Angle, and L. Jacobs (eds.) Beneficial co-utilization of agricultural, municipal and industrial by-products. Kluwer Academic Press, Dordrecht, The Netherlands.

Trevett, M.F. 1962. Nutrition and growth of the lowbush blueberry. Maine Agr. Expt. Stat Bull. 605.

# The Wild Blueberry Industry– Past

George W. Wood

**SUMMARY.** This paper, presented as part of a symposium at the 9th North American Blueberry Research and Extension Workers Conference, Halifax, Nova Scotia, Canada, August 18-21, 2002, traces the early history of the wild blueberry (*Vaccinium angustifolium*, Aiton, and *Vaccinium myrtilloides*, Michaux) in eastern North America. Wild blueberry production is traced from consumption of native blueberries by animals and native North Americans, through early commercial production in Canada and the United States, up to the present time. Different management methods adopted as the scientific knowledge increased are described. Early management consisted of burning fields one spring and harvesting the fruit in August of the following year. Over the last century, several new practices have been introduced to increase blueberry yields. These include the oil burner, the flail mower, the mechanical harvester, increased use of selective herbicides, increased use of managed pollinators, and extensive land leveling. *[Article copies available for a fee from The Haworth Document Delivery Service: 1-800-HAWORTH. E-mail address: <docdelivery@haworthpress.com> Website: <http://www.HaworthPress. com> © 2004 by The Haworth Press, Inc. All rights reserved.]*

George W. Wood is a retired research officer with Agriculture Canada, Fredericton, New Brunswick.

Address correspondence to: George W. Wood, 18 Floral Avenue, Fredericton, NB E3A 1K7, Canada (E-mail: gwood@nb.sympatico.ca).

[Haworth co-indexing entry note]: "The Wild Blueberry Industry–Past." Wood, George W. Co-published simultaneously in *Small Fruits Review* (Food Products Press, an imprint of The Haworth Press, Inc.) Vol. 3, No. 1/2, 2004, pp. 11-18; and: *Proceedings of the Ninth North American Blueberry Research and Extension Workers Conference* (ed: Charles F. Forney, and Leonard J. Eaton) Food Products Press, an imprint of The Haworth Press, Inc., 2004, pp. 11-18. Single or multiple copies of this article are available for a fee from The Haworth Document Delivery Service [1-800-HAWORTH, 9:00 a.m. - 5:00 p.m. (EST). E-mail address: docdelivery@haworthpress.com].

**KEYWORDS.** *Vaccinium angustifolium*, history, management practices

Blueberries grow wild in many parts of the world, but the wild blueberry discussed in this symposium refers to two species, *Vaccinium angustifolium*, Aiton, and *Vaccinium myrtilloides*, Michaux which are indigenous to North America. The commercial wild blueberry industry being reviewed is further restricted to Eastern Canada and Maine.

Wild berries are an important source of food for birds and other wild animals, and native people enjoyed them long before the first Europeans discovered North America. There are several references to their usage in the writings of early explorers. One early account states that the French explorer, Samuel de Champlain, found native people gathering wild blueberries for use during the winter months. They dried them in the sun, beat them into a powder which was added to parched meal, and the resulting product was used as seasoning for soups and stews and curing of meats (Munson, 1901). The fruit was also believed to have healing powers.

Native people originally harvested the berries where they grew naturally, that is in treeless barrens or where forested areas had been burned over after lightning strikes. Later, some tribes encouraged continuing production by deliberately setting fire to favorite picking areas, and this method of pruning resulted in improved growth and increased yield of fruit.

The early settlers began to harvest the fruit for their own use after the manner of the native people. Crop management was very primitive and focused on maintaining existing sites and establishing new areas through pruning by fire and cutting of competing bushes.

The wild blueberry industry as we know it began on the blueberry barrens of Washington County, Maine. There are several thousand hectares of these treeless sandy or gravelly tracts of land with acidic soil that favors the growth of blueberries and other heath-like plants. It appeared that native people burned parts of this area for many years before settlers took an interest around 1800 (Day, 1959). For the next hundred years, the barrens were treated as public lands and open for general picking. The pickers and their families consumed much of the fruit but they soon found a ready market in nearby communities.

In Canada, wild blueberries have a similar history. There are many parts of Quebec and the Atlantic Provinces where open barrens exist, and where burned over forests offered opportunities for this common native fruit to grow. In Yarmouth County, Nova Scotia, for example, there are large areas that were repeatedly burned by forest fires, and because the land is slow to regenerate forest, it filled in with lowbush blueberry and other plants that thrive in acidic soil. Early records of the harvesting and sale of wild blueberries date back to the 1800s. At this time, berries were handpicked and sold in baskets in nearby towns and shipped in barrels or cans to the Boston market (Kinsman, 1986).

As berry picking became more popular and the supply of fruit increased, interest in marketing the crop grew accordingly. The sale of fresh fruit expanded to major centres such as Boston and Montreal and more people were recruited to harvest berries. Although the pickers sometimes received less than three cents per quart, good seasonal wages were possible where there was lots of fruit, and many people, often including whole families, were attracted to carry out the harvest. In the beginning berries were handpicked into round boxes, but these were soon replaced by wooden (veneer) square-top quart boxes shipped in thirty-two quart crates. Transportation to distant markets was by rail or ship.

In 1866, a cannery in Milbridge, Maine began putting up blueberries for sale to the Union army. This was the beginning of the processing industry and successful movement of berries into wider markets. Other canneries soon took advantage of the new opportunities, and the demand for fruit resulted in the expansion of wild blueberry harvesting from its base in the barrens to include the full coastal area of Maine and beyond. Also, the establishment of canning factories along the Canada-US border encouraged the development of land for wild blueberry production in the neighbouring Province of New Brunswick.

Systematic management of blueberry land started with owners leasing some of their land to operators on the Maine barrens. Each year a section of land was burned by the leasee who used a piece of bent pipe closed at one end, filled with kerosene, and having a cloth plug at the other end which served as a wick. There was no crop during the burn year; harvesting began in the second year. The first ripened berries were handpicked for fresh fruit sales, and the rest were harvested at one time when the bulk of the crop had ripened, and then delivered to the canning factory. A hand-held blueberry rake was used to harvest the later berries and they were passed through a winnowing machine, which removed leaves and other trash before the fruit was sent on for processing.

Several types of blueberry rakes were tried in the different production areas. In Nova Scotia, native people devised a home made rake made of wood with teeth fashioned from bedsprings (Kinsman, 1986). In Quebec, some rakes were modeled after bear claws, and pickers beat the berries from the bushes backwards into a pail or basket–as opposed to the forward motion employed with the rakes used in Maine and the Maritime Provinces. But it was the rake developed by Abijah Talbot of Columbia Falls, Maine that became the standard for the industry. This rake was similar to those used for harvesting cranberries, but had metal teeth and a flat bottom like a dustpan. With only minor modifications it is still in use by many growers.

In the beginning, fields were pruned by simply setting fire to dried natural vegetation, but hay or straw was soon used to help carry the fire. It was spread on the fields in the fall of the year and burning was carried out in the spring, shortly after the snow had melted from the fields. Operators generally divided their land into three divisions, burning a third each year, but they later changed to a two-year cycle, which was found to provide greater yields in the long term.

During the 1950s, a blower-type oil burner was introduced and was soon used by most of the larger and more progressive growers. This equipment enabled burning to be carried out under a wider range of weather conditions. Propane gas was also tried as a fuel, but this option never became popular. Tests with a rotary mower to prune wild blueberries were carried out for several years, but it was not until the 1980s that the growers generally accepted it.

Another early innovation was to divide the fields into ten-foot wide lanes with string at harvest time. This increased efficiency by helping control the pickers and assured complete harvesting of the berries.

Management practices for wild blueberry changed gradually, but steadily during much of the twentieth century, one of the early changes being the introduction of pest control measures. This was prompted by complaints in the market of finding maggots in canned Maine blueberries in 1922. Government inspectors were placed in canning plants and by the following year, fruit was graded under the Maine Pure Food and Drug Law. The US Department of Agriculture staff undertook studies on the blueberry fruit fly, *Rhagoletis mendax Curran*, and recommended control procedures which included uniform burning of the fields, destruction of waste in the fields and at the factory, and dusting of fields with an insecticide.

Problems with the fruit fly have not been confined to Maine. It is prevalent throughout the Maritime Provinces of Canada, but has not been reported from Quebec or Newfoundland and Labrador. In 1930, fruit from Yarmouth County, NS, was returned from Boston and New York when shipments of fresh fruit were found infested with maggots.

This led to inspections by Canadian authorities and further research on blueberry insect pests and their control (Kinsman, 1986).

While the canning of wild blueberries was the common method of processing in Maine, it never became popular in Canada. Early sales of Canadian fruit went fresh to domestic and American markets, and later on most of it went to the canneries in Maine for processing. By the 1930s, however, some berries were being frozen, often in fish plants that already had freezing and storage facilities, and this soon became the preferred method for shipping fruit to distant markets. By the mid 1950s, technology was available for providing individually-quick-frozen fruit and this has remained as the most common method for processing wild blueberries.

As markets expanded, new blueberry fields were continually being developed in Maine and Canada, and industry leaders began looking for new technology and management practices that would further improve the efficiency and profitability of their industry. In 1945, two bills were passed in the Maine legislature. The first was to provide funds for the purchase of a farm and its equipment for blueberry research; the second, was to provide for an industry tax of one and one-quarter mills per pound of fresh fruit on all blueberries grown, purchased, sold, handled or processed in the State. A research program began at Blueberry Hill Farm, Jonesboro, Maine the following year, under the direction of the Maine Agriculture Experiment Station, Orono, Maine, and the Maine Blueberry Commission, composed of industry leaders, was named to advise on Research and Extension activities (Day, 1959).

At the same time, interest in wild blueberry production was increasing in Canada. Following major crop losses from feeding by the black army cutworm, *Actebia fennica* Tausch, an entomological field station was opened at Tower Hill, New Brunswick in 1946. In 1949, a Blueberry Substation was established at this location to conduct a full range of research on wild blueberries. Staff of Agriculture Canada, Fredericton, NB, and Kentville, NS directed these programs. Cooperation between Canadian and Maine research and extension teams already existed, but it became much closer with the addition of these new facilities and more staff.

Problems proposed by the industry and studied at these research establishments included plant propagation and breeding, the use of fertilizers and mulching, pollination, insect pests and diseases, the control of weeds, irrigation, and mechanical harvesting. Some components of these studies were conducted at Orono and Kentville where more sophisticated facilities were available.

Early experiments on preparing land, sowing seed, setting out sods and rhizomes, or planting stems and seedlings were discouraging. However, research on propagation at the Kentville Research Station eventually resulted in several varieties being named from superior clones selected from native stands in Eastern Canada and Maine. Select clone material was used to establish pick-your-own operations in a number of locations, but was never accepted as a method of expanding existing fields of wild blueberries.

Researchers ran into difficulty in demonstrating the value of fertilisers because of the high degree of variability in native stands, and because such amendments often increased the growth of grass and other weeds to the detriment of blueberry growth. The proven benefit of fertilizing had to wait until later introduction of selective herbicides.

Wild blueberries were shown to be dependent on insects, mostly bees, for pollination and fruit set, and many growers reported improved pollination when honeybees were used to supplement the native bee species. However, it was difficult to demonstrate the contribution of honeybees in field experiments because of the confounding effects of clonal variability and incompatibility, weather conditions, and other uncontrollable factors. Honeybees were recommended when native bee populations were considered to be unusually low, and many growers often used them to increase the probability of successful pollination. The need for more pollinators also increased when field size or plant density became greater.

In addition to the blueberry fruit fly, several species of insects were identified as occasional pests; among them were climbing cutworms, leaf tiers and caterpillars, flea beetles and casebeetles, sawflies, and thrips. Although damage from these insects could sometime cause significant crop loss, control measures were only applied when the problem was identified as warranting such action.

The blueberry fruit fly was the only insect that required chemical controls on a regular basis. Annual monitoring of insect populations, however, was encouraged as part of an integrated pest control program. Sweeping the fields with a stout insect net provided information on most species, and sticky traps were also used to determine the prevalence of fruit fly and other flying insects.

Insecticides were initially applied in dust form, but the use of wettable powders and liquid concentrates later became the norm. Small areas were sprayed with powered ground equipment, and larger areas were treated by using fixed-wing aircraft or helicopters.

At the conclusion of World War II several new chemicals became available for agricultural use, and many of them were tested and registered for use on wild blueberries. Plant protection and weed control guides were subsequently prepared and made available to all growers by extension workers in Maine and Canada. The four Atlantic Provinces of Canada also agreed to work together in preparing their guides, and set up committees that met on an annual basis to coordinate their activities.

Some of the new chemicals, such as DDT were used for a few years and later withdrawn because of their long residual properties. In the second half of the twentieth century, there was increasing concern for pesticide side effects on beneficial insects and the environment in general, and their usage never reached the levels common to many other agricultural crops.

Monilinia blight, *Monilinea vacciniicorymbosi* (Reade), and Botrytis blight, *Botrytis cinerea* Pers. were identified as the major diseases affecting wild blueberries, and control measures were frequently required. Red leaf disease, *Exobasidium vaccinii* (Wor.) and witches broom, *Pucciniastrum goeppertianum* (Kuhn) were controlled by eradicating diseased plants with a herbicide.

As new chemical herbicides became available they were tested by research workers and by growers, as well. Trials were also carried out with different methods of application. When small areas were treated, growers applied herbicides such as 2,4-D to stumps of woody plants and wiped foliage with a woolen mitten soaked with herbicide. In some areas, it was more practical to apply the chemical with a knapsack sprayer using a hand pump to apply pressure. Larger areas were most often treated with a power sprayer that employed a revolving brush or roller.

A review of available data from Maine clearly shows the impact of new technology and management practices on wild blueberry production. Initial harvesting on the Maine barrens probably yielded less than one hundred pounds per acre. In 1922, the Maine Extension Service reported average yields of about four hundred pounds per acre, and in 1955, it reported an average of over six hundred pounds. In 1976, the State average exceeded one thousand pounds per acre, with some fields producing over a ton per acre (Metzer and Ismail, 1976).

Through the twentieth century, sales of wild blueberries kept pace with increases in production and provided good returns to growers and processors. The success of the industry is the result of having a quality product, committed and innovative growers and processors, good research support, and aggressive marketing.

The bakery trade has always been the primary market for wild blueberries, but the popularity of blueberry in the preparation of muffins in coffee shops and home baking since World War II created an even larger market. In addition, expanded usage in jams, dairy products, beverages and cereals guaranteed strong demand.

The USA has always been the largest market for wild blueberries, but trade missions by industry leaders to Europe and Japan since the 1970s have resulted in major overseas markets, and berries are now exported to over twenty countries around the world.

In 1981, industry Leaders from Canada and Maine formed the Wild Blueberry Association of North America. This association brought together growers and processors from all production areas to extend awareness of wild blueberries and promote a quality product. The success of their activities soon gained them recognition as a model for cooperative international marketing of an agricultural product.

## LITERATURE CITED

Day, C.A. 1959. A history of the blueberry industry in Washington County. Presentations at Farm and Home Week in Orono, Maine.

Kinsman, G. 1986. The history of the lowbush industry in Nova Scotia 1880-1950. Rept. N. S. Dept. Agr. Mktg.

Metzer, H.B. and A.A. Ismail. 1976. Management practices and operating costs in lowbush blueberry production. Bull723 Life Sci. and Agr. Expt. Sta., Univ. Maine, Orono.

Munson, W.M. 1901. The horticultural status of the genus *Vaccinium*. Rpt. Maine Agr. Exp. Sta. Orono, Maine, pp. 113-158.

# Soil and Plant Response to MSW Compost Applications on Lowbush Blueberry Fields in 2000 and 2001

P. R. Warman
C. J. Murphy
J. C. Burnham
L. J. Eaton

**SUMMARY.** Field experiments were initiated in May 1999 to investigate the application of municipal solid waste (MSW) compost to lowbush blueberry (*Vaccinium angustifolium* Ait.) fields. Three sites were selected: Debert, NS (Truro sandy loam) and two sites near Musquodoboit, NS (both Rawdon gravely loamy sands). Treatments at each site consisted of a randomized complete block design with six treatments (Control [no fertilizer], NK fertilizer, NPK fertilizer, and three rates of MSW

P. R. Warman is Professor, C. J. Murphy is former MSc Student, J. C. Burnham is Research Associate, and L. J. Eaton is Research Professor, Department of Environmental Sciences, Nova Scotia Agricultural College, Truro, NS B2N 5E3, Canada.

The authors acknowledge the financial support provided by the NSERC-AAFC Research Partnership Support Program and the NS-AAFC Technology Development Program. The authors are indebted to Ryan Ring, Jason Murphy, and Mike Munroe who helped set up, maintain, and harvest the plots and perform the lab analyses of the soils and plant tissue.

[Haworth co-indexing entry note]: "Soil and Plant Response to MSW Compost Applications on Lowbush Blueberry Fields in 2000 and 2001." Warman, P. R. et al. Co-published simultaneously in *Small Fruits Review* (Food Products Press, an imprint of The Haworth Press, Inc.) Vol. 3, No. 1/2, 2004, pp. 19-31; and: *Proceedings of the Ninth North American Blueberry Research and Extension Workers Conference* (ed: Charles F. Forney, and Leonard J. Eaton) Food Products Press, an imprint of The Haworth Press, Inc., 2004, pp. 19-31. Single or multiple copies of this article are available for a fee from The Haworth Document Delivery Service [1-800-HAWORTH, 9:00 a.m. - 5:00 p.m. (EST). E-mail address: docdelivery@haworthpress.com].

compost) blocked four times. Compost treatments provided the equivalent of 100, 200, and 400 kg ha$^{-1}$ of total N, respectively. The experimental objectives were to evaluate soil and plant response to the compost and to determine whether the organic amendment could be used as an alternative to chemical fertilizers. Yield, soil fertility, and plant nutrients were evaluated in blueberry leaf tissue and fruit over two years. The MSW compost had a strong (K) and a mild effect (P, Ca, Mg, S, Cu, Zn) on extractable soil nutrients, while a strong effect (Mn) and a mild effect (N, K) was observed on leaf tissue nutrients. The fruit yield was not affected by the treatments. Therefore, the compost treatments provided equivalent amounts of plant essential nutrients without negatively influencing trace element absorption. *[Article copies available for a fee from The Haworth Document Delivery Service: 1-800-HAWORTH. E-mail address: <docdelivery@haworthpress.com> Website: <http://www.HaworthPress. com> © 2004 by The Haworth Press, Inc. All rights reserved.]*

**KEYWORDS.** Municipal solid waste, soil and plant response, *Vaccinium angustifolium* Ait.

## *INTRODUCTION*

The lowbush blueberry (*Vaccinium angustifolium* Ait.) is a perennial shrub that is commercially produced in Maine (USA), and the Atlantic Provinces (Nova Scotia, New Brunswick, Prince Edward Island and Newfoundland) and Eastern Quebec in Canada. The plant grows well in Nova Scotia's acidic Podzolic soils with a reported optimum growth in the pH range of 4 to 5.5 (Hall et al., 1979). The blueberry is unique because it occurs naturally in forests and abandoned farmlands (Hall et al., 1964), exhibits large genetic variability (Vander Kloet, 1978), and is managed by regular pruning back to ground level, usually every second year (Hall et al., 1979). Pruning forces the lowbush blueberry into a biennial production cycle in which new and vigorous vegetative stems grow and initiate flower buds in the first (sprout) year, followed by flowering, pollination and fruit development in the second (crop) year. Selective herbicides and chemical fertilizers are typically applied in the spring of the sprout year. The fruit is harvested in late summer of the crop year. The cycle is repeated over the next two growing seasons.

Lowbush blueberries typically receive inputs of chemical fertilizers to stimulate plant growth (Eaton, 1994); however, researchers have been testing alternative nutrient sources to replace fertilizers. One such amendment is source separated municipal solid waste (MSW) compost that has been found to improve soil properties (He et al., 1992) and increase yields (Maynard, 1995). One possible benefit of composts, such as MSW compost, is to provide increased organic matter onto lowbush blueberry soils, which helps to reduce soil erosion by holding soil particles (Eaton and Jensen, 1996). As yet, few studies have evaluated the effects of MSW compost applied to crops such as the lowbush blueberry. Warman (1987) applied organic amendments, such as animal manures and sawdust, to lowbush blueberries, with the manure treatments providing the highest levels of plant macronutrients. In another study by Warman (1988), it was found that green manure and dairy manure increased the organic carbon, $K_2O$ (green manure treatment only), and Ca and Mg content (dairy manure treatment only) of a blueberry soil compared with fertilizer and control plots. Lafond et al. (1999) investigated the effects of paper mill sludge on soil chemical properties and yield of lowbush blueberries and found that yields with sludge were increased compared to a fertilizer treatment.

Municipal solid waste composting is gaining popularity in Nova Scotia as a way to divert refuse from landfills. As more communities in the province produce MSW compost, there is an increased pressure to apply the compost to agricultural land. Few studies have investigated compost additions to a crop grown under low pH conditions or one, such as the lowbush blueberry, that requires the surface application of material. The objectives of this study have been to: (1) evaluate changes in soil fertility associated with MSW compost application; and (2) evaluate the application of MSW compost on yield, plant nutrients, and trace metal content of lowbush blueberry leaf tissue and fruit. The fulfillment of these objectives will help to determine whether MSW compost is a suitable fertilizer for lowbush blueberries.

## MATERIALS AND METHODS

*Site selection and field management.* Three Nova Scotia sites were selected for the study. The Debert site (N 45° 26', W 63° 27') is located on a well-drained Truro sandy loam soil. The South Branch and White Field sites (N 45° 06', W 63° 09') are located on heavier Rawdon sandy loam soils, near Middle Musquodoboit, NS. In 2000, both the Debert and White Field locations were in the crop year of their two-year management cycles, whereas South Branch was in the sprout year. The Debert and White Field locations were pruned by flail mowing in the spring of 1999 and spring of 2001. The South Branch field was pruned in the fall of 1997 and fall of 1999. Selective herbicides were applied in the spring of the prune year at the three sites. Terbacil or 3-tert-butyl-5-chloro-6-methyluracil was applied at a rate of 2.0 kg ha$^{-1}$ to the White Field and Debert sites in May 1999. Hexazinone or 3-cyclo-hexyl-6-(dimethylamino)-1-methyl-1, 3, 5-triazine-2, 4-(1H, 3H)-dione at a rate of 2.0 kg ha$^{-1}$ was applied to South Branch in May 2000 and May 2001 by the owner. The owner monitored the fields on a regular basis for pests and diseases and no major outbreaks were noted throughout the 2000 and 2001 growing seasons.

*Compost and fertilizer applications, experimental design.* Treatments at each site were in a randomized block design with six treatments: Control, NK fertilizer, NPK fertilizer, MSW (municipal solid waste compost) rate 1, MSW rate 2, and MSW rate 3 blocked four times. Plots were 3 m by 3 m in size with 1 m walkways. Fertilizer blends were made to match the MSW compost with respect to total N, $P_2O_5$, and $K_2O$ in the compost, with the assumption of 50% (MSW 1), 25% (MSW 2), and 12.5% (MSW 3) total N availability from the compost, respectively, and were based on a lowbush blueberry nitrogen requirement of 50 kg ha$^{-1}$. Inorganic fertilizer blends were made by mixing the proper amounts of ammonium nitrate for N (34-0-0), triple superphosphate for $P_2O_5$ (0-46-0), and muriate of potash for $K_2O$ (0-0-60). Compost treatments provided the equivalent of 100, 200 and 400 kg of total N ha$^{-1}$, respectively.

The MSW compost was obtained from the Lunenburg Regional Recycling and Composting Facility, where it was produced under aerobic condition using an in-vessel system and placed in windrows for final maturing. The entire process took about three months. All fertilizer and compost were hand broadcast on each plot in April. In 1999, the first year of the experiment (results not discussed), fertility treatments were applied to all three sites, even though only one site was in its crop year (South Branch). In the second (2000) and third year (2001) of the experiment only sites that were in the sprout year received fertility treatments (South Branch in 2000, Debert and White Field in 2001).

*Soil and tissue sampling.* All soil samples were taken with a soil sample probe at a depth of 0 to 15 cm and the leaf/stem litter was discarded. Five soil cores were taken in each plot and mixed to give a representative sample. Leaf samples were acquired by removing all the leaves from the top 10 cm of 50 randomly chosen blueberry stems within the plot, and bulking them into one sample. In 2000, both Debert and White Field (crop year) soil and leaf samples were taken at harvest in mid-August, after the blueberry fruit were ripe and harvested. In 2000, South Branch (sprout year) soil and leaf samples were taken at tip dieback, the point at which a small black tip appears when the terminal bud dies (Hall et al., 1979) in mid-July. In 2001, Debert and White Field (sprout year) sites were sampled for leaves and soil at tip dieback, and South Branch (crop year) was sampled at harvest. Debert and White Field were harvested by hand raking in mid-August 2000, whereas South Branch was hand raked on 21 August 2001. All fruit yields were weighed and recorded on site.

*Laboratory procedures.* Soil samples were air dried and sieved through a number 10 mesh (2 mm) sieve. The Mehlich 3 (M3) soil extractant (Mehlich, 1984) was used to evaluate soil extractable elements including P, K, Ca, Mg, S, Fe, Mn, B, Mo, Cu, Zn, Cd, Cr, Ni, and Pb. The process consisted of adding 100 mL of M3 to 10 mL of soil followed by shaking for 15 minutes at 200 oscillations per minute, then filtered through Whatman # 5 filter paper. All plant tissue samples were oven dried at 65°C before grinding through a Wiley micromill, using a 40 mm mesh stainless steel screen. Then, 2.00 g of sample was placed in a 250 mL digestion tube and digested with 10 mL of concentrated nitric acid ($HNO_3$), a modification of the method of Zarcinas et al. (1987). The solution was analyzed for P, K, Ca, Mg, S, Fe, Cu, Mn, Zn, B, Cd, Cr, Ni, and Pb. Dried tissue were analysed for C and N using a CNS-1000 Analyzer (Leco Corporation, St. Joseph, MI).

Both the 2000 and 2001 MSW composts were analyzed for their nutrient and trace element contents (Table 1). Four replications of a previously obtained composite sample were air dried and ground with a mortar and pestle, and were digested and analysed as above, except 1.00 g of sample was used and filtered into 100 mL volumetric flasks.

*Statistical analysis and interpretations.* The data was not pooled because of the numerical difference among the sites and between the years. Statistical analysis was completed on the SAS System software

TABLE 1. Dry weight analysis of compost macronutrients (g kg$^{-1}$), micronutrients, and trace elements (mg kg$^{-1}$) in 2000 and 2001

| Year | N | P | K | Ca | Mg | S | Na | Fe |
|------|------|-----|-----|----|-----|-----|------|------|
| 2000 | 23.2 | 6.0 | 7.8 | 25 | 2.6 | 3.1 | 5330 | 6800 |
| 2001 | 25.5 | 6.3 | 8.1 | 26 | 2.9 | 6.4 | 5401 | 7900 |

| Year | Cu | Mn | Zn | B | Cd | Cr | Ni | Pb |
|------|-----|------|-----|----|------|----|----|----|
| 2000 | 99 | 1910 | 206 | 12 | 0.37 | 15 | 9 | 68 |
| 2001 | 104 | 884 | 213 | 18 | 0.19 | 14 | 12 | 70 |

package (SAS Institute, 1999). After assumptions were verified, a one-way ANOVA (P ≤ 0.05) was used to determine if the treatments produced a significant effect for that year and location; if significant, treatment means were compared using a Tukey's Means Comparison Test (P = 0.05).

For each nutrient the results from the three sites and two years are presented for both the soil and leaf samples. The term 'site-year combination' (SYC) will refer to the six possible site-year combinations for each nutrient (three sites × two years). The fertility treatments were deemed to have a 'strong' effect on a nutrient if six or five SYC had a significant difference. If four or three SYC were different, it was called a 'mild' effect, while a 'weak' effect was when only two, one or zero SYC had a difference among the five fertility treatments.

## RESULTS AND DISCUSSION

*Soil nutrients.* Levels of Mehlich-3 extractable soil elements for the fertility treatments of each site in 2000 ('00) and 2001 ('01) are listed in Table 2 (macronutrients) and Table 3 (micronutrients). Generally, except for P, the South Branch site had the highest levels of extractable soil nutrients in both 2000 and 2001. Fertility treatments had a strong effect on extractable K soil levels. The MSW 3 treatment produced higher values of extractable K then the control, NK and/or NPK. In a study by Warman and Cooper (2000), high application rates of poultry manure compost showed the highest M3 extractable K in the top 15 cm of two different soil types (silty clay and a sandy loam) compared to other manure and inorganic fertilized forage plots.

The treatments had a mild effect on soil extractable values of P, Ca, and Mg and a weak effect on S levels (Table 2). Overall, the soil macronutrients were numerically higher in 2001 compared to 2000, and levels in the MSW3 and MSW2 treated plots were generally higher then those in the controls. The increase P, Ca, and Mg can be attributed to the fact that the compost had relative high values of these nutrients. The NK treated soil did not receive P, hence the low M3 P levels, especially in 2001; we suspect that effective mycorrhizae must be helping these plants absorb and translocate P from the soil solution in order to meet their nutritional P requirements.

Treatments did not produce a strong effect in any soil extractable micronutrients and trace elements, with only a mild effect with Cu and Zn (Tables 3 and 4). There were weak effects of treatments on Mehlich-3

**TABLE 2.** Analysis of Mehlich-3 soil macronutrients from the blueberry plots in 2000 ('00) and 2001 ('01) (mg kg$^{-1}$)[z]

| Location | Treatment | P | | K | | Ca | | Mg | | S | |
|---|---|---|---|---|---|---|---|---|---|---|---|
| Year | | '00 | '01 | '00 | '01 | '00 | '01 | '00 | '01 | '00 | '01 |
| Debert | Control | 11a | 32a | 24ab | 32a | 3a | 113a | 3a | 16a | 19a | 3a |
| | NK | 11a | 35a | 20ab | 35a | 4a | 49a | 3a | 14a | 19a | 4a |
| | NPK | 12a | 33a | 14a | 26a | 3a | 18a | 2a | 8a | 20a | 2a |
| | MSW1 | 11a | 36a | 21ab | 41a | 14a | 63a | 4a | 13a | 20a | 4a |
| | MSW2 | 12a | 55a | 26ab | 70b | 22a | 125a | 4a | 18a | 19a | 6a |
| | MSW3 | 12a | 54a | 34b | 128c | 14a | 363b | 5a | 35b | 21a | 9a |
| White | Control | 5a | 2a | 28a | 28a | 35ab | 75a | 9ab | 11a | 28a | 30b |
| Field | NK | 4a | 2a | 25a | 34a | 27a | 74a | 7a | 11a | 23a | 27b |
| | NPK | 4a | 3ab | 30a | 32a | 31ab | 58a | 7a | 8a | 25a | 10ab |
| | MSW1 | 5a | 4abc | 27a | 56b | 45ab | 146ab | 8ab | 19ab | 27a | 4a |
| | MSW2 | 4a | 7c | 34a | 85c | 75ab | 261c | 11ab | 32c | 25a | 6a |
| | MSW3 | 6a | 6bc | 51b | 92c | 76b | 225bc | 13b | 29bc | 29a | 4a |
| South | Control | 7a | 5ab | 33a | 40a | 259a | 587a | 97a | 108a | 28a | 44b |
| Branch | NK | 6a | 2a | 39ab | 29a | 285a | 344a | 109a | 98a | 24a | 10a |
| | NPK | 7a | 5ab | 34ab | 40a | 262a | 411a | 71a | 92a | 32a | 21ab |
| | MSW1 | 8a | 7ab | 59bc | 60b | 353ab | 622a | 94a | 130a | 36a | 37ab |
| | MSW2 | 9a | 11ab | 73c | 89c | 328a | 637a | 91a | 111a | 32a | 39b |
| | MSW3 | 16b | 17b | 133d | 114c | 548b | 660a | 128a | 135a | 32a | 32b |

[z]Treatment means in a column (each site) followed by the same letter are not significantly different at p ≤ 0.05

**TABLE 3.** Analysis of Mehlich-3 soil micronutrients from 2000 ('00) and 2001 ('01) (mg kg$^{-1}$)[z]

| Location | Treatment | B | | Fe | | Cu | | Mn | | Zn | |
|---|---|---|---|---|---|---|---|---|---|---|---|
| Year | | '00 | '01 | '00 | '01 | '00 | '01 | '00 | '01 | '00 | '01 |
| Debert | Control | < 0.05a | < 0.05a | 67a | 65a | 1.0a | 1.2a | 5a | 8ab | 1.2a | 2.4a |
| | NK | < 0.05a | < 0.05a | 60a | 62a | 1.0a | 1.6b | 4a | 10b | 1.2a | 1.8a |
| | NPK | < 0.05a | < 0.05a | 86a | 57a | 1.2a | 1.4ab | 4a | 5a | 1.1a | 1.0a |
| | MSW 1 | < 0.05a | < 0.05a | 85a | 62a | 1.4a | 1.0a | 4a | 5a | 1.1a | 1.4a |
| | MSW 2 | < 0.05a | < 0.05a | 74a | 68a | 1.5a | 0.8a | 4a | 5a | 1.4a | 1.8a |
| | MSW 3 | < 0.05a | < 0.05a | 91a | 62a | 1.5a | 1.1a | 3a | 10b | 1.3a | 3.6b |
| White | Control | < 0.05a | 0.06a | 128a | 70b | 2.6a | 2.1a | 43a | 33a | 3.3a | 3.1ab |
| Field | NK | < 0.05a | 0.46b | 127a | 67ab | 2.5a | 1.8a | 37a | 27a | 3.1a | 2.1ab |
| | NPK | < 0.05a | < 0.05a | 108a | 58ab | 2.7a | 1.7a | 33a | 23a | 3.1a | 1.5a |
| | MSW 1 | < 0.05a | < 0.05a | 108a | 57ab | 2.1a | 1.7a | 30a | 25a | 2.9a | 2.2ab |
| | MSW 2 | < 0.05a | 0.13ab | 141a | 58ab | 2.2a | 1.3a | 47ab | 28a | 3.1a | 2.7b |
| | MSW 3 | < 0.05a | 0.09ab | 153a | 55a | 2.7a | 1.3a | 61b | 26a | 3.8a | 3.0b |
| South | Control | < 0.05a | 0.17ab | 139a | 53ab | 2.1ab | 1.8b | 52a | 41a | 2.8ab | 4.5bc |
| Branch | NK | < 0.05a | 0.07a | 138a | 34a | 1.8a | 0.9ab | 41a | 33a | 2.4a | 1.4a |
| | NPK | < 0.05a | 0.11ab | 135a | 56ab | 2.1ab | 0.8a | 47a | 37a | 2.6ab | 1.8ab |
| | MSW 1 | < 0.05a | 0.40b | 142a | 67b | 2.3ab | 1.9ab | 47a | 51a | 3.5b | 3.4abc |
| | MSW 2 | < 0.05a | 0.19ab | 130a | 62ab | 2.4ab | 1.8ab | 49a | 44a | 3.4ab | 3.5abc |
| | MSW 3 | < 0.05a | 0.31b | 158a | 59ab | 3.0b | 1.7ab | 62a | 44a | 4.7c | 4.8bc |

[z]Treatment means in a column (each site) followed by the same letter are not significantly different at P ≤ 0.05

TABLE 4. Fruit yields (kg m$^2$) and leaf macronutrient (g kg$^{-1}$) analysis for 2000 ('00) and 2001 ('01)[z]

| Location | Treatment | Yield | | N | | P | | K | |
|---|---|---|---|---|---|---|---|---|---|
| Year | | '00 | '01 | '00 | '01 | '00 | '01 | '00 | '01 |
| Debert | Control | 0.31a | na | 14a | 13a | 1.1a | 1.0a | 3.6b | 3.3a |
| | NK | 0.33a | na | 13a | 13a | 1.1a | 1.0a | 3.8b | 3.6ab |
| | NPK | 0.38a | na | 13a | 13ab | 1.1a | 1.0a | 3.1a | 3.4ab |
| | MSW 1 | 0.30a | na | 13a | 14ab | 1.1a | 1.1a | 3.1a | 3.8abc |
| | MSW 2 | 0.32a | na | 14a | 14ab | 1.1a | 1.1a | 3.5ab | 4.0bc |
| | MSW 3 | 0.35a | na | 13a | 15b | 1.1a | 1.1a | 3.5ab | 4.3c |
| White | Control | 0.47a | na | 15a | 11a | 1.0a | 1.1abc | 3.6a | 4.1a |
| Field | NK | 0.64a | na | 17b | 12ab | 0.9a | 0.9a | 4.1a | 4.0a |
| | NPK | 0.71a | na | 16ab | 11ab | 0.9a | 1.0ab | 3.8a | 4.1a |
| | MSW 1 | 0.60a | na | 15ab | 11ab | 0.9a | 1.1abc | 3.7a | 4.5ab |
| | MSW 2 | 0.66a | na | 16ab | 11ab | 0.9a | 1.1bc | 3.8a | 4.8ab |
| | MSW 3 | 0.68a | na | 16ab | 13b | 1.0a | 1.2c | 3.9a | 5.5b |
| South | Control | na | 0.42a | 15a | 9a | 0.9a | 0.9a | 3.8a | 2.9a |
| Branch | NK | na | 0.38a | 17b | 10a | 1.0ab | 0.8a | 4.4abc | 3.1a |
| | NPK | na | 0.57a | 16ab | 10a | 1.0ab | 0.9a | 4.2ab | 3.3a |
| | MSW 1 | na | 0.57a | 16ab | 10a | 1.0ab | 0.9a | 4.3ab | 3.4a |
| | MSW 2 | na | 0.65a | 16ab | 11a | 1.1b | 0.9a | 4.9bc | 3.4a |
| | MSW 3 | na | 0.52a | 17b | 10a | 0.9ab | 0.9a | 5.1c | 3.4a |

[z]Treatment means in a column (each site) followed by the same letter are not significantly different at P ≤ 0.05
na–not applicable

levels of B, Fe, and Mn. Levels of Cd, Cr, Ni, and Pb (data not shown) did not differ among treatments, and values were below 0.10 mg kg$^{-1}$, except for Ni, which was 0.97 mg kg$^{-1}$ for the NK treatment at White Field in 2000. Overall, micronutrients and trace elements in the MSW2 and MSW3 amended plots were numerically higher then the control and the fertilizer treatments, but were statistically similar.

*Fruit yield.* Blueberry fruit yield data at all three sites indicated there was no effect of the five fertilizer or compost treatments on fruit yields (Table 4). The yields were extremely variable at these sites, which could account for the lack of significance. White Field had the highest yield of the three sites, which ranged from 0.47-0.71 kg m$^2$; however, yields at the three sites did not reflect the M3 extractable nutrient levels. The Control treatment produced the lowest and the NPK the highest yielding plots for Debert and White Field. The NPK plots induced higher yields than the NK fertilized plots at all three sites. As well, at Debert and White Field, yield increased with increasing compost application rates. It should be noted that it is common to find no yield response to fertility applications to lowbush blueberries; for example, Warman (1987) found no effect of inorganic and organic fertilizers on lowbush blueberry yield compared to control plots, while Eaton (1994) reported an inconsistent yield response of lowbush blueberry stands to NPK fertilization.

*Leaf nutrients.* For leaf extractable nutrients, only 19 of the possible 96 element-site-year combinations (16 elements × 3 sites × 2 years) resulted in a significant difference among the five fertility treatments (Tables 4, 5, and 6). These differences were equally distributed in the two years, ten in 2000 and nine in 2001. The fertility treatments produced a mild and a strong effect on leaf N and K, respectively. The NK treatment had the highest N values in 2000 but in 2001 the MSW3 treatment produced the highest levels at Debert and White Field. In general, the inorganic fertilizers supplied the most N to the plant when compared to the control and compost-treated plots, except for Debert in 2001. Using MSW compost from the same source, Warman and Rodd (1998) reported a similar response to vegetables. In our experiment, the plant levels of K increased with the higher compost rate at all SYC, except for South Branch in 2001.

The treatments produced a weak effect on the remaining macronutrients P, Mg, and S, while no effect was observed with Ca. The relationship between P levels in the soil and the ability of a plant to uptake P is still unclear in lowbush blueberries. For example, throughout the en-

TABLE 5. Macronutrient (g kg$^{-1}$) and micronutrient (mg kg$^{-1}$) analysis of 2000 ('00) and 2001('01) leaf tissue[z]

| Location | Treatment | Ca | | Mg | | S | | B | |
|---|---|---|---|---|---|---|---|---|---|
| Year | | '00 | '01 | '00 | '01 | '00 | '01 | '00 | '01 |
| Debert | Control | 6.2a | 3.4a | 2.4a | 1.5a | 1.0a | 2.4a | 39a | 17a |
| | NK | 5.6a | 3.3a | 2.1a | 1.4a | 1.0a | 2.3a | 38a | 20a |
| | NPK | 6.1a | 3.5a | 2.1a | 1.4a | 0.9a | 2.3a | 38a | 20a |
| | MSW 1 | 6.2a | 3.6a | 2.2a | 1.4a | 0.9a | 2.5a | 38a | 18a |
| | MSW 2 | 6.2a | 3.5a | 2.1a | 1.4a | 0.9a | 2.5a | 37a | 20a |
| | MSW 3 | 6.0a | 3.4a | 2.0a | 1.3a | 0.9a | 2.5a | 40a | 16a |
| White | Control | 6.3a | 3.6a | 2.0ab | 1.3a | 1.1a | 3.0a | 33a | 17a |
| Field | NK | 6.5a | 3.8a | 1.8a | 1.4a | 1.1a | 3.1ab | 31a | 15a |
| | NPK | 6.3a | 4.0a | 1.9ab | 1.4a | 1.0a | 3.0a | 32a | 16a |
| | MSW 1 | 6.3a | 3.8a | 2.1ab | 1.4a | 1.1a | 3.0ab | 32a | 16a |
| | MSW 2 | 6.7a | 4.3a | 2.3b | 1.5a | 1.1a | 3.3b | 28a | 16a |
| | MSW 3 | 6.6a | 4.0a | 2.1ab | 1.5a | 1.1a | 3.7c | 33a | 17a |
| South | Control | 3.3a | 6.0a | 2.0a | 3.3a | 0.9a | 2.6a | 19a | 32a |
| Branch | NK | 3.1a | 5.4a | 2.2a | 3.0a | 1.0a | 2.6a | 18a | 29a |
| | NPK | 3.5a | 6.0a | 2.1a | 3.1a | 1.0a | 2.6a | 18a | 29a |
| | MSW 1 | 3.5a | 5.7a | 2.1a | 3.0a | 1.0a | 2.6a | 20a | 31a |
| | MSW 2 | 3.3a | 6.0a | 2.0a | 3.0a | 1.0a | 2.8a | 23a | 33a |
| | MSW 3 | 3.0a | 5.6a | 2.0a | 2.8a | 1.0a | 2.8a | 18a | 35a |

[z]Treatment means in a column (each site) followed by the same letter are not significantly different at P ≤ 0.05

TABLE 6. Micronutrient (mg kg$^{-1}$) analysis of 2000 and 2001 leaf tissue[z]

| Location | Treatment | Fe | | Cu | | Mn | | Zn | |
|---|---|---|---|---|---|---|---|---|---|
| Year | | '00 | '01 | '00 | '01 | '00 | '01 | '00 | '01 |
| Debert | Control | 28a | 51a | 4b | 4a | 2190b | 1377b | 9a | 11ab |
| | NK | 29a | 26a | 3ab | 3a | 2070b | 1421b | 8a | 9a |
| | NPK | 24a | 18a | 3a | 3a | 1760ab | 1260b | 7a | 12ab |
| | MSW 1 | 29a | 72a | 3a | 3a | 1890ab | 840ab | 7a | 22b |
| | MSW 2 | 30a | 26a | 3ab | 3a | 1910ab | 678a | 8a | 10a |
| | MSW 3 | 25a | 18a | 3ab | 4a | 1300a | 383a | 7a | 10ab |
| White | Control | 133a | 184a | 5a | 4a | 2420ab | 1566ab | 11a | 19a |
| Field | NK | 156a | 164a | 5a | 5a | 2890b | 1513ab | 11a | 18a |
| | NPK | 172a | 178a | 7a | 5a | 2690ab | 1555b | 10a | 18a |
| | MSW 1 | 165a | 195a | 5a | 5a | 2340ab | 1149ab | 10a | 12a |
| | MSW 2 | 157a | 166a | 6a | 4a | 1990a | 1017a | 11a | 11a |
| | MSW 3 | 177a | 160a | 6a | 4a | 2420ab | 1059ab | 10a | 19a |
| South | Control | 56a | 115a | 3a | 3a | 323a | 317a | 8a | 6a |
| Branch | NK | 60a | 118a | 3a | 3a | 338a | 278a | 10a | 7a |
| | NPK | 77a | 106a | 3a | 3a | 417a | 269a | 10a | 7a |
| | MSW 1 | 93a | 128a | 3a | 4a | 354a | 251a | 11a | 6a |
| | MSW 2 | 70a | 131a | 3a | 3a | 402a | 257a | 9a | 7a |
| | MSW 3 | 77a | 139a | 4a | 3a | 402a | 278a | 10a | 6a |

[z]Treatment means in a column (each site) followed by the same letter are not significantly different at P $\leq$ 0.05

tire experiment, only twice was there a significant treatment effect on leaf tissue P. Even the NPK treatment was never statistically higher than the control plots with regards to leaf P. As indicated previously, the plant must be acquiring P from sources other than directly applied P.

The leaf micronutrient values indicated that the treatments resulted in a strong effect for Mn but weak or no effects for the other nutrients (Table 6). Leaf Mn was different in four of the six SYC, which was caused by the control and the fertilizer treatments. Generally, the MSW2 and MSW3 treatments had lower Mn values then the control or NK values. Manganese availability is directly related to soil pH; as pH increases, Mn availability decreases. Acidic soils have larger amounts of available Mn$^{2+}$ for plant uptake than do neutral or alkaline soils. Plants in the compost plots had lower Mn in their tissue simply due to the pH effect of the compost. Warman (1988) discovered that conventional fertilizers applied to lowbush blueberry fields significantly increased Mn levels in both the below and the above ground portions of the plant when compared to organic amendments. This was probably due to the acidifying effect of fertilizers coupled with the liming effect of organic amendments.

There were no differences in levels of B and Fe among treatments at all six SYC (Tables 4 and 5) and the treatments resulted in only a weak effect on Cu and Zn. Cadmium, Cr, Ni, and Pb were analyzed, but the results are not shown because the treatments did not produce a significant effect on the content of leaf elements, except for Ni at South Branch in 2000, in which the MSW3 treatment resulted in a higher value than the other five treatments. To date, therefore, MSW compost has not increased these potentially toxic elements in lowbush blueberry leaves. It is to be noted that none of the compost treatments resulted in increased trace element levels in blueberry fruit in 2000, the only year we have data (data not shown) (Murphy, 2001).

As the level of Mehlich-3 extractable K increased in the soil, at all three sites and years, the corresponding leaf K values also increased (Tables 2 and 4). A Pearson correlation coefficient of 0.526 between K soil and leaf values indicated a linear relationship between the two variables (P = 0.001). None of the other elements showed a significant positive correlation between M3 soil and leaf samples. Therefore, the Mehlich-3 extractant may not be appropriate to evaluate all "plant available" nutrients and trace elements in lowbush blueberry soils. Warman (1987), however, found that soil extractable P, K, Ca, and Mg were highly correlated to plant tissue contents of these elements in a lowbush blueberry study using Bray II for P and 1 M $NH_4OAc$ for K, Ca, and Mg. Furthermore, it has been suggested that plant roots or rhizomes may be a more appropriate plant organ to sample; however, removing root structures are destructive to blueberry production.

The leaf tissue content of K, Ca, Mg, B, and Zn reflect differences in the two years of the production cycle, in that the concentration of these five nutrients alternated between crop and sprout year regardless of location or treatment (Tables 4, 5, and 6). For example, Ca values for Debert and White Field in 2000 ranged between 5.6 and 6.5 mg $kg^{-1}$ while South Branch had lower values, 3.0 to 3.5 mg $kg^{-1}$. In the next season, leaf Ca at Debert and White Field decreased to a value similar to those at South Branch in 2000; correspondingly, the 2001 South Branch values increased to values similar to the 2000 Debert and White Field.

## CONCLUSION

The results of this study, to date, demonstrate some effects of MSW applications on lowbush blueberry soil and leaf nutrient levels, particularly in the increased levels of N and K. It does appear that compost treatments are providing equivalent amounts of plant essential nutrients without negatively influencing trace element absorption. Many of these results are similar to those demonstrated by Warman (1987; 1988). The data also illustrate the influence of soil properties associated with the differences between the sandy Debert soil and the heavier Rawdon soils at the South Branch and White Field sites.

It appears that long-term (several production cycles) studies are required to determine if MSW compost applications provide significant benefits to lowbush blueberry production systems. In order to do that, this study will be continued for at least one more production cycle.

## GROWER BENEFITS

Medium and high rates of municipal solid waste compost increased the levels of some macro- and micronutrients in the blueberry soils. The compost had an equivalent effect on leaf tissue nutrient content as the fertilizer treatments. The fruit yields were similar among the control, the two fertilizer treatments and the three compost treatments. We predict that the increase in extractable soil nutrients will positively increase fruit yields and plant nutrition over the long term.

## LITERATURE CITED

Eaton, L.J. 1994. Long-term effects of herbicide and fertilizers on lowbush blueberry growth and production. Can. J. Plant Sci. 74:341-345.

Eaton, L.J. and K. Jensen. 1996. Protection of lowbush blueberry soils from erosion. Lowbush Blueberry Fact Sheet. Nova Scotia Dept. Agr. Mktg.

Hall, I.V., L.E. Aalders, N.L. Nickerson, and S.P. Vander Kloet. 1979. The biological flora of Canada 1. *Vaccinium angustifolium* Ait., sweet lowbush blueberry. Can. Field Nat. 93:415-430.

Hall, I.V., L.E. Aalders, and L.R. Townsend. 1964. The effects of soil pH on the mineral composition and growth of the lowbush blueberry. Can. J. Plant Sci. 44:433-438.

He, X.T., S.J. Traina, and T.J. Logan. 1992. Chemical properties of municipal solid waste composts. J. Environ. Qual. 21:318-329.

Lafond, J., R.R. Simard, and M. Roy. 1999. Application of paper mill sludge on blueberry: effects on soil chemical properties and yield. 1999 AIC Meeting, Charlottetown, PEI. (poster).

Maynard, A.A. 1995. Cumulative effect of annual additions of MSW compost on the yield of field grown tomatoes. Comp. Sci. Util. 3:47-54.

Mehlich, A. 1984. Mehlich 3 soil test extractant: A modification of Mehlich 2 extractant. Commun. Soil Sci. Plant Anal. 15:1409-1416.

Murphy, C. 2001. Soil and plant response to MSW compost applications on lowbush blueberry fields. MSc. Thesis, Dept. of Env. Sciences, NSAC, Truro, N.S. Dalhousie.

SAS Institute Inc. 1999. Release 8.00, Version 4.10, Cary, NC.

Vander Kloet, S.P. 1978. Systematics, distribution and nomenclature of the polymorphic *Vaccinium angustifolium* Ait. Rhodora. 80:358-376.

Warman, P.R. 1987. The effects of pruning, fertilizers and organic amendments on lowbush blueberry production. Plant Soil. 101:67-72.

Warman, P.R. 1988. A lysimeter study directed at nutrient uptake by the acid tolerant lowbush blueberry. Commun. Soil Sci. Plant Anal. 19:1031-1040.

Warman, P.R. and J.M. Cooper. 2000. Fertilization of a mixed forage crop with fresh and composted chicken manure and NPK fertilizer: Effects of dry matter yield and soil and tissue N, P, and K. Can. J. Soil Sci. 80:337-344.

Warman, P.R. and V. Rodd. 1998. Influence of source-separated MSW compost on vegetable crop growth and soil properties: Year 3. Proc. Eighth Annu. Conf. Composting Council Canada, Toronto, Ont., pp. 263-273.

Zarcinas, B.A., B. Cartwright, and L.R. Spouncer, 1987. Nitric acid digestion and multi-element analysis of plant material by inductively coupled spectrometry. Commun. Soil Sci. Plant Anal. 18:131-146.

# Factors Contributing
# to the Increase in Productivity
# in the Wild Blueberry Industry

David E. Yarborough

**SUMMARY.** Wild blueberries (*Vaccinium angustifolium* Ait.) are unique
in that they are commercially grown in the State of Maine in the United
States and in Quebec and the Atlantic Provinces of Canada. Approximately half the crop is grown in Maine and the remainder from Canada.
The wild blueberry crop has increased by an average of 2.3 million kg
each year over the last 20 years and now averages over 68.2 million kg
per year. Since 1980, approximately 4050 new ha added in Maine, and
over 6073 ha have been added in New Brunswick and Nova Scotia; PEI
has also developed more than 1620 new ha. However, most of the gains
in yield have come from improved management of the fields. Pre-emergence weed control with the herbicides terbacil and hexazinone in
the 1980s provided a release from the weed competition, and immediately doubled yields on many fields. It also allowed for improved fertility management and the increased use of bees for pollination which
resulted in even more production. Good disease and pest control using
integrated pest management keeps crop losses from pests to a minimum.
Recently, in Maine, there has been an investment in use of irrigation,

David E. Yarborough is Professor of Horticulture, Department of Plant, Soil and
Environmental Sciences, University of Maine, 5722 Deering Hall, Orono, ME
04469-5722 (E-mail: DavidY@maine.edu). Maine Agricultural Forest and Experiment Station Publication # 2569.

[Haworth co-indexing entry note]: "Factors Contributing to the Increase in Productivity in the Wild Blueberry Industry." Yarborough, David E. Co-published simultaneously in *Small Fruits Review* (Food Products
Press, an imprint of The Haworth Press, Inc.) Vol. 3, No. 1/2, 2004, pp. 33-43; and: *Proceedings of the Ninth
North American Blueberry Research and Extension Workers Conference* (ed: Charles F. Forney, and Leonard
J. Eaton) Food Products Press, an imprint of The Haworth Press, Inc., 2004, pp. 33-43. Single or multiple copies of this article are available for a fee from The Haworth Document Delivery Service [1-800-HAWORTH,
9:00 a.m. - 5:00 p.m. (EST). E-mail address: docdelivery@haworthpress.com].

with approximately 3,000 ha of in-ground irrigation and 850 ha of above-ground irrigation now in use.

Recent research has shown a 43% increase in yield with irrigation. All of these factors have combined to produce a three fold increase in the wild blueberry crop over the past 20 years. Mechanical harvesting has increased but is only used on a small percentage of the Maine crop, whereas more than half of the Canadian crop is harvested mechanically. This contributes to the efficiency of production since it reduces the cost of the most expensive production practice. *[Article copies available for a fee from The Haworth Document Delivery Service: 1-800-HAWORTH. E-mail address: <docdelivery@haworthpress.com> Website: <http://www.HaworthPress. com> © 2004 by The Haworth Press, Inc. All rights reserved.]*

**KEYWORDS.** Lowbush blueberry, weed management, fertility management, pollination, irrigation

## INTRODUCTION

Cultivated blueberries (*V. corymbosum* L. and *V. ashei* Reade) grow on 19,500 ha and account for nearly 60% of the blueberry production in North America. They are propagated and planted through out the world. Cultivated production doubled from 1980 from just over 45.5 million kg to an average of over 91 million kg a year (Holbeine, 1991; Yarborough, 2002b). Most of the gains in production from cultivated blueberries has come from increases in the number of acres planted, with the largest expansion occurring in the Pacific Northwest and Southeast, but improved varieties and high density plantings have also contributed to the increases (Moore, 1994; Trinka, 1997).

Wild blueberries are confined to a much smaller geographical area. Maine has 36,316 ha; Quebec has 16,194 ha of improved fields plus hundreds of thousands of ha of Provincial Crown land with wild blueberries that are harvested when prices are high. They are grown on 9,717 ha in New Brunswick, on 14,980 ha in Nova Scotia, and in PEI there will be nearly 2,429 ha in production soon (Yarborough, unpublished data). There are also minor areas of commercial production in New Hampshire, approximately 400 ha; Massachusetts, about 200 ha; and Michigan about 80 ha, but most of these berries are sold on the fresh market (Yarborough, unpublished data). The wild producing acreage greatly exceeds the cultivated area but production is considerably less (Figure 1); since only half of the fields are harvested annually because

FIGURE 1. North American production and area, 160 million kg on 86 thousand ha.

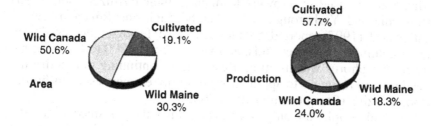

of the two-year pruning cycle and the wild clones within the field have a wide range of yielding potential (Hepler and Yarborough, 1991) the yield per acre from the wild stands is considerably less than the cultivated plantings.

Wild blueberry production has increased on the average by 3.5 fold over the past 20 years (ranged from 1.5 to 5 fold increase), but the land base has only increased by 1.5 fold over the past 20 years, (ranged from 1.2 to 5.7 fold) (Yarborough, unpublished data). Much of the improvements in the productivity of wild blueberries has come from improvements in management (Yarborough, 1997).

## FACTORS CONTRIBUTING TO INCREASED PRODUCTIVITY

*Weed management.* The catalyst for the improvement of wild stands has been improved weed management. Terbacil was the first selective preemergence herbicide registered in 1979 in Maine for wild blueberries, and two fold increase in yields was reported with its use (Ismail et al., 1981). A second herbicide, hexazinone, provided a much wider spectrum of weed control and increases of two fold or more in yield were reported with its use (Yarborough and Bhowmik, 1988; Yarborough and Ismail, 1985) and net income was maximized at 1.8 kg/ha hexazinone (Yarborough et al., 1986).

*Fertility management.* The use of herbicides allowed the addition of fertilizer to improve wild blueberry growth with out stimulating the weeds and reducing yield as fertilizer will do without adequate weed control (Hepler and Ismail, 1985; Yarborough et al., 1986). A method of testing leaves to determine the levels of N and P and recommendations for the use of diammonium phosphate (DAP) or monoammonium phosphate (MAP) fertilizers has been developed at the University of Maine and is used by growers to manage their fertilizer applications (Smagula and Yarborough, 1999; Yarborough and Smagula, 1993). Litten et al. (1997) reported that the use of DAP fertilizer increased stem length, number of flower buds, and yields, from 4,900 to 6,235 kg/ha, when phosphorus was limiting. Research is continuing to identify the need for and response to applications of minor elements such as boron, iron, and zinc (Smagula and Litten, 2002).

*pH*. Soil samples are only taken to determine the pH, since no significant correlations with blueberry yield have been found with nutrient levels in the soil. A reduction in soil pH reduces available nutrients and increases metals which makes the environment more favorable for blueberries and less so for weeds (Buckman and Brady, 1969). Patten and Wang (1994) indicated a reduction in soil pH could reduce weeds not controlled by herbicides in cranberry (*V. macrocarpon* Ait.) beds. Since a reduction in the pH from 6.0 to 4.0 had no effect on the yield of wild blueberries (Smagula, unpublished data), sulfur is being used as a cultural method to reduce the pH, thereby reducing the herbicide inputs.

*Pollination*. As early as 1961, George Wood (1961) indicated increased fruit set in wild blueberries could be obtained by using honeybees. Karmo (1974) promoted the use of honeybees to improve pollination of wild blueberries in Nova Scotia, and in Maine, Ismail (1987) recommended the use of two to four hives per 0.4 ha. Data from a survey of growers in Maine (Dill, unpublished data) shows a positive increase of 785 kg/ha for each hive used (Figure 2). All of the wild blueberry growing areas (Figures 3-6), with the exception of Quebec, have substantially increase their use of honey bees over the last 20 years (Figure 7).

Drummond (2002) has published methods to determine colony strength of the hive and a method of determining if there is an adequate pollination force in the field (Yarborough and Drummond, 2001). The use of alternate pollinators has been explored (Stubbs et al., 2000; 2001; 2002) and although some growers are using other pollinators to diversify, honeybees still remains the dominant pollinator because of price and availability.

*Irrigation*. When water becomes limiting, interest in irrigation increases. Research on wild blueberry irrigation on the crop year in the 1950s indicated a positive response (Struchtemeyer, 1956) but no response was seen for irrigating on the burn year (Trevett, 1967). A more controlled greenhouse and field study conducted in 1978-79 showed a strong correlation between the total water used and the number of flow-

FIGURE 2. Honeybee hives per acre versus yield in pounds per acre × 1000.

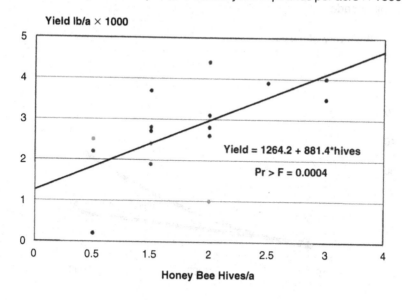

FIGURE 3. Maine growing area × 1000 ha, yield × 1 million kg, hives × 1000 and irrigated ha × 1000 by decade.

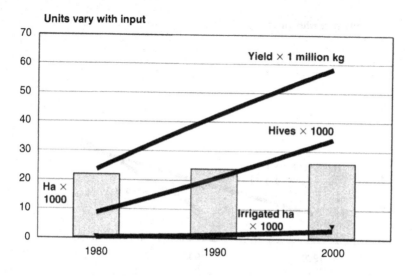

FIGURE 4. Nova Scotia growing area × 1000 ha, yield × 1 million kg, hives × 1000 by decade.

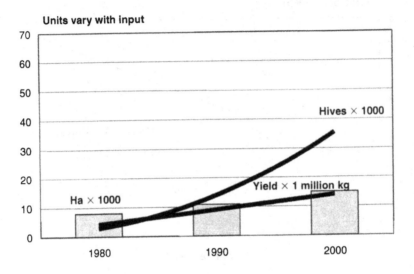

FIGURE 5. New Brunswick growing area × 1000 ha, yield × 1 million kg, hives × 1000 by decade.

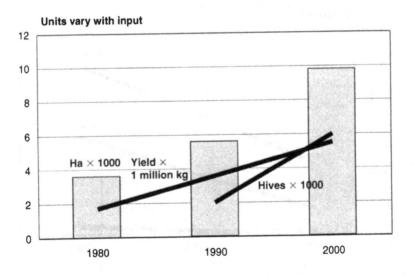

FIGURE 6. PEI growing area × 1000 ha, yield × 1 million kg, hives × 1000 by decade.

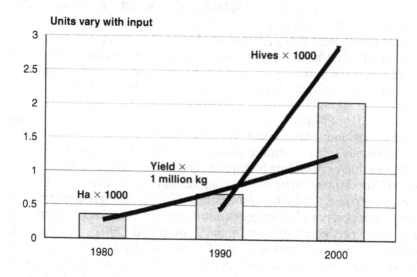

FIGURE 7. Quebec growing area × 1000 ha, yield × 1 million kg, hives × 1000 by decade.

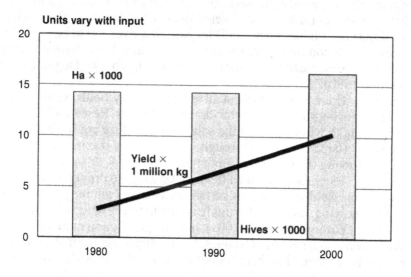

ers produced by the plant (Benoit et al., 1984). Studies done on irrigation in Nova Scotia found only a weak response to irrigation (Gordon et al., 1994). Research conducted in Maine in 2000-2001 has shown a 43% increase in yield with irrigation in both the burn and crop year (Seymour, unpublished data). Dalton et al. (2002) determined that with a yield of 2,050 kg/ha that irrigation is profitable. They also looked at the rainfall over the past 35 years and found less than a 20% probability of have one inch of rainfall per week in August. With 1,012 ha aboveground and 3036 ha of in-ground irrigation systems (Yarborough, unpublished data). Maine is the only wild blueberry growing area that has made a substantial investment in irrigation (Figure 2).

*Pest management.* Periodic outbreaks of chewing insects, fruit fly infestations and/or Monilina blight can cause substantial crop losses. Integrated Crop Management monitoring methods for insects include using traps, sweep nets, and action thresholds, which enable growers to determine when pest populations reach damaging levels and apply pesticides only when needed (Yarborough and Drummond, 2001). A method to predict the incidence and severity of *Monilina* (Delbridge et al., 1998) allowed growers in Maine to reduce the use of fungicide application by half in 2001 (Yarborough, unpublished data). The net result of this management is to reduce costs and to prevent loss in yield.

*Land improvement.* Land leveling with excavators removes rocks and smooths out knolls, which is required for in-ground irrigation. This procedure also allows for mechanical pruning and harvesting, thereby reducing the two most expensive costs of production (Yarborough et al., 2001). Nova Scotia has land leveled over 8,100 ha, New Brunswick, over 4,000 ha, Maine, over 3,000 ha, and PEI over 2,000 ha, and this process is still continuing. A surface mulch is also being used on some fields to increase rhizome spread and reduce soil erosion (Degomez and Smagula, 1990).

*Pruning.* Prior to 1980 almost all wild blueberry fields were pruned by burning using straw or a tractor-drawn oil burner. Research indicated that mowing to within a cm of the soil would produce equivalent yields to burning (Ismail and Yarborough, 1979). After the increase in oil prices following the OPEC oil embargo in the 1974, this pruning technique was widely adopted on all fields flat enough to properly mow. Although some field sanitation of surface insects and mummy berries are lost with mowing, the cost of pruning is substantially reduced compared to bruning (Yarborough et al., 2001), making production more efficient.

*Mechanical harvesting.* The quest to develop a mechanical harvester for wild blueberries has been going on since the 1950s (Dale et al.,

1994). A successful harvester was developed by the Bragg Lumber Company in Collingwood, NS in 1979 using an open reel head developed by the University of Maine. According to Doug Bragg (personal communication) there are 750 operating harvester heads. Sibley (1993) indicated a 92% recovery could be obtained using a properly adjusted and operated Bragg harvester. A recent wild blueberry production acreage analysis by Kevin Sibley (personal communication) revealed that only 15% of the acres in Maine are mechanically harvested, whereas 77% in Nova Scotia, 52% in PEI, 25% in New Brunswick, 5% in Quebec, and < 2% in Newfoundland are harvested mechanically. There is an ongoing effort to develop a smaller, more efficient harvester that produces fruit quality equivalent to hand-raking (Yarborough, 2002a). Since harvesting is the largest cost of production, the increased use of mechanical harvesters will reduce production costs and improve efficiency.

## CONCLUSIONS AND GROWER BENEFITS

Most of the increases in wild blueberry production did not come from the addition of newly developed acres, although New Brunswick, Nova Scotia, and PEI have significantly increased their producing acreage. Improved weed management allowed for the use of fertilizers, and substantial increases in the number of pollinators that have contributed to produce larger crops. More recently, improved pest management has reduced losses and the use of irrigation in Maine has produced substantially higher yields on irrigated fields. Land improvement has allowed for mechanical pruning and harvesting, thereby reducing the cost and improving the profitability of production. It is expected that as more growers adopt a more intensive management approach, yield and efficiency will continue to improve.

## LITERATURE CITED

Benoit, G.R., W.J. Grant, A.A. Ismail, and D.E. Yarborough. 1984. Effect of soil moisture and fertilizer on the potential and actual yield of lowbush blueberries. Can. J. Plant Sci. 64:683-689.
Buckman, H.O. and N.C. Brady. 1969. The nature and property of soils. 7th ed. The Macmillian Co., New York.

Dale, A., E. J. Hanson, D.E. Yarborough, R.J. McNicol, E.J. Stang, R. Brenan, J. Morris, and G.B. Hergert. 1994. Mechanical harvesting of berry crops. Hort. Rev. 16:255-382.

Dalton, T., A. Files, and D. Yarborough. 2002. Investment, ownership and operating costs of supplemental irrigation systems for Maine wild blueberries. Maine. Agr. For. Agr. Expt. Sta. Tech. Bul. 183.

Degomez, T. and J. Smagula. 1990. Mulching for improved plant cover. Wild Blueberry Fact Sheet No. 228. Univ. Maine Coop. Ext. Orono, Maine.

Delbridge, R. P. Hildebrand, and D. E.Yarborough. 1998. A method to control Monilina blight. Wild Blueberry Fact Sheet No. 217. Univ. Maine Coop. Ext. Orono, Maine.

Drummond, F.A. 2001. Honey bees and blueberry pollination. Wild Blueberry Fact Sheet No. 629. Univ. Maine Coop. Ext. Orono, Maine.

Gordon, R., L. Eaton, and D. McIsaac. 1994. The effect of a simple irrigation procedure on lowbush blueberry production. N.S. Dept. Agr. Fish. 94-35.

Hepler, P.R. and A.A. Ismail. 1985. The split block design: A useful design for extension and research in lowbush blueberries. HortScience. 20:735-737.

Hepler, P.R. and D.E. Yarborough. 1991. Natural variability in yield of lowbush blueberries. HortScience 26:245-246.

Holbein, P.J. 1991. NABC statistical record, 1990. North American Blueberry Council. Folsom, CA.

Ismail A.A. 1987. Honeybees and blueberry pollination. Coop. Ext. Serv. Bul. 629. Univ. Maine, Orono, Maine.

Ismail, A.A., J.M. Smagula, and D.E. Yarborough. 1981. Influence of pruning method, fertilizer, and terbacil on the growth and yield of the lowbush blueberry. Can. J. Plant Sci. 61:61-71.

Ismail, A.A. and D.E. Yarborough. 1979. Pruning lowbush blueberries–A review and update. Pp. 87-95. In: J.N. Moore (ed.) Proc. Fourth North Amer. Blueberry Res. Workers Conf. Univ. Ark. Fayeteville, AR.

Karmo, E.A. 1974. Blueberry pollination-problems, possibilities. N.S. Dept. Agr. Mkt. Hort. Bio. Serv. No. 109. Truro, N.S.

Litten, W., J.M. Smagula, and S. Dunham. 1977. Blueberry surprise from phosphorus. Maine Agr. Forest Expt. Sta. Misc. Rpt. 404. Univ. Maine, Orono, Maine.

Moore, J.N. 1994. The blueberry industry of North America. HortTechnology 4:96-102.

Patten, K.D. and J. Wang. 1994. Cranberry yield and fruit quality reduction caused by weed competition. HortScience 29:1127-1130.

Sibley, K.J. 1993. Effect of head-speed-ground-speed ratio on the picking effectiveness of a lowbush blueberry harvester. Can. Agr. Eng. 35(1):33-39.

Smagula, J.M. and W. Litten. 2002. Correcting lowbush blueberry boron deficiency with soil or foliar application. Acta Hort. 574:363-371.

Smagula, J.M. and D.E. Yarborough. 1999. Leaf and soil sampling procedures. Wild Blueberry Fact Sheet No. 222. Univ. Maine Coop. Ext. Orono, Maine.

Struchtemeyer, R.A. 1956. For larger yields irrigate lowbush blueberries. Maine Farm Res. Maine Agr. Expt. Sta. 4(2):7-18.

Stubbs, C.S., F.A. Drummond, and D.E. Yarborough. 2000. Field conservation management of native leafcutting and mason osmia bees. Wild Blueberry Fact Sheet No. 301. Univ. Maine Coop. Ext. Orono, Maine.

Stubbs, C.S., F.A. Drummond, and D.E. Yarborough. 2001. Commercial bumble bee (*Bombus impatiens*) management for wild blueberry pollination. Wild Blueberry Fact Sheet No. 302. Univ. Maine Coop. Ext. Orono, Maine.

Stubbs, C.S., F.A. Drummond, and D.E. Yarborough. 2002. How to manage alfalfa leaf cutting bees for wild blueberry pollination. Wild Blueberry Fact Sheet No. 300. Univ. Maine Coop. Ext. Orono, Maine.

Trevett, M.F. 1967. Irrigating lowbush blueberries the burn year. Maine. Farm Res. Maine Agr. Expt. Sta. 15(2):1-4.

Trinka, D. 1997. Production trends in the cultivated blueberry industry of North America. Acta Hort. 446:37-39.

Wood, G.W. 1961. The influence of honeybee pollination on fruit set of the lowbush blueberry. Can. J. Plant Sci. 41:332-335.

Yarborough, D.E. 1997. Production trends in the wild blueberry industry of North America. Acta Hort. 446:33-35.

Yarborough, D.E. 2002a. Progress towards the development of a mechanical harvester for wild blueberries. Acta Hort. 574:329-334.

Yarborough, D.E. 2002b. North American Blueberry Production. Wild Blueberry Newsletter, July 2002. Univ. Maine Coop. Ext. Orono, Maine.

Yarborough, D.E. and P.C. Bhowmik. 1988. Effect of hexazinone on weed populations and on lowbush blueberries in Maine. Acta Hort. 241:344-349.

Yarborough, D.E., T. Degomez, and A.L. Hoepler. 2001. Blueberry enterprise budget. Wild Blueberry Fact Sheet No. 260. Univ. Maine Coop. Ext. Orono, Maine.

Yarborough, D.E. and F.A. Drummond 2001. Integrated crop management field scouting guide for lowbush blueberries. Wild Blueberry Fact Sheet No. 204. Univ. Maine Coop. Ext. Orono, Maine.

Yarborough, D.E., J.J. Hanchar, S.P. Skinner, and A.A. Ismail. 1986. Weed response, yield and economics of hexazinone and nitrogen use in lowbush blueberry production. Weed Sci. 34:723-729.

Yarborough, D.E. and A.A. Ismail. 1985. Hexazinone on weeds on lowbush blueberry growth and yield. HortScience 20:406-407.

Yarborough D.E. and J.M. Smagula. 1993. Fertilizing with nitrogen and phosphorus. Wild Blueberry Fact Sheet No. 225. Univ. Maine Coop. Ext. Orono, Maine.

# Lowbush Blueberry
# (*Vaccinium angustifolium*)
# with Irrigated and Rain-Fed Conditions

R. M. Seymour
Gordon Starr
David E. Yarborough

**SUMMARY.** Growers desire a better understanding of the effect of irrigation on lowbush blueberry (*Vaccinium angustifolium*) crop yield and quality. A randomized block design of irrigated and rain-fed plots was established in a lowbush blueberry field near the coast of Maine. The irrigated plots averaged 43% higher yield than rain-fed plots for the 2000-2001 production cycle. Irrigated plots had the highest yield three weeks after the highest yield for the rain-fed plots. Rain-fed berries had greater firmness, higher Brix solids and lower moisture content than irrigated berries for three different harvest dates. Berry size distribution was

R. M. Seymour is Public Service Assistant, University of Georgia, Griffin Campus, Griffin, GA. Gordon Starr is Research Scientist with the USDA-ARS in Orono, ME. David E. Yarborough is Professor of Horticulture, University of Maine, Orono, ME.

Address correspondence to: Dr. R. M. Seymour, 1109 Experiment Street, Griffin, GA 30223 (E-mail: rseymour@griffin.peachnet.edu).

This study was funded by the Maine Agricultural Experiment Station, USDA-Cooperative State Research, Education and Extension Service and the Maine Wild Blueberry Commission.

[Haworth co-indexing entry note]: "Lowbush Blueberry (*Vaccinium angustifolium*) with Irrigated and Rain-Fed Conditions." Seymour, R. M., Gordon Starr, and David E. Yarborough. Co-published simultaneously in *Small Fruits Review* (Food Products Press, an imprint of The Haworth Press, Inc.) Vol. 3, No. 1/2, 2004, pp. 45-56; and: *Proceedings of the Ninth North American Blueberry Research and Extension Workers Conference* (ed: Charles F. Forney, and Leonard J. Eaton) Food Products Press, an imprint of The Haworth Press, Inc., 2004, pp. 45-56. Single or multiple copies of this article are available for a fee from The Haworth Document Delivery Service [1-800-HAWORTH, 9:00 a.m. - 5:00 p.m. (EST). E-mail address: docdelivery@haworthpress.com].

not significantly different for treatments, but for different harvest dates the berry size distributions were significantly different. *[Article copies available for a fee from The Haworth Document Delivery Service: 1-800-HAWORTH. E-mail address: <docdelivery@haworthpress.com> Website: <http://www. HaworthPress.com> © 2004 by The Haworth Press, Inc. All rights reserved.]*

**KEYWORDS.** Blueberry quality, yield, evapotranspiration

## *INTRODUCTION*

Lowbush blueberry growers in Maine are adopting many cultivation practices to enhance yields, prevent weather related losses and improve quality of their harvest. Since 1990 irrigated land in lowbush blueberry (*Vaccinium angustifolium*) production in Maine has increased 400% (Yarborough, 2002). At the same time that irrigated area is increasing, additional water supplies have not been easy to obtain. Competing water needs and environmental impacts have made acquiring suitable water supply for irrigation much more difficult. A better understanding of the difference irrigation can make in yields is desired by growers utilizing irrigation and those considering irrigation for their operations.

Struchtemeyer's (1953) results indicated that supplemental irrigation during berry filling in dry years increased yields and improved quality of the blueberries. Benoit et al. (1984) determined that irrigation during the bud formation year could result in increased yield during the crop year.

In highbush blueberries (*Vaccinium corymbosum* L.), Storlie and Eck (1996), and Haman et al. (1997) developed crop coefficients for irrigation scheduling. Because of the difference in cultural practices and morphology of the plants, highbush blueberry crop coefficients are not appropriate as estimates of the water needs of lowbush blueberries. For lowbush blueberries, the recommendation of 2.54 cm (1 in) per week was provided through a Maine farm research report by Trevett (1967). However, little information on the yield and quality effects of supplemental irrigation under improved production practices for lowbush blueberry in Maine is available.

This study reports measured yield and quality differences between lowbush blueberries with and without supplemental irrigation in Maine during the 2000-2001 production cycle.

## MATERIALS AND METHODS

The irrigation study was conducted at the Blueberry Hill Farm at Jonesboro, ME where the soil is a Colton sandy loam. The blueberry field where plots were established had been in production for over 60 years being harvested and burned regularly at 2-year intervals. The project began with spring 2000, after a fall burn pruning in 1999, and was carried on through the harvesting of berries in summer 2001.

The irrigated treatment had scheduling based on soil moisture measurements during the prune year and supplemental irrigation up to 2.54 cm (1 in) per week during the harvest year compared to a treatment without irrigation. All other cultural practices were the same for the entire field. Plots were established in a random block design with six replicated plots for each treatment. Each plot area was 12.2 m (40 ft) × 12.2 m (40 ft). The plot positions within the field were determined by the layout of the hand-move irrigation system pipes. The pipes were laid out at 12.2 m (40 ft) spacing with sprinkler risers spaced 12.2 m (40 ft) along the pipes, so four sprinkler risers delineated the corners of each square plot.

In the first year, coefficients of uniformity of irrigation in plots ranged from 67 to 80%. To compensate for the non-uniformity of the irrigation patterns, the gross application rate of water was 25% higher than the desired net application. For example, to apply 13 mm (0.50 in) irrigation net depth, 18 mm (0.64 in) gross amount was applied.

In the first season, irrigation scheduling began June 25. An irrigation application was initiated when soil moisture was approximately 50-70% of field capacity in the three inches of mineral soil just below the organic mat. This was determined by taking two soil samples from one randomly chosen irrigation plot and measuring soil moisture. When the average of the two samples fell below 70% of field capacity, irrigation would be initiated the next day. For each irrigation 13 mm (0.50 in) net was applied in the morning.

For the second year of the study, treatment plots from the previous year continued with the same treatment either irrigated or rain-fed. However, irrigations were scheduled by a different protocol than during the first year. The biomass in the spring 2001 was much larger than the previous year, so irrigation scheduling began earlier in this year, on June 1. An individual irrigation application would not exceed 13 mm (0.50 in) net to prevent excessive leaching below the root zone. Rainfall was the only weather parameter needed to decide whether or not to irrigate. The decision to irrigate was made on each Monday and Friday throughout the season. Therefore, if there was no rainfall between any two decision days, there would be one irrigation application of 13 mm (0.5 in). If there were rainfall events between any two decision days, then a needed irrigation amount was determined. The net irrigation amount would be the difference in 13 mm (0.5 in) and the rainfall amount since the previous decision day. If more than 13 mm (0.5 in) of rain fell between decision days, there would be no irrigation applied that day. This ensured that each week no less than one inch of water per week would be applied to the irrigated plots.

Complete weather data at the farm was collected from June 23 in the harvest year. Wind speed, wind direction, relative humidity, temperature, rainfall and solar radiation were measured. The data were collected every 15 minutes by a Campbell Scientific CR21X datalogger. The raw data were summarized to hourly and daily data. After the growing season, the computer program REF-ET (Allen, 2000) was used to calculate daily reference evapotranspiration (Ref-ET) from June 23 until the last harvest. The cumulative Ref-ET was compared with water applications and rainfall amounts to determine the fraction of the Ref-ET provided for both treatments.

Harvest samples were taken August 1st, 8th and 15th. A 1 m (3.3 ft) × 1 m (3.3 ft) area randomly chosen within each plot was harvested. The harvest samples were winnowed in the field immediately after being raked. Yield of berries was determined by weighing the winnowed harvest samples. From the total sample of each plot, a random sampling of 150 ml (2.5 oz) of berries was taken to determine size distribution and firmness of berries. All of the berries in the 150 ml sample were divided into 4 sizing groups (< 6 mm (0.2 in), 6-9 mm (0.2-0.3 in), 9-12 mm (0.3-0.5 in), and > 12 mm (0.5 in)). For each sub-sample, the number of berries in each size group was counted. For the first and second harvest dates, 30 berries from each of the three larger size groups for each treatment were tested for the maximum compression force required to break the skin of the berry. The tests for firmness of the sample berries were carried out according to ASAE Standard No. S368.4 (2001). The average maximum force to break the skin for each size group in each treatment was calculated. Two 10 g (0.35 oz) sub-samples were taken from each plot harvest sample to determine moisture content. These sub-samples were pureed and placed in a vacuum oven to remove the water and then the dry mass was measured. Brix solids test was also carried out on two additional pureed sub-samples from each plot to characterize solids content.

Statistical analysis consisted of ANOVA to determine if differences were significant ($\alpha = 0.05$) for yield and quality data. Pair wise comparison of treatment means for all harvest dates was carried out using Tukey's procedure (Steel and Torrie, 1980).

## RESULTS AND DISCUSSION

*Evapotranspiration and Water Applied.* Between May 15 to the first frost for the 2000 season (prune year), 21 cm (8.3 in) of rain fell and 2.9 cm (1.1 in) of additional water was applied to the irrigated plots. In the 2001 season, 16.5 cm (6.5 in) of rain fell and 13.1 cm (5.2 in) of additional water was applied to the irrigated treatment from May 15 to the final harvest date. Total rainfall in the harvest year was low because of dry weather with almost no rainfall after July 18 at the farm. Therefore, the difference in water available for the two treatments was greater in the harvest year.

In the harvest season, the average weekly Ref-ET was 3.3 cm (1.3 in) for the eight weeks calculated (June 23-August 18). The weekly Ref-ET ranged from 4.6 to 2.3 cm (1.8 to 0.9 in) during this time. For the rain-fed plots, the average water applied weekly was 0.8 cm (0.3 in) with a range of weekly total amounts from 0 to 2.9 cm (0 to 1.1 in). The irrigated plots averaged 2.4 cm (0.9 in) per week with a range of 1.3 to 3.4 cm (0.5 to 1.3 in) per week. Figure 1 shows cumulative weekly water applied for the two treatments with the calculated weekly Ref-ET for the 2001 season. Figure 2 shows the water applied to the treatments as a fractional amount of the Ref-ET. The irrigated plots received 73% of the total Ref-ET from irrigation and rain while the rain-fed plots received 24% of the total Ref-ET as rain.

*Harvest.* When the harvests for the three dates were averaged for each treatment, the average irrigated yield was 5921 kg/ha (5282 lb/ac) while the yield for rain-fed plots was 4132 kg/ha (3686 lb/ac) (Table 1). The average irrigated yield was 43% greater than the average rain-fed yields. While the irrigated yields were larger than the rain-fed yields on each of the three harvest dates, the difference in yields between the two treatments increased with each later harvest. Average yield for treatments on August 1 was 4909 kg/ha (4379 lb/ac) for the irrigated plots and 4826 kg/ha (4305 lb/ac) for the rain-fed plots, a difference that was not significant. A week later on August 8 the difference in irrigated and rain-fed yield was 1421 kg/ha (1268 lb/ac). While it was a greater dif-

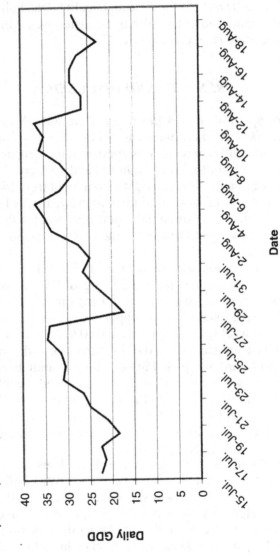

FIGURE 1. Cumulative irrigation and rain applied for the two treatments along with the reference evapotranspiration (Ref-ET) calculated for the Blueberry Hill Farm location.

FIGURE 2. Fractional amount of rain and irrigation applied for the two treatments in 2001.

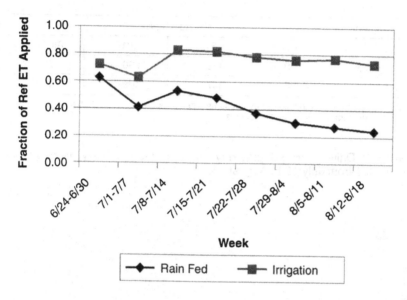

ference, it was not a significant difference. However, the third harvest on August 15 did have significantly different yields between treatments, with a difference of 3864 kg/ha (3447 lb/ac). While the irrigated yields increased dramatically for the third harvest date compared to the first two harvest dates, the rain-fed yields did not change significantly over the three harvest dates.

The ten days from August 1-10 had the highest ten-day accumulation of growing degree days (GDD) for the 2001 season. Consequently, the harvest on August 8 occurred at the end of the most stressful conditions of the season and had the lowest yields for both treatments. During these days the average daily GDD accumulation was 33 while the average daily accumulation for the month of July was 24. Figure 3 shows daily GDD for mid-July through mid-August.

Quality of the berries taken from the three harvest dates was also variable from harvest to harvest and between treatments. The berry size distribution is given in tabular form in Tables 2 and 3. The majority of the berries were always the 6-9 mm (0.2-0.3 in) berry size. In this size category, there was significant difference in the number of berries from harvest to harvest. Differences between treatments were not significant for this size category.

TABLE 1. Average yields for each of the harvest date and the average for all harvest dates for each of the treatments.

| Treatment | Yield kg/ha | | | |
|---|---|---|---|---|
| | 8/1 | 8/8 | 8/15 | All Dates |
| Irrigated | 4909a[z] | 4893a | 7961a | 5921 |
| Rain-fed | 4826a | 3472a | 4097b | 4132 |

[z]Values in the same column followed by the same letter do not differ at $\alpha \leq 0.05$ significance level. Different letters indicate significant difference.

FIGURE 3. Daily growing degree day values for the Blueberry Hill Farm location in 2001 from July 15 to August 18.

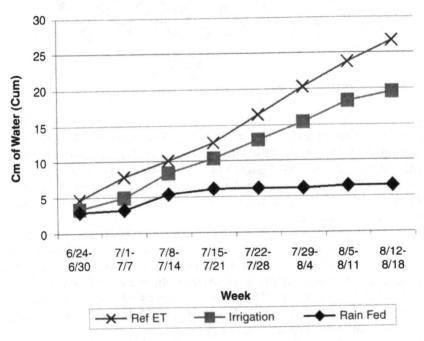

The number of berries 9-12 mm (0.3-0.5 in) size were less than the 6-9 mm (0.2-0.3 in) size number of berries for all treatments and harvest dates. The number of berries in the 9-12 mm (0.3-0.5 in) size category was not significantly different for the treatments or harvest dates. The number of berries > 12 mm (0.5 in) in diameter was so few that their contribution to yields was negligible.

TABLE 2. Berry size distribution for each of the harvest dates with treatments averaged together.

| Treatment | Number of Berries | | | |
|---|---|---|---|---|
| | < 6 mm | 6-9 mm | 9-12 mm | > 12 mm |
| 8/1 | 353a[z] | 742a | 524a | 9a |
| 8/8 | 157b | 990b | 521a | 17a |
| 8/15 | 367a | 909ab | 487a | 22a |

[z]Values in the same column followed by the same letter do not differ at α ≤ 0.05 significance level. Different letters indicate significant difference.

TABLE 3. Berry size distribution for the two treatments with harvests averaged together.

| Treatment | Number of Berries | | | |
|---|---|---|---|---|
| | < 6 mm | 6-9 mm | 9-12 mm | > 12 mm |
| Irrigated | 317a[z] | 860a | 512 | 16 |
| Rain-fed | 269a | 900a | 510 | 15 |

[z]Values in the same column followed by the same letter do not differ at α ≤ 0.05 significance level. Different letters indicate significant difference.

The berries < 6 mm (0.2 in) in size may have been unripe, diseased, damaged or just small ripe berries. There was no separation of the smallest berry sizes into the above categories. It was assumed that the majority of berries < 6 mm (0.2 in) would be removed in the sorting line. Treatment differences were not significant for this smallest size range, but the number of smallest size berries among harvests was significantly different. The August 8 harvest had the least number of berries in the smallest size range for both treatments. This would suggest that in the excessive heat in the week before this harvest more small berries were dropped from plants or removed with winnowing because they were shriveled or their density was very low.

Table 4 shows the results of blueberry firmness tests. The firmness was significantly different between the two harvest dates tested and between treatments. Berries from the first harvest required a lower maximum force to break the skin than the berries from the second harvest. Berries from the rain-fed treatment required a higher maximum force to break the skin than the irrigated berries for both harvests. The firmness results confirm expectations that irrigation will in general produce softer berries.

Moisture content of the berries harvested varied significantly from one harvest to the next and between treatments. Tables 5 and 6 show the results of moisture content measurement and Brix solids. The irrigated berries were higher in moisture content than the rain-fed berries for all harvests. Both treatments had a steady decline in moisture content of berries from the first harvest date to the last. The higher the average moisture content of berries from a treatment or harvest date, the lower the firmness values indicating that higher moisture content berries are

TABLE 4. Blueberry firmness measured by the force, in Newtons, to break the skin for he harvests of August 1 and 8.

| Treatment/Harvest Date | 8/1 | 8/8 |
|---|---|---|
| Irrigated | 2.55a$^z$ | 2.8a |
| Rain-fed | 3.77b | 4.42b |

$^z$Values followed by the same letter do not differ at P ≥ 0.05. Different letters indicate significant difference.

TABLE 5. Summary of average Brix and moisture content of blueberry fruit from irrigated and rain-fed plots.

| Treatment | Brix | Moisture Content |
|---|---|---|
| Irrigated | 5.70a$^z$ | 0.80a |
| Rain-fed | 7.42b | 0.71b |

$^z$Values in columns followed by the same letter do not differ at P ≥ 0.05. Different letters indicate significant difference.

TABLE 6. Summary of average Brix and moisture content measurements for the two treatments.

| Harvest Date | Brix | Moisture Content |
|---|---|---|
| 8/1 | 5.83a$^z$ | 0.80a |
| 8/8 | 6.67b | 0.77ab |
| 8/15 | 7.17b | 0.71b |

$^z$Values in columns followed by the same letter do not differ at α ≤ 0.05 significance. Different letters indicate significant difference.

more easily damaged from handling. However, the sample with the highest moisture content did not have the highest number of larger berries, so berry size distribution was not definitively related to firmness or moisture content.

The irrigated berries had lower Brix solids than rain-fed berries. The average Brix solids was 5.7 for the irrigated berries and 7.4 for the rain-fed berries. The harvest samples were significantly different, and the difference between treatments was significant as well. The first harvest had significantly lower solids content compared to the last two harvest dates which were not significantly different from each other.

## CONCLUSIONS

Because the difference in total water applied was much greater in the harvest year than the prune year for this study, the yield and quality differences result primarily from water availability differences in the harvest season. On average, the irrigated yields were 43% greater than the rain-fed yields. The yield differences between rain-fed and irrigated treatments became more pronounced as the harvest date was extended. The yields for the rain-fed plots were highest for the earliest harvest while the yields for the irrigated plots were highest for the latest harvest. The berry size distributions along with yield results indicate that for the irrigated plots a later harvest date was preferred, while the earliest harvest date was best for the rain-fed plots. Although yield was greater for the irrigated treatment, the rain-fed berries had greater firmness, higher Brix solids and lower moisture content than irrigated berries indicating a higher quality of berries for the rain-fed treatment. The berry size distributions were similar for both treatments although there was significant variation among harvest dates. For all harvest dates and both treatments, the majority of berries were 6-9 mm (0.2-0.3 in) in diameter.

Further studies must be conducted measuring crop water use directly and comparing the crop water use to Ref-ET to determine more accurate irrigation scheduling guidelines than was achieved in this project.

## GROWER BENEFITS

This study indicates that harvest timing should be managed differently for irrigated blueberry fields than rain-fed blueberries, particularly in dry years. Growers have a better understanding of yield differences and quality differences between irrigated and non-irrigated lowbush blueberries.

# LITERATURE CITED

Allen, Richard G. 2000. REF-ET evapotranspiration software. *www.kimberly.uidaho. edu* University of Idaho.

American Society of Agricultural Engineers. 2001. Compression test of food materials of convex shape. ASAE Standard No. 368.4. ASAE Standards. pp. 580-587.

Benoit, G.R., W.J. Grant, A.A. Ismail and D.E. Yarborough. 1984. Effect of soil moisture and fertilizer on the potential and actual yield of lowbush blueberries. Can. J. Plant Sci. 64:683-689.

Haman, D.Z., R.T. Pritchard, A.G. Smajstrla, F.S. Zazeuta and P.M. Lyrene. 1997. Evapotranspiration and crop coefficients for young blueberries in Florida. App. Eng. In Agric. ASAE. St. Joseph MI. 13:209-216.

Steel, R.G.D. and J.H. Torrie. 1980. Principles and Procedures of Statistics A Biometrical Approach. McGraw Hill, Inc. New York.

Storlie, C.A. and P. Eck. 1996. Lysimeter-based crop coefficients for young highbush blueberries. Hortscience 31:819-822.

Struchtemeyer, R.A. 1953. Operation rainfall plus. Univ. Maine Farm Res. 1:7-9.

Trevett, M.F. 1967. Irrigating lowbush blueberries the burn year. Univ. Maine Farm Res. 15:1-4.

Yarborough, D. 2002. Personal communication.

# Gypsum–
# An Alternative to Chemical Fertilizers
# in Lowbush Blueberry Production

Kevin R. Sanderson
Leonard J. Eaton

**SUMMARY.** Five experimental trials were established in commercial lowbush blueberry (*Vaccinium angustifolium* Ait.) fields in eastern Canada and monitored over two cropping cycles to determine the value of gypsum as a soil amendment. Lowbush blueberry plants were treated each cropping cycle with either: (1) no application (control), (2) 10-10-10 fertilizer applied at 300 kg/ha (268 lb/acre), (3) gypsum applied at 4 t/ha (3572 lb/acre), or (4) the combination of (2) and (3). In the first cropping

---

Kevin R. Sanderson is Research Scientist, Agriculture and Agri-Food Canada, Crops and Livestock Research Centre, 440 University Avenue, Charlottetown, PE, Canada, C1A 4N6. Leonard J. Eaton is Research Professor, Department of Environmental Sciences, Nova Scotia Agricultural College, P.O. Box 550, Truro, NS, Canada, B2N 5E3.

Address corresponding to: Kevin R. Sanderson (E-mail: sandersonk@agr.gc.ca).

The authors would like to thank Gordon Dickie, Shaw Resources Ltd., P.O. Box 60, Shubenacadie, NS, BON 2H0, Gary Brown, Bragg Lumber Co. Ltd., Collingwood, NS, B0M 1E0, and Agriculture and Agri-Food Canada Matching Investment Initiative for cooperation and financial support.

In addition, the authors gratefully acknowledge the five commercial blueberry growers in Prince Edward Island and Nova Scotia who provided research sites and management input and Thomas Gallant for his assistance in the manuscript preparation.

[Haworth co-indexing entry note]: "Gypsum–An Alternative to Chemical Fertilizers in Lowbush Blueberry Production." Sanderson, Kevin R., and Leonard J. Eaton. Co-published simultaneously in *Small Fruits Review* (Food Products Press, an imprint of The Haworth Press, Inc.) Vol. 3, No. 1/2, 2004, pp. 57-71; and: *Proceedings of the Ninth North American Blueberry Research and Extension Workers Conference* (ed: Charles F. Forney, and Leonard J. Eaton) Food Products Press, an imprint of The Haworth Press, Inc., 2004, pp. 57-71. Single or multiple copies of this article are available for a fee from The Haworth Document Delivery Service [1-800-HAWORTH, 9:00 a.m. - 5:00 p.m. (EST). E-mail address: docdelivery@haworthpress.com].

cycle, the application of gypsum and gypsum with fertilizer significantly increased tissue concentrations of N, P, K, Ca, Mn, and S in comparison to the control and fertilizer only treatments. Tissue K, Ca, Mn, and S were significantly increased in the second cropping cycle with gypsum application. Soil pH was reduced in the second cropping cycle with gypsum application. Gypsum with fertilizer increased stem length, live buds, total buds, and total blossoms in the first cropping cycle in comparison to the control. The treatments had no significant effect on marketable yield. *[Article copies available for a fee from The Haworth Document Delivery Service: 1-800-HAWORTH. E-mail address: <docdelivery@haworthpress. com> Website: <http://www.HaworthPress.com> © 2004 by The Haworth Press, Inc. All rights reserved.]*

**KEYWORDS.** *Vaccinium augustifolium* Ait., organic fertilizer, nutrient content, growth, yield

## INTRODUCTION

The lowbush blueberry (*Vaccinium augustifolium* Ait.) is native to North America (Vander Kloet, 1978) and is typically found on Podzolic soils (Canada Soil Survey Committee, 1978). This species prefers soils that are generally acidic (pH 3.5-5.5), infertile, and have well developed organic layers (Trevett, 1962). Natural genetic variability within fields appears to exert the greatest influence on fruit yields (Hepler and Yarborough, 1991). The cultural practices of pruning and selective herbicides are used to maintain the blueberry as the dominant plant in fields (Yarborough et al., 1986). Regular pruning forces the plant into a biennial production cycle, with maximum yields occurring in the second year after pruning (Jordan and Eaton, 1995).

The introduction of selective herbicides has encouraged producers to apply chemical fertilizers in the hopes of increasing production (Yarborough et al., 1986), even though lowbush blueberry plant and yield responses to fertilizer applications have been inconsistent in short-term research studies (Eaton, 1988; Eaton and Patriquin, 1988; Warman, 1987). Sanderson (1983) and Cutcliffe and Sanderson (1984) demonstrated substantially increased yields of lowbush blueberry with a complete fertilizer (10-10-10). Sanderson and Ivany (1994) and Smagula and Dunham (1997) produced significant effects on lowbush blueberry yield with diammonium phosphate. In contrast, Warman (1987) demonstrated no beneficial effects on yield or growth characteristics of the lowbush blueberry treated with various manures and fertilizers, and Eaton et al. (1997) reported no increased yields with several rates of a P fertilizer.

A long-term study in Nova Scotia demonstrated increases in stem length and fruit bud numbers after three or more cycles in which an N-P-K fertilizer was applied in conjunction with herbicide (Eaton, 1994). Eaton (1994) stated, however, that any increases in vegetative and reproductive characteristics of the lowbush blueberry can be more affected by herbicide application than from fertilizer amendments, although the latter did produce some beneficial results over a 12 year period. Long-term, repeated applications of fertilizer appear to be of limited benefit to the crop and result in a buildup of P and K which may be detrimental to wild blueberry soils (Eaton et al., 1998). The overall indication is that the application of fertilizer to the lowbush blueberry often produces inconsistent results in terms of yield.

In recent years, there has been an increasing interest in adopting organic farm practices, with the use of natural elements being favored over chemical fertilizers (Browne et al., 2000; Doing, 1997). One such substance, gypsum, a material on the Permitted Materials List for organic crop production (Canadian General Standards Board, 1999), is a common soil amendment used to supply Ca to the soil (Singh et al., 1989). This soil management practice has been applied to a number of agricultural crops (Carter and Cutcliffe, 1990; Shainberg et al., 1989), including the lowbush blueberry (Barber, 1984). Sanderson et al. (1996a) reported that a single application of gypsum can increase blueberry yield.

It has been reported that yield increases of the lowbush blueberry are possible with gypsum application, and further research is warranted to explore the potential benefits of this product (Chaisson and Argall, 1996). Sanderson et al. (1996a) suggest that Ca may be a limiting factor in blueberry proliferation because soils where this species grows is characteristically low in exchangeable ions.

The objective of this study was to assess the effect of gypsum on nutrient uptake, plant growth, and production of the lowbush blueberry, and to compare similarities and differences between gypsum (a material on the Permitted Materials List for organic crop production) and standard chemical fertilizers in terms of lowbush blueberry growth and production.

## MATERIALS AND METHODS

Five experimental sites representative of commercial *V. angustifolium* stands were evaluated for two cropping cycles (1998 to 2001). Experimental sites were located in Belle River (62°48'W, 46°02'N) and Mount Stewart (62°53'W, 46°24'N), Prince Edward Island (PE) and Tower (62°59'W, 45°28'N; Mt. Thom), Fern Walker (62°59'W, 45°30'N; Mt. Thom), and Debert (63°27'W, 45°26'N), Nova Scotia (NS). Both of the Prince Edward Island sites have Culloden soils that are well drained and consist of loamy sand to sandy loam (MacDougall et al., 1988). The soils at the Nova Scotia sites are: at Debert, a Truro sandy loam to loamy sand, at Fern Walker, a Thom gravelly sandy loam, and at Tower, a Kirk Hill gravelly sandy loam (Webb et al., 1991).

The treatments applied were: (1) control (no fertilizer or gypsum); (2) 10-10-10 fertilizer at 300 kg/ha (268 lb/acre); (3) gypsum at 4 t/ha (3572 lb/acre); and (4) 10-10-10 fertilizer at 300 kg/ha (268 lb/acre) plus gypsum at 4 t/ha (3572 lb/acre). The 10-10-10 fertilizer was commercial grade and the gypsum was agricultural grade (Ca = 22%; S = 18%; 100% pass-through with a 2 mm screen; 60% pass-through a 0.15 mm screen). The treatment plots were 3 m × 8 m (9.8 ft × 26.2 ft) in size with a 1 m (3.3 ft) border and arranged in a randomized complete block design with four replicates at each site. The fertilizer and gypsum were broadcast-applied by hand on the soil surface in early to mid-May of the sprout year in 1998 and reapplied to the same plots in the sprout year of the second cropping cycle in the spring of 2000. All experimental sites were managed according to a commercial two-year production cycle and pesticides were applied by individual growers as necessary.

Vegetative growth was estimated from samples collected from each site in early spring (April/May) of the crop year (1999 and 2001). Blueberry stems were obtained by clipping 50 stems per plot at ground level along a 10 m line transect; total stem length, buds and blossoms were assessed. During August of the crop year, yield data was obtained by hand raking the entire plots when fruit were 95-100% mature.

Soil and leaf samples were taken in the sprout year of each cropping cycle at all sites. Leaf samples of *V. angustifolium* plants were randomly collected from all plots when approximately 90% of the stems had reached the terminal abortion stage (Hall et al., 1979), and were obtained by taking the top 5 to 10 leaves from approximately 50 stems in each plot. Samples were dried at 80°C, ground to 1 mm, ashed at 500°C, and extracted in 2 M HCl. The extract was centrifuged and concentrations of P, K, Ca, Mg, Cu, Zn, Mn, Fe, B, and S were determined in the supernatant using ICAP 1100 (Thermo Jarrell Ash Corp., Watham, MA). The N content of plant tissue was determined by gas analysis of the combustion stream using a LECO 2000 CNS analyzer (LECO Corp., St. Joseph, MI) (Windham, 1997).

Soil samples were taken with a hand held soil probe to a depth of 15 cm from all plots at timing similar to leaf samples. Samples were air dried and passed through a 2 mm sieve prior to analysis. Soil pH was determined with a 1:1 soil to water ratio. Soil nutrient concentration of P, K, Ca, Mg, Cu, Zn, B, Fe, Mn, and S were extracted with Mehlich III (Tran and Simard, 1993) and the supernatant analyzed using ICAP 1100.

The data were statistically analyzed according to the randomized complete block design. The treatment sums of squares were partitioned into contrasts (Steel and Torrie, 1960) of the mean of the no gypsum (1, 2) vs. gypsum (3, 4) treatments, and mean of the no fertilizer (1, 3) vs. fertilizer (2, 4) treatments. All calculations were done within the ANOVA directive of the statistical programming language of GENSTAT 5 (Genstat 5 Committee, 1987). Statistical tests were evaluated at $P = 0.05$. The five sites were representative of commercial wild blueberry fields, and therefore considered to be random. Analysis of preliminary experiments determined that there were no significant site by treatment effects and negligible provincial effects (data not shown), therefore results are presented as combined data across experimental locations (Steel and Torrie, 1960).

## RESULTS

### Tissue Nutrient Levels

*Cycle 1–1998 Sprout Year.* Tissue K was the only leaf nutrient significantly increased by all treatments in comparison to the control (Table 1). The gypsum and gypsum with fertilizer treatments significantly increased tissue N, P, K, Ca, Mn, and S, and significantly decreased Mg and Fe in comparison to the control and fertilizer treatments; Ca was not significantly different from the control (Table 1). Fertilizer and gypsum with fertilizer significantly reduced tissue B concentration in comparison to the control and gypsum treatments. Applying gypsum with fertilizer significantly increased tissue P in comparison to the other treatments. The tissue concentration of samples taken from plots with fertilizer

TABLE 1. Nutrient content of wild blueberry tissue samples as affected by gypsum and/or fertilizer application with data combined from all experimental sites

| Treatment | N | P | K | Ca | Mg | S | Mn | Zn | Cu | Fe | B |
|---|---|---|---|---|---|---|---|---|---|---|---|
| | % | | | | | | ppm | | | | |
| **Cycle 1–1998 Sprout Year** | | | | | | | | | | | |
| 1. Control[z] | 1.59 | 0.121 | 0.47 | 0.47 | 0.14 | 0.13 | 321 | 6.7 | 3.1 | 20.0 | 30.0 |
| 2. Fertilizer[y] | 1.61 | 0.123 | 0.50 | 0.46 | 0.14 | 0.14 | 301 | 7.4 | 2.8 | 22.2 | 26.1 |
| 3. Gypsum[x] | 1.73 | 0.134 | 0.57 | 0.53 | 0.12 | 0.37 | 435 | 7.0 | 3.0 | 15.8 | 29.5 |
| 4. Gypsum with Fertilizer[w] | 1.74 | 0.140 | 0.58 | 0.49 | 0.12 | 0.34 | 400 | 6.9 | 2.6 | 17.7 | 27.4 |
| Significant Contrasts[v] (n = 40; df = 12) | g | f, g | g | g | g | g | g | NS | f | f, g | f |
| LSD (n = 20; df = 12) | 0.06 | 0.005 | 0.03 | 0.03 | 0.01 | 0.04 | 57 | NS | 0.3 | 2.4 | 2.4 |
| **Cycle 2–2000 Sprout Year** | | | | | | | | | | | |
| 1. Control[z] | 1.71 | 0.122 | 0.47 | 0.41 | 0.16 | 0.12 | 438 | 12.4 | 4.3 | 23.3 | 28.2 |
| 2. Fertilizer[y] | 1.82 | 0.132 | 0.52 | 0.41 | 0.16 | 0.12 | 436 | 13.5 | 4.1 | 44.9 | 26.2 |
| 3. Gypsum[x] | 1.77 | 0.132 | 0.58 | 0.50 | 0.15 | 0.36 | 632 | 13.9 | 4.4 | 21.5 | 28.2 |
| 4. Gypsum with Fertilizer[w] | 1.88 | 0.142 | 0.62 | 0.46 | 0.15 | 0.35 | 587 | 13.7 | 4.2 | 22.2 | 26.8 |
| Significant Contrasts[v] (n = 40; df = 12) | f, g | f, g | f, g | g | g | g | g | g | NS | NS | f |
| LSD (n = 20; df = 12) | 0.07 | 0.006 | 0.05 | 0.04 | 0.01 | 0.09 | 59 | NS | NS | NS | NS |

[z]Control = no fertilizer or gypsum. [y]Fertilizer = 300 kg ha⁻¹ 10-10-10. [x]Gypsum = applied at 4 t ha⁻¹. [w]Gypsum with fertilizer = 4 t ha⁻¹ gypsum + 300 kg ha⁻¹ 10-10-10. [v]f = significant no fertilizer (1,3) vs. fertilizer (2,4) contrast; g = significant no gypsum (1,2) vs. gypsum (3,4) contrast; $P = 0.05$; NS = not significant.

were increased in P (0.003 ± 0.0010%) and Fe (2.1 ± 0.55 ppm) and decreased in Cu (0.3 ± 0.07 ppm) and B (3.0 ± 0.55 ppm) in comparison to those which did not receive fertilizer (Table 1). Samples taken from the plots amended with gypsum were increased in N (0.14 ± 0.014%), P (0.015 ± 0.0010%), K (0.09 ± 0.008%), Ca (0.04 ± 0.008%), Mn (0.02 ± 0.003%), and S (0.22 ± 0.009%), while decreased in Mg (0.02 ± 0.003%) and Fe (4.3 ± 0.55 ppm) in comparison to those without fertilizer.

*Cycle 2–2000 Sprout Year.* Tissue P and K concentrations were significantly increased by all treatments when compared to the control; but the highest concentrations were found in the gypsum + fertilizer treated plants (Table 1). Gypsum and gypsum with fertilizer significantly increased tissue Ca, Mn, and S; tissue Mg was decreased compared to the control and fertilizer treatment. The tissue concentration of samples taken from plots with fertilizer were increased in N (0.11 ± 0.017%), P (0.010 ± 0.0014%), and K (0.04 ± 0.011%), while decreased in B (1.8 ± 0.46 ppm). The tissue concentration of samples taken from plots with gypsum were increased in N (0.06 ± 0.017%), P (0.010 ± 0.0014%), K (0.10 ± 0.011%), Ca (0.07 ± 0.008%), Mg (0.01 ± 0.003%), Mn (173 ± 13.5 ppm), Zn (0.9 ± 0.29 ppm), and S (0.23 ± 0.021%) in comparison to plots that did not receive gypsum.

Tissue nutrient concentrations were affected in the year of application, with the exception of Zn in 1998 and Fe and Cu in 2000. In the year of application, gypsum applied alone significantly increased tissue N, P, K, Ca, Mn, and S in comparison to the control, except for N in 2000. When gypsum was applied with fertilizer, tissue N, P, Ca, Mn, and S were increased in both sprout year samples when compared to the fertilizer alone treatment. The gypsum with fertilizer treatment combination significantly reduced tissue Ca compared to the application of gypsum alone.

### Soil Nutrient Levels

*Cycle 1–1998 Sprout Year.* Soil P was significantly increased with application of gypsum with fertilizer in comparison to the control (Table 2). Soil K was significantly increased by the fertilizer treatment in comparison to the gypsum treatment. Gypsum and gypsum with fertilizer significantly increased soil Ca and S concentration and decreased soil Mg when compared to the control and fertilizer treatments. Gypsum with fertilizer treatments decreased soil Zn in comparison to the

TABLE 2. Effect of gypsum and fertilizer on wild blueberry soil nutrients (ppm) and pH with data combined from all experimental sites

| Treatment | P | K | Ca | Mg | Zn | Mn | S | pH |
|---|---|---|---|---|---|---|---|---|
| | | | ---------- | ppm | ---------- | | | |
| Cycle 1–1998 Sprout Year | | | | | | | | |
| 1. Control[z] | 50 | 51 | 260 | 51 | 3.1 | 32 | 38 | 4.57 |
| 2. Fertilizer[y] | 56 | 56 | 256 | 47 | 2.9 | 34 | 41 | 4.56 |
| 3. Gypsum[x] | 50 | 47 | 400 | 33 | 2.4 | 27 | 61 | 4.56 |
| 4. Gypsum with Fertilizer[w] | 61 | 54 | 439 | 37 | 2.4 | 27 | 57 | 4.53 |
| Significant Contrasts[v] (n = 40; df = 12) | f | f | g | g | g | g | g | NS |
| LSD (n = 20; df = 12) | 7 | 6 | 86 | 9 | 0.6 | 7 | 11 | NS |
| | | | | | | | | |
| Cycle 2–2000 Sprout Year | | | | | | | | |
| 1. Control[z] | 44 | 51 | 228 | 53 | 3.3 | 30 | 43 | 4.54 |
| 2. Fertilizer[y] | 53 | 70 | 242 | 52 | 3.0 | 33 | 42 | 4.52 |
| 3. Gypsum[x] | 45 | 49 | 692 | 43 | 2.2 | 23 | 271 | 4.31 |
| 4. Gypsum with Fertilizer[w] | 55 | 58 | 645 | 42 | 2.0 | 25 | 222 | 4.28 |
| Significant Contrasts[v] (n = 40; df = 12) | f | f, g | g | g | g | g | g | g |
| LSD (n = 20; df = 12) | 9 | 7 | 201 | 9 | 0.7 | NS | 101 | 0.07 |

[z]Control = no fertilizer or gypsum. [y]Fertilizer = 300 kg ha$^{-1}$ 10-10-10. [x]Gypsum = applied at 4 t ha$^{-1}$. [w]Gypsum with fertilizer = 4 t ha$^{-1}$ gypsum + 300 kg ha$^{-1}$ 10-10-10. [v]f = significant no fertilizer (1,3) vs. fertilizer (2,4) contrast; g = significant no gypsum (1,2) vs. gypsum (3,4) contrast; $P$ = 0.05; NS = not significant.

control. Soil Cu, B, and Fe ranged from 0.2 to 0.7, 0.1 to 0.3, and 163 to 315 ppm, respectively (data not shown), while pH (Table 2) was not affected by treatment. The plots with fertilizer showed an increase in P (8.4 ± 1.70 ppm) and K (6.3 ± 1.40 ppm). The plots with gypsum showed an increase in Ca (162 ± 19.7 ppm), and S (19.1 ± 2.63 ppm), while decreases were noted for Mg (14.1 ± 2.12 ppm), Zn (0.6 ± 0.13 ppm), and Mn (6.1 ± 1.68 ppm) (Table 2).

*Cycle 2–2000 Sprout Year.* Soil P and K were significantly increased by the fertilizer and gypsum with fertilizer treatments in comparison to the control and gypsum treatment (Table 2). Gypsum and gypsum with fertilizer significantly increased soil Ca and S, and significantly decreased Mg and pH; both in comparison to the control and fertilizer treatments. The plots with fertilizer were increased in soil P (9.8 ± 2.01 ppm) and K (13.8 ± 1.54 ppm) levels in comparison to those which did not receive fertilizer. The plots that received gypsum were increased in Ca (433 ± 46.0 ppm) and S (204 ± 23.2 ppm), while decreased in K (6.8 ± 1.54 ppm), Mg (9.7 ± 2.06 ppm), Zn (1.1 ± 0.16 ppm), Mn (7.5 ± 2.00 ppm), and pH (0.24 ± 0.016) (Table 2). Overall, fertilizer application increased soil P and K, while gypsum application increased soil Ca and S, but decreased soil Mg, Zn, Mn, and pH (Table 2).

## Yield and Growth

*Cycle 1–1999 Crop.* Marketable yield in the first cropping cycle ranged from 6.03 to 6.51 t ha$^{-1}$ (2.69 to 2.90 t acre$^{-1}$) and was not affected by treatment (Table 3). The gypsum with fertilizer treatment significantly increased stem length compared to all other treatments, and the number of live buds, total buds, and total blossoms per stem in the first cropping cycle in comparison to the control (Table 3). All treatments with fertilizer significantly increased total number of buds per stem compared to the control. The contrasts of no fertilizer vs. fertilizer and no gypsum vs. gypsum were significant for stem length in the first cropping cycle; the former contrast significant for number of live, total buds, and total blossoms per stem. When comparing plots with fertilizer in the first cropping cycle, there was an increase in stem length (2.2 ± 0.56 cm) (0.9 ± 0.22 in), live buds per stem (0.9 ± 0.22), total buds per

TABLE 3. Effect of gypsum and fertilizer treatments on blueberry yield (t ha$^{-1}$) and plant growth parameters (mean amounts per stem) during the first and second cropping cycles

| Treatment | Total Yield | Stem Length | Live Buds | Dead Buds | Total Buds | Total Blossoms |
|---|---|---|---|---|---|---|
| | t ha$^{-1}$ | cm | ------------------------ # ------------------------ | | | |
| Cycle 1–1998 Sprout Year | | | | | | |
| 1. Control[z] | 6.03 | 18.3 | 4.4 | 0.2 | 4.5 | 23.3 |
| 2. Fertilizer[y] | 6.51 | 19.9 | 5.3 | 0.4 | 5.7 | 27.4 |
| 3. Gypsum[x] | 6.03 | 19.6 | 5.0 | 0.4 | 5.4 | 23.9 |
| 4. Gypsum with Fertilizer[w] | 6.49 | 22.4 | 5.8 | 0.3 | 6.2 | 31.7 |
| Significant Contrasts[v] | | | | | | |
| (n = 40; df = 12) | NS | f, g | f | NS | f | f |
| LSD (n = 20; df = 12) | NS | 2.5 | 1.0 | NS | 1.1 | 6.0 |
| | | | | | | |
| Cycle 2–2000 Sprout Year | | | | | | |
| 1. Control[z] | 4.69 | 13.9 | 3.3 | 0.1 | 3.4 | 17.0 |
| 2. Fertilizer[y] | 4.37 | 15.9 | 3.3 | 0.2 | 3.5 | 17.8 |
| 3. Gypsum[x] | 4.51 | 14.3 | 4.7 | 0.2 | 4.8 | 18.1 |
| 4. Gypsum with Fertilizer[w] | 4.36 | 17.5 | 3.5 | 0.2 | 3.7 | 19.3 |
| Significant Contrasts[v] | | | | | | |
| (n = 40; df = 12) | NS | f, g | NS | NS | NS | NS |
| LSD (n = 20; df = 12) | NS | 1.1 | NS | NS | NS | NS |

[z]Control = no fertilizer or gypsum. [y]Fertilizer = 300 kg ha$^{-1}$ 10-10-10. [x]Gypsum = applied at 4 t ha$^{-1}$. [w]Gypsum with fertilizer = 4 t ha$^{-1}$ gypsum + 300 kg ha$^{-1}$ 10-10-10. [v] f = significant no fertilizer (1,3) vs. fertilizer (2,4) contrast; g = significant no gypsum (1,2) vs. gypsum (3,4) contrast; $P$ = 0.05; NS = not significant.

stem (0.9 ± 0.24) and total blossoms per stem (5.9 ± 1.37) compared to plots without fertilizer. With gypsum application, there was an increased stem length (1.9 ± 0.56 cm) (0.7 ± 0.22 in) in comparison to plots without gypsum.

*Cycle 2–2001 Crop.* In the second cropping cycle, stem length was the only parameter significantly affected by treatment; significantly increased by 2.6 ± 0.25 cm (1.0 ± 0.10 in) and 1.0 ± 0.25 cm (0.4 ± 0.10 in) by the two contrasts, respectively (Table 3). There were no significant treatment effects on total number of buds per stem in the second cropping cycle. Yield in the second cropping cycle was not significantly affected by treatment and ranged from 4.36 to 4.69 t/ha (1.94 to 2.09 t/ acre) (Table 3).

## DISCUSSION

Gypsum has been demonstrated to increase leaf nutrient levels in a number of agricultural crops (Evanylo, 1989; Shainberg et al., 1989), including Brussel sprouts (Sanderson and Carter, 1993) and lowbush blueberries (Sanderson et al., 1996a; Sanderson and Ivany, 1996; 2000). In the present study, applications of gypsum and gypsum + fertilizer produced similar results. Gypsum application increased leaf tissue P, K, Ca, Mn, and S in both production cycles, and decreased leaf Mg. Gypsum + fertilizer increased N, P, K, Mn, and S and reduced leaf Mg in both production cycles.

Shainberg et al. (1989) postulated that tissue nutrient benefits from gypsum application may be due to antagonistic/agonist effects of Ca in the leaf tissue, while Souza and Ritchey (1986) suggest that increased root development from gypsum application results in increased nutrient uptake from the soil. The former explanation has been supported by Nakamura et al. (1990), who indicate that adequate amounts of Ca can improve plant growth by acting as an antagonist in the uptake of K and P; thus improving the nutritional status of the plant. The alteration of soil pH with Ca can improve P nutrition in the low pH soils of lowbush blueberry fields (Warman, 1987). Increases in nutrient availability from gypsum application is also possible because decomposition of organic matter is enhanced with gypsum application through heightened microbiological activity in the soil (Subramani and Scifres, 1988). Increased breakdown of organic matter may be a substantial factor with the lowbush blueberry as their soils characteristically exhibit a slow rate of organic material decomposition (Stevenson, 1986).

Gypsum application in the first sprout year (1998) significantly decreased soil Mg and Zn; Ca and S were increased when applied in both sprout years (1998 and 2000). The application of gypsum with fertilizer in the second sprout year (2000) significantly increased soil K in comparison to the control and gypsum applied alone. Other authors have shown that, in comparison to soil amendment with lime, gypsum releases a substantial quantity of Ca and evenly distributes it throughout the soil profile due to its high solubility (Carter et al., 1986; Evanylo, 1989). Gypsum and gypsum with fertilizer increased soil S when applied in both sprout years (1998 and 2000). Farina and Channon (1988) reported enhanced S nutrition following gypsum application to maize and alfalfa, and proposed that this elevation could result in significant crop yield increases. Similarly, Sanderson et al. (1996b) reported enhanced S nutrition in cabbage following gypsum application.

Our work has shown that gypsum application, alone or combined with fertilizer, decreases soil Mg and Zn. This result is in accordance with research by Evanylo (1989), Farina and Channon (1988), and McCray and Sumner (1990) who found that applying gypsum under acidic soil conditions can alleviate the phytotoxic effects of elevated levels of elements such as Zn. It has been shown that reducing the amount of phytotoxic elements in the soil, in combination with greater Ca availability, can induce greater root growth (McCray and Sumner, 1990).

The application of gypsum decreased soil pH after the second application in 2000 in our study, similar to the results of others (Carter et al., 1986). Hall et al. (1964) proposed that soil amendments that lower pH would provide a more productive soil environment for *V. angustifolium*. The effects of gypsum on soil pH, however, can be variable as Sanderson et al. (1996a) and Sanderson and Ivany (2000) report that gypsum application only slightly reduces pH.

In this study, fertilizer applications to lowbush blueberry significantly increased stem length, live and total buds per stem, and total blossoms per stem. With gypsum application, increases were noted for stem length only. Statistically significant results were not shown from treatments of fertilizer or gypsum in terms of marketable yield. These results contrast with those of Sanderson et al. (1996a), who reported that the yield of lowbush blueberry increased in the first cropping cycle (47%) with gypsum applied at a rate of 4 t/ha broadcast-applied in the spring of the sprout year. Other crop species, such as cabbage, also show increased yield from gypsum application (Sanderson et al., 1996b). The lowbush blueberry may have exhibited increased soil and tissue nutrient content as a result of gypsum-induced improvements in soil structure. Generally, the use of gypsum as a soil treatment benefits the crop through more efficient water infiltration and stand development (Shainberg et al., 1989). With gypsum present in the soil horizon, roots increase in proliferation and depth and are able to extract an increased quantity of water from the soil; thus they are better able to withstand periods of drought (Evanylo, 1989; Shainberg et al., 1989). However, the effects of water availability on the lowbush blueberry may be small as the water requirement of this species is low in comparison to other crops, and as such, is well adapted to dry conditions (Chaisson and Argall, 1996).

## CONCLUSION AND GROWER BENEFITS

The results of this study indicate that gypsum application is an effective method to increase nutrient uptake in the wild blueberry. When used in conjunction with fertilizer, additional levels of N and P were observed in tissue samples. While the present study demonstrated little benefit to fruit production, increased nutrient levels should provide the plant with a higher yield potential, and possibly greater yields over the long term.

As a common practice, lowbush blueberry producers traditionally use chemical fertilizers, although research has indicated inconsistent yield response from this practice. Previous authors have reported varied yield response to gypsum application; supported by the current work. Nevertheless, if the objective of producers is to improve plant health, gypsum amendments can induce more efficient uptake of nutrients with or without fertilizer application. In addition, gypsum is a permitted materials listed product for organic crop production and would be an useful addition to organic management of the lowbush blueberry.

## LITERATURE CITED

Barber, S. A. 1984. Soil nutrient bioavailability, A mechanistic approach. Wiley, New York.

Browne, A. W., P. J. C. Harris, A. H. Hofny-Collins, N. Pasiecznik, and R. R. Wallace. 2000. Organic production and ethical trade: Definition, practice and links. Food Policy 25:69-89.

Canada Soil Survey Committee. 1978. The Canadian system of soil classification. Can. Dept. of Agr. Publ., Supply and Services Canada 1646.

Canadian General Standards Board. 1999. Organic Agriculture. Natl. Stnd. of Canada CAN/CGSB-32.310-99.

Carter, M. R. and J. A. Cutcliffe. 1990. Effects of gypsum on growth and mineral content of Brussels sprouts, and soil properties of Orthic Podzols. Fert. Res. 24:77-84.

Carter, M. R., J. R. Pearen, P. G. Karkansis, R. R. Cairns, and D. W. McAndrew. 1986. Improvements of soil properties and plant growth on a brown solonetzic soil using irrigation, calcium amendments and nitrogen. Can. J. Soil Sci. 66:581-589.

Chaisson, G. and J. Argall. 1996. Growth and development of the wild blueberry. Wild blueberry fact sheet A.2.0. Wild blueberry production guide. N. B. Dept. Agri. Rural Dev., Fredericton, N.B.

Cutcliffe, J. A. and K. R. Sanderson. 1984. Effect of Velpar and fertilizer on yields of native lowbush blueberries in Prince Edward Island. Agr. Agri-Food Can. Hort. Crops Canadex 541:235.641.

Doing, H. 1997. The landscape as an ecosystem. Agr., Ecosystems Environ. 63: 221-225.

Eaton, L. J. 1988. Nitrogen cycling in lowbush blueberry stands. PhD Diss., Dalhousie Univ., Halifax, N.S.

Eaton, L. J. 1994. Long term effects of herbicides and fertilizers on lowbush blueberry growth and production. Can. J. Plant Sci. 74:341-345.

Eaton, L. J. and D. G. Patriquin. 1988. Inorganic nitrogen levels and nitrification potential in lowbush blueberry. Can. J. Soil Sci. 68:63-75.

Eaton, L. J., G. W. Stratton, and K. R. Sanderson. 1997. Fertilizer phosphorus in lowbush blueberries: Effects and fate. Acta. Hortic. 446: 447-486.

Eaton, L. J., K. R. Sanderson, and G. W. Stratton. 1998. Short- and long-term consequences of repeated fertilizer applications to wild blueberry fields. Proc. VIII. North Amer. Blueberry Res. Ext. Workers Conf. Wilmington, NC. pp. 161-168.

Evanylo, G. K. 1989. Amelioration of subsoil acidity by gypsum. Veg. Growers News 44:3-4.

Farina, M. P. W. and P. Channon. 1988. Acid-subsoil amelioration. II. Gypsum effects on growth and subsoil chemical properties. Soil Sci. Soc. Amer. J. 52:175-180.

Genstat 5 Committee. 1987. Genstat 5 reference manual. Oxford Univ. Press, New York.

Hall, I. V., L. E. Aalders, and L. R. Townsend. 1964. The effects of pH on the mineral composition and growth of the lowbush blueberry. Can. J. Plant. Sci. 44:433-438.

Hall, I. V., L. E. Aalders, N. L. Nickerson, and S. P. van der Kloet. 1979. The biological flora of Canada. 1. *Vaccinium angustifolium* Ait., sweet lowbush blueberry. Can. Field Nat. 93:415-430.

Hepler, P. R. and D. E. Yarborough. 1991. Natural variability in yield of lowbush blueberries. HortScience 26:245-246.

Jordan, W. C. and L. J. Eaton. 1995. A comparison of first and second cropping years of Nova Scotia blueberries (*Vaccinium angustifolium* Ait.). Can. J. Plant Sci. 75: 703-707.

MacDougall, J. I., C. Veer, and F. Wilson. 1988. Soils of Prince Edward Island: Prince Edward Island Soil Survey. Agr. Can. LRRC Contrib., No. 84-54.

McCray, J. M. and M. E. Sumner. 1990. Assessing and modifying Ca and Al levels in acid subsoils. Pp. 47-75. In: B.A. Stewart (ed.). Adv. Soil Sci., Springer-Verlag, New York.

Nakamura, Y., K. Tanaka, E. Ohta, and M. Sakata. 1990. Protective effects of external $Ca^{2+}$ on elongation and intracellular concentration of $K^+$ in intact mung bean roots under high NaCl stress. Plant Cell Physiol. 31:815-821.

Sanderson, K. R. 1983. Native lowbush blueberry management trial 1982-83. Agr. Can. Prince Edward Island Regional Dev. Branch Productivity Enhancement Program Project Rpt.

Sanderson, K. R. and M. R. Carter. 1993. Effect of gypsum on yield of Brussels sprouts. Agr. Agri-Food Can. Agri-Info Factsheet, 93-12: Agdex:252.540.

Sanderson, K. R., M. R. Carter, and J. A. Ivany. 1996a. Effects of gypsum on yield and nutrient status of native lowbush blueberry. Can. J. Plant Sci. 76:361-366.

Sanderson, K. R. and J. A. Ivany. 1994. Crop year applied fertilizer in wild blueberry management. Agr. Agri-Food Can. Agri-Info Factsheet, 94-29: Agdex:235.23.

Sanderson, K. R. and J. A. Ivany. 1996. Effect of gypsum on wild lowbush blueberry yield. Agr. Agri-Food Can. Agri-Info Factsheet, 96-02: Agdex:235.540.

Sanderson, K. R. and J. A. Ivany. 2000. Effect of dolomitic limestone, calcitic limestone and gypsum applied in the sprout year on nutrient status of wild blueberry. Agr. Agri-Food Can. Agri-Info Factsheet, 00-03: Agdex:235.23.

Sanderson, K. R., J. B. Sanderson, and J. A. Ivany. 1996b. Supplemental soil sulphur increases cabbage yield. Can. J. Plant Sci. 76:857-859.

Shainberg, I., M. E. Sumner, W. P. Miller, M. P. W. Farina, M. A. Pavan, and M. V. Fey. 1989. Use of gypsum on soils: A review. Adv. Soil Sci. 9:1-111.

Singh, G., I. P. Abrol, and S. S. Cheena. 1989. Effects of gypsum application on mesquite (*Prosopis juliflora*) and soil properties in an abandoned sodic soil. Forest Ecol. Mgt. 29:1-14.

Smagula, J. M. and S. Dunham. 1997. Blueberry surprise from phosphorus. Maine Agr. Forest Expt. Sta. Miscellaneous Rpt. No. 404. Univ. Maine, Dept. Biosystems Sci. Eng., Orono, Maine.

Souza, D. M. G. and K. D. Ritchey. 1986. Uso do gesso no solo de cerrado. Pp. 199-244 In: D. F. Brasilia (ed.). An. Sem. Uso Fosfogesso Agricultura. EMBRAPA, Brazil.

Steel, R. G. D. and J. H. Torrie. 1960. The principles and procedures of statistics. McGraw-Hill, New York.

Stevenson, F. J. 1986. Cycles of soil: Carbon, nitrogen, phosphorus, sulphur, micronutrients. Wiley, New York.

Subramani, S. and J. Scifres. 1988. Effect of gypsum and lime application and *Azolla pinnata* inoculation on grain yield of rice. Trop. Agr. 65:226-228.

Tran T. S. and R. R. Simard. 1993. Mehlich III extractable elements. Pp. 43-49 In: M. R. Carter (ed.). Soil sampling and methods of analysis. Can. Soc. Soil Sci., Lewis Publishers, London.

Trevett, M. F. 1962. Nutrition and growth of the lowbush blueberry. Maine Agr. Expt. Sta. Bul. 605.

Vander Kloet, S. P. 1978. Systematics, distribution and nomenclature of the polymorphic *Vaccinium angustifolium* Ait. Rhodora 80:358-376.

Warman, P. R. 1987. The effects of pruning, fertilizers, and organic amendments on lowbush blueberry production. Plant Soil 101:67-72.

Webb, K. T., R. L. Thompson, G. J. Beke, and J. L. Nowland. 1991. Soils of Colchester County, Nova Scotia. Nova Scotia Soil Survey. Res. Branch, Agr. Can., Ottawa, Ont., Rpt. No. 19.

Windham, W. R., 1997. Animal feed. (Method 990.03). Pp. 26-27. In: P. Cunnif (ed.), Official methods of analysis of AOAC Intl. (2000) 17th Edition. AOAC Intl., Gaithersburg, MD.

Yarborough, L. E., J. J. Hancher, S. P. Skinner, and A. A. Ismail. 1986. Weed response, yield and economics of hexazinone and nitrogen use in lowbush blueberry production. Weed Sci. 34:723-729.

# The Economics
# of Supplemental Irrigation
# on Wild Blueberries:
# A Stochastic Cost Assessment

Timothy J. Dalton
David E. Yarborough

**SUMMARY.** Wild blueberry growers face numerous production perils of which drought and heat produce the greatest losses. Using partial budgets, this research derives stochastic cost estimates of four alternative irrigation technologies that apply supplemental irrigation to wild blueberries. Uncertain parameters related to irrigation water demand are subjected to Monte Carlo simulation to derive the expected distribution of total annual irrigation costs. This study finds that labor and energy intensive systems supply water more cost effectively to Maine producers due to limited overall demand for irrigation water. In addition, the total an-

Timothy J. Dalton is Assistant Professor, Maine Agricultural Center, Department of Resource Economics and Policy, 5782 Winslow Hall, University of Maine, Orono, ME 04469-5782.

David E. Yarborough is Professor of Horticulture, Maine Agricultural Center, 414 Deering Hall, University of Maine, Orono, ME 04469-5782.

Maine Agricultural and Forest Experiment Station publication 2571. This research was supported by a grant from the Maine Agricultural Center.

[Haworth co-indexing entry note]: "The Economics of Supplemental Irrigation on Wild Blueberries: A Stochastic Cost Assessment." Dalton, Timothy J., and David E. Yarborough. Co-published simultaneously in *Small Fruits Review* (Food Products Press, an imprint of The Haworth Press, Inc.) Vol. 3, No. 1/2, 2004, pp. 73-86; and: *Proceedings of the Ninth North American Blueberry Research and Extension Workers Conference* (ed: Charles F. Forney, and Leonard J. Eaton) Food Products Press, an imprint of The Haworth Press, Inc., 2004, pp. 73-86. Single or multiple copies of this article are available for a fee from The Haworth Document Delivery Service [1-800-HAWORTH, 9:00 a.m. - 5:00 p.m. (EST). E-mail address: docdelivery@haworthpress.com].

nual cost of irrigation is highly site specific and necessitates spatially disaggregated estimation of costs. *[Article copies available for a fee from The Haworth Document Delivery Service: 1-800-HAWORTH. E-mail address: <docdelivery@haworthpress.com> Website: <http://www.HaworthPress.com> © 2004 by The Haworth Press, Inc. All rights reserved.]*

**KEYWORDS.** Blueberries, irrigation, stochastic, cost, *Vaccinium augustifolium*

## INTRODUCTION

Stochastic weather events can impact the profitability and risk of wild blueberry (*Vaccinium angustifolium* Ait.) production in humid regions such as Maine. The use of irrigation for agricultural crops has different implications depending on the amount of rainfall needed by the crop and average annual rainfall. In arid areas, where irrigation is essential, the issue devolves to a comparison between yields with, and without, irrigation. In temperate regions, where in some years there is enough rainfall to meet crop needs, the cost analysis is contingent upon usage. In these regions, the range of use for supplemental irrigation can be from "not at all" in wet years to "frequently" in dry years and as such, the economic costs of an irrigation system are highly variable from year to year. Epperson et al. (1992) found that irrigation is needed in much of the humid regions of the United States because uneven rainfall creates uncertain net returns and lower overall profits. It has also been identified as an important risk management strategy (Boggess and Ritchies, 1988; Hatch et al., 1991; Vandeveer et al., 1989).

Since demand for irrigation is dependent upon deficient natural rainfall, annual operating costs are conditional upon usage. Caswell and Zilberman (1985) determined that cost savings have a significant impact upon the choice of, and tendency to, adopt new irrigation technologies. While cost savings play an important role in triggering the adoption of more efficient irrigation systems for the grower already versed in technology usage, cost uncertainty can act as a significant impediment to the uninitiated grower. This source of uncertainty can be a critical factor in delaying a producer's decisions to adopt new technology (Purvis et al., 1995). Boggess and Amerling (1983) simulated irrigation investment decisions in humid areas and found that weather pattern variability had a significant effect upon costs, revenue, and investment viability.

Annual irrigation costs are uncertain due to stochastic weather events and the demand for irrigation water. This source of uncertainty largely determines the annual cost of irrigation. During the production season, annual operating costs will accumulate depending upon the labor intensity of the system, the number of irrigation sets per season, fuel and oil requirements, maintenance, and financing charges. Using an economic-engineering simulation modeling approach, limited information on investment costs and the technical requirements were determined as key impediments to producer adoption in the case of zero runoff sub-irrigation systems (Uva et al., 2000). Bernardo (1988) also used a simulation approach to evaluate the impact of spatial variability of irrigation application and determined that non-uniform water application increased water usage and the cost of irrigation. As Schneekloth et al. (1995) state, one of the most important problems facing decision makers is the ordering of alternative investment decisions with different risky outcomes in order to determine which option is the least costly given operator conditions.

As the lack of information on investment and operating cost can deter producer investment in new production technologies, the objective of this paper is to determine which irrigation systems are the least costly for wild blueberry production under growing conditions facing Maine producers. An ex ante stochastic cost function approach is used to derive the expected distribution of economic costs for alternative technologies to irrigate blueberries for three regions in Maine. Comparison of the results can be used to determine whether a spatially uniform irrigation recommendation can be made or if site-specific recommendations are required.

## MATERIALS AND METHODS

In order to derive stochastic cost estimates of irrigation systems, a three step procedure was employed. First, a representative partial cost budget was developed to capture the economic costs associated with irrigation. Secondly, the stochastic factors associated with irrigation investment and annual usage were identified. Thirdly, after identifying the stochastic elements affecting economic costs, the budgets were repeatedly calculated by varying the stochastic factors. The resulting cost estimates were analyzed to determine which technology is the most cost-efficient applicator of water.

To quantify the tradeoffs between initial investment and annual operating expense, four different irrigation systems were evaluated: (1) moveable large gun systems; (2) handline small sprinkler systems; (3) hose reel traveler systems; and (4) permanent set small sprinkler systems. Since irrigation is not an essential tool for crop production in the study region, many growers consider adopting irrigation only partially over their operation. As a result, the three technology alternatives were analyzed at 10, 20, and 40 ha (25, 50, and 100 acres).

*Stochastic weather factors.* The microclimate surrounding the farm and rainfall are two of the most important factors to consider when making the irrigation investment decision. As a general planning rule, lowbush blueberries need about 25 mm (one inch) of water a week from April to October, and irrigation is commonly applied during the months of June, July, and August for proper plant and fruit development under Maine growing conditions. Irrigation cost analysis is conducted for three growing regions in Maine in order to determine whether site specific cost estimates are required or whether one single estimate can be used to represent all growing areas. Forty years of weather data are examined for each of the three growing sites.

*Economic analysis of irrigation costs.* This study focuses on calculating the ownership and operating costs of the four systems at three different acreage sizes for three different regions. Ownership costs consist of depreciation and interest charges for the irrigation system plus taxes and insurance charges. Depreciation and interest charges are calculated on an annual cost basis using the Capital Recovery Method cited in similar economic evaluations (Collier and Glagola, 1998). Operating costs accumulate from annual variable costs of irrigation.

*Ownership expenses.* Total investment costs are presented in Table 1. These amounts represent the approximate cost to establish the four different systems including water source development, diesel engine, pump, mainline, lateral delivery lines, sprinkler system, and installation labor. These costs were derived from interviews with irrigation equipment dealers conducted during 2001. Sales tax is not added to the total cost under the assumption that the grower holds a commercial agricultural production sales tax exemption certificate.

To derive the annual cost of these systems, each piece of equipment is depreciated over its estimated lifespan. Interest charges are added to the depreciation calculation and any salvage value at the end of the lifespan deducted from the annual charge. Long-term interest rates on irrigation equipment range between 7.5% to 10% depending upon the loan amount and buyer qualifications (FCS). A fixed nominal investment in-

TABLE 1. Total investment costs for irrigation system establishment ($/field).

| Irrigation Systems | 10 ha | 20 ha | 40 ha |
|---|---|---|---|
| Handline Moveable Large Gun | $41,000 | $56,780 | $80,760 |
| Handline Small Sprinkler | $45,038 | $61,615 | $93,315 |
| Hose Reel Large Gun | $39,900 | $53,759 | $54,630 |
| Permanent Set Small Sprinkler | $63,564 | $103,329 | $185,750 |

Source: Authors' calculation from model results

terest rate of 9%, representative of "average" credit for loans written over a five- to ten-year period, is inflation adjusted to a real rate of 5.6%. Inflation averaged 3.2% annually for the 15-year period between 1987 and 2001.

In addition to depreciation and interest, ownership costs include insurance charges and property tax adjustments. Insurance rates range from $0.0056 to $0.0075 per $100 dollars of coverage for irrigation equipment (Diversified Agrinsurance). The baseline analysis assumes a rate of $0.0068 per $100 dollars of coverage, calculated over the replacement value of the piece of equipment. Annual ownership cost is a fixed component of the total annual cost of irrigation. Ownership costs must be repaid yearly irrespective of the demand for irrigation water.

*Operating expenses.* Operating costs to run and maintain the irrigation systems are calculated in a partial budget format; that is only costs associated with the operation of the irrigation system are captured. Each of the models calculates operating costs contingent upon the demand for irrigation water. There are four primary components of the operating cost budgets: labor, power, maintenance, and interest charges.

Labor costs accumulate from two different sources: initial set up and end-of-season take-down of the system and variable labor usage per irrigation. Managerial time is not included in the labor cost calculation. These per acre coefficients are applied uniformly across the three different acreage examples. A $9.40 hourly wage rate is applied in the calculations. This wage rate is based upon the 2001 Adverse Effect Wage Rate of $8.17 and inflated by 15% to account for meals and other benefits entitled to immigrant workers (USDA, 2001a; USDA, 2001b). Alternatively, it can be seen as the benefits premium (Social Security, Unemployment Compensation, and Workers Compensation Insurance) attached to attract local workers from non-agricultural employment alternatives. Since managerial labor is not included in the calculation, a constant cost-per-acre labor charge is calculated for the four different systems.

Power costs are calculated by determining the number of hours that the pumping unit operates to apply the required amount of water. Irrigation water demand is determined by comparing observed weekly rainfall amounts with the recommended threshold of 25 mm per week. Once a deficit of 13 mm occurs, the model applies 19 mm of irrigation water to ensure that 13 mm of irrigation water reaches the plant. Total pumping time is inflated by 10% to account for flushing, system testing, and mistakes. Total pumping time is then multiplied by hourly fuel-consumption rates of the different diesel motors and then by the price of diesel fuel ($0.33/L). This diesel price is based upon sales-tax-free prices from summer 2001.

Maintenance and upkeep charges are calculated for these systems as a fixed coefficient of initial purchase price. Maintenance and upkeep coefficients are derived from Patterson et al. (1996). These coefficients represent an average charge that should be incurred over the life of the irrigation component, not one representing a new piece of equipment with little or no maintenance nor an old one with high upkeep costs. Pieces of equipment with moving parts require higher maintenance costs than fittings for example. Maintenance and upkeep on tubing represents limited unforeseen breakage.

The final component of the operating budget is an interest charge on working capital used during the production season. The interest charge represents the financial cost of a short-term operating loan or the opportunity cost of producer capital used to pay for these expenses before blueberry receipts are received. A short-run nominal interest rate of 8%, inflation adjusted to 4.7%, is applied over a seven-month period of time, e.g., April through October on the balance of labor, fuel, and maintenance charges. This rate is a representative rate provided to producers by the Farm Credit Service for short-term operating loans.

Ownership and operating costs are totaled for the different irrigation systems and the three field sizes in a partial budget format. Key bioeconomic parameters of the model are then varied using Monte Carlo techniques to derive the probability distribution of total annual costs. There are several uncertain parameters in the model of which annual rainfall is one of the key parameters. The number of irrigations applied per season is derived from the empirical rainfall distributions described in Figures 1 to 3.

An irrigation scheduling rule was created and then compared against historical weather patterns. The rule states that once a cumulative rain-

FIGURE 1. Weekly probabilities of rainfall exceeding limit of 12.5, 19, or 25 mm, Jonesboro, Maine 1959-1998

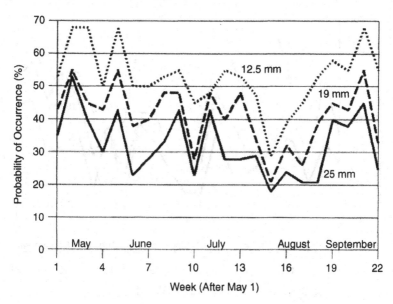

fall deficit of 13 mm occurs, 19 mm of irrigation water is applied. Based upon a 75% application efficiency, this ensures that 13 mm of irrigation water compensates for the rainfall deficit. Using this rule, the annual number of seasonal irrigations was calculated from historical rainfall data for each season. These historical requirements were then fit to a discrete probability density distribution. For each of the sites, the number of irrigations is represented by a negative binomial density function with a mean and variance of 4.3 and 3.8 applications for Jonesboro, 6.2 and 3.5 for Ellsworth, and 5.4 and 3.4 for Belfast. In addition to the biological uncertainty represented in the demand for irrigation water, several key uncertain economic parameters are modeled.

Three input costs are varied: the price of diesel fuel, the wage rate, and the nominal interest rate. These three prices are systematically varied by 10% above and below their stated values derived in 2001.

Each irrigation system model and acreage combination was simulated 1500 times by randomly selecting values from the irrigation demand, fuel, wage, and interest rate distributions. The 1500 combinations of these variables are used to derive a probability density distribution of total annual costs. The same base economic models were then simulated

FIGURE 2. Weekly probabilities of rainfall exceeding limit of 12.5, 19, or 25 mm, Ellsworth, Maine 1959-1998

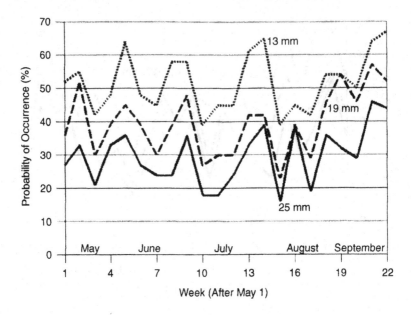

for each of the three locations in the blueberry production region in order to test the hypothesis of spatially homogenous irrigation costs.

## RESULTS AND DISCUSSION

Total irrigation cost is highly dependent upon the demand for irrigation water. As demand for irrigation water increases, operating cost begin to accumulate. Given that the overall demand for irrigation is limited for a large percentage of the years, total annual cost distributions are skewed towards the annual ownership component of the total.

Figures 1 through 3 summarize rainfall histories along an east to west transect of the blueberry-growing region in Maine. Forty years of weather data (1959-98) from the National Climatic Data Center from three sites were selected for illustration (National Climatic Data Center). These sites are Jonesboro, Ellsworth, and Belfast. In the case of Ellsworth, data collection ceased in 1994, so the weather series is reduced to 35 years. Daily weather events were aggregated into weeks

FIGURE 3. Weekly probabilities of rainfall exceeding limit of 12.5, 19, or 25 mm, Belfast, Maine 1959-1998

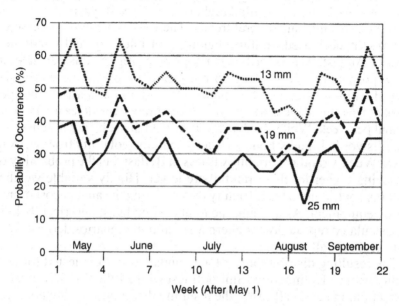

and presented over the period from the beginning of May until the end of September.

One graph for each site is presented that depicts the probability that rainfall in a particular week (starting on May 1) will exceed the threshold level of 25 mm, 19 mm, and 13 mm (Figures 1, 2, and 3 for Jonesboro, Ellsworth, and Belfast, respectively). For nearly all of the growing season, the probability of receiving 25 mm of rainfall per week is less than 50%. The graphs also indicate a distinct decrease in the probability of rainfall during the critical fruiting period from mid-July (week 11) through August (week 17). This valley is more apparent in Figure 1 (Jonesboro) and Figure 3 (Ellsworth). During this period, the probability of receiving any rainfall reaches its lowest in the season. There is less than a 20% chance of 25 mm of rainfall during this period.

Overall, by mid-July, less than 33% of all years exceeded 25 mm cumulative per week in Jonesboro, only 21% in Ellsworth, and 25% in Belfast, indicating that early season rainfall deficits were not made up during mid-season. During the same period, Jonesboro and Belfast received 19 mm of rainfall in nearly two-thirds of the 40 years, while Ellsworth received the same amount approximately 50% of the time.

Jonesboro and Belfast are both virtually assured of at least 13 mm (approximately nine out of every 10 years) while Ellsworth is likely to receive at least 13 mm of rainfall eight out of every 10 years.

The figures indicate several trends. There are distinct intra-seasonal patterns of decreased rainfall beginning in mid-July and continuing through August. Across the blueberry-growing region, this translates to a low probability of receiving 25 mm per week of rainfall during the critical fruiting stage, a 50% chance of receiving at least 19 mm per week during this stage, and a high chance of receiving at least 13 mm of rainfall per week. While absolute levels of rainfall are higher in Jonesboro and Ellsworth, Figures 1 and 2 indicate extreme volatility from week to week. Weekly rainfall variability is less in Belfast, but the probability of achieving higher levels of rainfall are lower. Highly variable weather patterns contribute to the difficulty in calculating the annual cost of irrigation equipment. As a result, the total cost budgets, described below, are calculated repeatedly for alternative rainfall scenarios derived from historical rainfall patterns.

The resulting distributions of total annual costs were tested for normality using a Kolmogorov-Smirnoff test with a Lilliefors significance correction. In nearly all cases, the resulting distributions were non-normally distributed when evaluated at $\alpha = 0.01$. The cost distributions were normally distributed at the $\alpha = 0.01$ level in only two cases: the permanent set small sprinkler system on 40 ha in Ellsworth and Belfast. Normality was rejected, however, at the $\alpha = 0.05$ level. As a result, all distributions were assessed as non-normally distributed.

Given that all distributions of total cost are non-normally distributed, median estimates of total annual cost are presented in Table 2. A breakdown of a static budget into cost centers was reported by Dalton et al. (2002). In many cases, the median statistics indicate little difference between the expected cost of irrigation across location. In general, systems that require lower initial investment in equipment (the handline large gun and hose reels) are less expensive to operate on an overall total cost basis than high investment cost systems. This is due in part to the limited amount of time that the systems operate during the year and the limited acreage over which ownership costs are distributed.

Despite little difference in the median estimate of total annual cost for each technology and acreage combination, significant differences exists among sites. Using a Wilcoxon test, differences between sites were significant at $\alpha = 0.01$. This result is largely due to differing vari-

TABLE 2. Median total annual cost estimates ($/ha)

| | Jonesboro | Ellsworth | Belfast |
|---|---|---|---|
| Handline Moveable Large Gun | | | |
| 10 ha | 318.57 | 335.15 | 327.32 |
| 20 ha | 226.20 | 239.87 | 233.11 |
| 40 ha | 171.27 | 184.38 | 177.64 |
| Handline Small Sprinkler | | | |
| 10 ha | 468.92 | 490.27 | 479.92 |
| 20 ha | 336.78 | 355.70 | 346.00 |
| 40 ha | 257.46 | 276.49 | 266.40 |
| Hose Reel Large Gun | | | |
| 10 ha | 315.65 | 331.71 | 324.45 |
| 20 ha | 225.51 | 245.25 | 239.13 |
| 40 ha | 133.82 | 153.07 | 143.27 |
| Permanent Set Small Sprinkler | | | |
| 10 ha | 523.58 | 525.92 | 523.86 |
| 20 ha | 466.59 | 468.86 | 467.01 |
| 40 ha | 436.45 | 439.05 | 437.10 |

Source: Authors' calculation from model results

ances in cost estimates which is reflective of the variability of irrigation water demand, consistent with rainfall estimates.

Significant differences are most visible in a comparison of results between the least expensive and most expensive systems on the 20 ha field in Figure 4. Figure 4 presents the cumulative probability distributions, in $/ha, of the hose reel and permanent set small sprinkler systems for the three locations. Each technology is grouped together and curves lying to the right are more expensive than curves lying to the left.

A larger share of the total annual cost of irrigation is tied to annual operating costs for hose reel systems and the shapes of the cost curves are dependent upon the demand for irrigation water. Water demand increases as the cumulative probability increases. By comparison, ownership costs compose a larger share of total annual cost for the permanent set small sprinkler systems. Since operating costs are a smaller share of total annual costs, the curves rise more rapidly and the differences between locations is less pronounced.

## CONCLUSIONS

Stochastic weather and economic factors make it difficult to assess the total annual cost of supplemental irrigation systems and this diffi-

FIGURE 4. Cumulative probability distributions of total annual irrigation costs for two systems and three locations in Maine

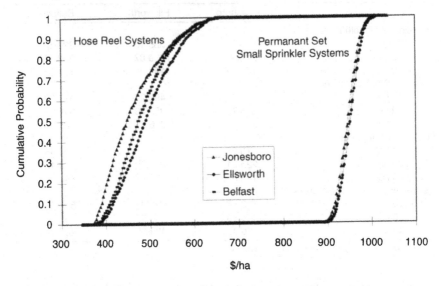

culty can act as an impediment to the adoption of irrigation technology. Using an economic-engineering methodology combined with Monte Carlo simulation, this research has derived probability density distributions of total annual costs of irrigation for four technologies, three acreages, and for three important areas in the wild blueberry growing region of Maine.

The resulting distributions of total annual cost are non-normally distributed thereby requiring the usage of non-parametric statistics to test for differences. Overall, hose reel and handline large gun systems are less expensive to operate than the handline small sprinklers or permanent set small sprinklers. This result is consistent across the three different locations examined in the study.

A second important finding is that the total annual cost of an irrigation system is dependent upon location. A Wilcoxon test was used to test the null hypothesis of equal annual costs across locations for each technology and acreage combination. In all cases, the null hypothesis of equal total annual costs was rejected. As a result, estimates of total annual irrigation costs must be spatially disaggregated to prevent biased producer recommendations. Future research will integrate the economic benefits of blueberry yield response to irrigation with the cost analysis in order to derive the profitability and risk reduction benefits once accurate yield response functions are derived.

## GROWER IMPLICATIONS

Annual irrigation costs are highly variable due to physical differences between irrigation equipment and the number of applications that the equipment is used during the cropping system. As a result, the total annual cost of alternative irrigation systems should be calculated under conditions of low, medium and high use in order to understand the averages and extremes of irrigation costs. Since irrigation is supplemental, the total annual cost of irrigation is dependent upon the frequency of its usage. Rainfall patterns across the wild blueberry producing region of Maine are dramatically different, and cost estimation should be conducted at a local level in relation to local rainfall patterns. Since the total annual cost of irrigation has been demonstrated to be significantly different between the three study sites, no single estimate can accurately represent total annual irrigation costs for the entire state.

## LITERATURE CITED

Bernardo, D. J. 1988. The effect of spatial variability of irrigation applications on risk-efficient irrigation strategies. Southern J. Agr. Econ. 20: 77-86.

Boggess, W. G. and C. B. Amerling. 1983. A bioeconomic simulation analysis of irrigation investments. Southern J. Agr. Econ. 15:85-91.

Boggess, W. G. and J. T. Ritchie. 1988. Economic and risk analysis of irrigation decisions in humid regions. J. Production Agr. 1: 116-122.

Caswell, M. and D. Zilberman. 1985. The choices of irrigation technologies in California. Amer. J. Agr. Econ. 67: 224-234.

Collier, C. A. and C. R. Glagola. 1998. Eng. Econ. Cost Analysis. Addison Wesley Longman, Inc.: Menlo Park, CA.

Dalton, T. J., A. Files, and D. Yarborough. 2002. Investment, ownership and operating costs of supplemental irrigation systems for wild blueberries. MAFES Tech. Bul. 183, Univ. Maine, Orono, Maine.

Diversified Agrinsurance. 2001. <http://www.dfsfin.com/directory/default.asp>.

Epperson, J. E., H. E. Hook, and Y. R. Mustafa. 1992. Stochastic dominance analysis for more profitable and less risky irrigation of corn. J. Production Agr. 5: 243-247.

Farm Credit Service (FCS). 2001. Personal interview. 26 Sept.

Hatch, L. U., W. E. Hardy Jr., E. W. Rochester, and R. L. Pickren. 1991. A farm management analysis of supplemental center-pivot irrigation in humid regions. J. Production Agr. 4:442-447.

National Climatic Data Center (NCDC). 2001. <http://www4.ncdc.noaa.gov/cgi-win/wwcgi.dll?wwDI~SelectStation~USA~ME>.

Patterson, P. E., B. R. King, and R. L. Smathers. 1996. Economics of sprinkler irrigation systems: Handline, solid set, and wheelline. Univ. Idaho Coop. Ext. Ser. Bul. No. 788.

Purvis, A., W. Boggess, C. Moss, and J. Holt. 1995. Technology adoption decisions under irreversibility and uncertainty: An ex ante approach. Amer. J. Agr. Econ. 77: 541-551.

Schneekloth, J., R. Clark, S. Coadym, N. L. Clocke, and G. W. Hergert. 1995. Influence of wheat-feed grain programs on riskiness of crop rotations under alternative irrigation levels. J. of Production Agr. 8: 415-423.

USDA. 2001a. 1 Oct. 2001. <affairs/aewr01.htm>.

USDA. 2001b. 1 Oct. 2001. <http://www.usda.gov/oce/oce/labor-affairs/aewr2001.htm>.

Uva, W. L., T. C. Weiler, R. A. Milligan, L. Albright, and D. Haith. 2000. Risk analysis of adopting zero runoff subirrigation systems in greenhouse operation: A Monte Carlo simulation approach. Agr. Resource Econ. Rev. 29: 229-239.

Vandeveer, L. R., K. W. Paxton, and D. R. Lavergne. 1989. Irrigation and potential diversification benefits in humid climates. Southern J. Agr. Econ. 21: 167-174.

# Diammonium Phosphate Application Date Affects *Vaccinium angustifolium* Ait. Nutrient Uptake and Yield

John M. Smagula
W. Litten
K. Loennecker

**SUMMARY.** A commercial lowbush blueberry field with a Colton sandy soil characteristic of the blueberry barrens of Washington County, Maine was the site for this study. Diammonium phosphate (DAP, 18-46-0) was applied at 448 kg/ha on 17 May (preemergent), 31 May, 14 June, 28 June, or 12 July 2000 for comparison with an unfertilized control plot. Stem height was measured on 20 tagged stems in each control plot of blocks 1, 2, and 3 at the time of fertilizer application. Composite leaf tissue samples taken for nutrient analysis indicated that there was a positive linear and quadratic relationship between leaf N concentration and

John M. Smagula is Professor of Horticulture, W. Litten is Faculty Associate, and K. Loennecker is Scientific Technician; all at the Department of Plant, Soil and Environmental Sciences, University of Maine, Orono, ME 04469-5722.

Address correspondence to: John M. Smagula (E-mail: smagula@mail.maine.edu).

The authors would like to thank the wild blueberry industry for their support of this research project through grower taxes administered by the Maine Wild Blueberry Commission.

This manuscript is publication number 2564 of the Maine Agriculture and Forestry Experiment Station.

[Haworth co-indexing entry note]: "Diammonium Phosphate Application Date Affects *Vaccinium angustifolium* Ait. Nutrient Uptake and Yield." Smagula. John M., W. Litten, and K. Loennecker. Co-published simultaneously in *Small Fruits Review* (Food Products Press, an imprint of The Haworth Press, Inc.) Vol. 3, No. 1/2, 2004, pp. 87-94; and: *Proceedings of the Ninth North American Blueberry Research and Extension Workers Conference* (ed: Charles F. Forney, and Leonard J. Eaton) Food Products Press, an imprint of The Haworth Press, Inc., 2004, pp. 87-94. Single or multiple copies of this article are available for a fee from The Haworth Document Delivery Service [1-800-HAWORTH, 9:00 a.m. - 5:00 p.m. (EST). E-mail address: docdelivery@haworthpress.com].

Digital Object Identifer: 10.1300/J301v03n01_09

fertilizer application date. Leaf P concentration had a quadratic relationship with application date. Applying fertilizer on 14 June resulted in the greatest uptake of N and P as indicated by leaf nutrient concentrations. Soil samples taken at the time of leaf sampling indicated no effect of any fertilizer application on soil P concentration. Fertilizer application did not affect stem density, stem height, or number of branches. Branch length was increased only by the 17 May preemergent application of DAP. Flower bud density was increased by fertilization on 17 May, 31 May, 14 June, or 28 June, compared to the control. Yield was increased by fertilizing with DAP applied preemergent (May 17) or on May 31 but not at the later dates. This quadratic relationship was significant at the 0.1% level. Apparently, N and P from DAP accumulated to high levels in leaf tissue when applied on 14 June but there was inadequate time to influence flower formation in flower buds. *[Article copies available for a fee from The Haworth Document Delivery Service: 1-800-HAWORTH. E-mail address: <docdelivery@haworthpress.com> Website: <http://www.HaworthPress. com> © 2004 by The Haworth Press, Inc. All rights reserved.]*

**KEYWORDS.** Fertilizer timing, lowbush blueberry, nitrogen, phosphorus

## INTRODUCTION

Lowbush blueberry (*Vaccinium angustifolium* Ait.) growers have relied primarily on the management of existing fields after clearing of forest land to encourage development of natural stands. A management practice unique to lowbush blueberry production is pruning above ground stems by fire or flail mowing to stimulate vigorous stem growth from rhizomes the following season. This practice results in an alternate year production cycle. Lowbush blueberry fields are managed on a two-year cycle with herbicide and fertilizer applications usually applied before shoots emerge from rhizomes the first year of the production cycle. Throop and Hanson (1997) reported that absorption of fertilizer-derived N was greatest for highbush blueberry plants from late bloom (May) until after fruit harvest (mid-August). An increase in absorption efficiency with increased growth rates was also reported in plum (Weinbaum et al., 1978) and grape (Conradie, 1986). Previous studies on lowbush blueberry suggest that date of fertilizer application may be more important on sandy textured than on heavier soils (Smagula and Litten, 2001). In this study, we determined the effect of time of fertilizer application on nutrient uptake, soil nutrient availability, plant growth, and yield as related to stage of plant development in the prune year.

## MATERIALS AND METHODS

A commercial lowbush blueberry field with a sandy soil, characteristic of the blueberry barrens of Washington County, Maine, was used in this study. A randomized complete block design with six blocks and six treatments was used. Treatment plots (1.5 m × 15 m) received a pre-emergent treatment of 448 kg DAP (18-46-0)/ha (80.6 kg N and 89.6 lbs P/acre) on 17 May or a post emergent treatment on 31 May, 14 June, 28 June, or 12 July. A control plot received no fertilizer. Fertilizer application was related to stem growth by measuring stem height of 20 tagged stems in each control plot of block 1, 2, and 3 at the time of fertilizer application. On 12 July 2000 after stems had stopped elongating, composite leaf samples were taken from 50 randomly cut stems in each treatment plot (Trevett et al., 1968). Soil samples (7.6 cm depth) were also taken at this time. Composite leaf samples were prepared according to the methods of Kalra and Maynard (1991). Solution analysis was by plasma emission spectroscopy. Stem samples from four randomly placed 0.03 m$^2$ quadrats within each treatment plot were collected in October 2000 and measured for stem length, branching, and flower bud formation. Yield was measured 30 August 2001 by handraking a strip 43 cm wide the length of each plot. Analysis of variance (ANOVA) was performed using the general linear models procedure (PROC GLM, SAS Institute, Cary, NC). Duncan's Multiple Range Test was used for means separation.

## RESULTS AND DISCUSSION

Leaf N concentrations were increased more by application of DAP on 31 May, 14 June, or 28 June, compared to the control or the 17 May application date (Figure 1). Stem height at each application date is also plotted in Figure 1, indicating growth was linear from 17 May to 12 July. Leaf N concentrations in control plots were at a level of sufficiency (> 1.6%), according to the standard proposed by Trevett (1972). Soil P concentration was not affected by date of fertilizer application but leaf P concentrations (sufficiency > 1.25%) were higher when fertilizer was applied on 31 May or 14 June, compared to the control (Fig-

FIGURE 1. Leaf N concentrations as influenced by time of DAP fertilization and the height of stems at fertilization. Leaf N Significance level = 0.01%

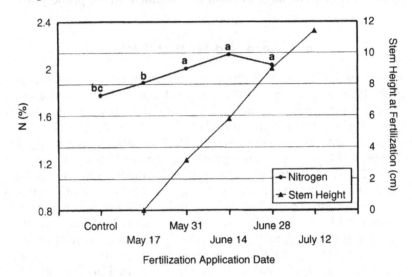

Fertilization Application Date

ure 2). Fertilizer applied when shoots were between 3.2 and 5.8 cm tall was more effective in raising leaf P than when shoots were shorter or taller than this height. Average soil pH in treatment plots ranged from 4.2 to 4.4 and was not influenced by fertilizer treatments (data not shown). Soil nutrient concentrations were not affected by fertilizer treatments. Stem density (number of stems per unit area) and stem length were not affected by date of fertilizer application (Figure 3). The average number of branches per stem was also not affected by treatment date but the average branch length was greater in plots receiving fertilizer at the earliest application date (Figure 4). Flower bud density was increased at all fertilizer treatment dates except the last, 12 July, compared to the control (Figure 5). August 2001 fruit yield was greatest for plots receiving the fertilizer on 17 May or 31 May, compared the control or to later application dates.

## CONCLUSIONS AND GROWER BENEFITS

It appears that timing may be more important on sandy textured soils than on heavier soils for maximizing lowbush blueberry nutrient uptake

FIGURE 2. Leaf and soil P as influenced by time of DAP fertilization. Four hundred forty-eight kilogram DAP/ha applied on indicated dates. Leaf P Significance level = 0.01%. Soil P not significant at 5% level.

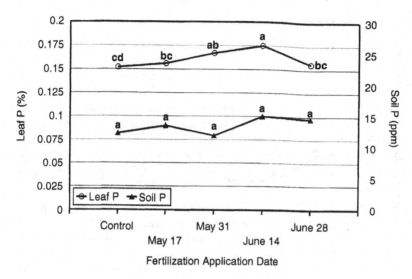

Fertilization Application Date

FIGURE 3. Stem length and density as influenced by time of DAP application. Four hundred forty-eight kilogram DAP/ha applied on indicated dates. Significance levels = 5% for stem number and stem length.

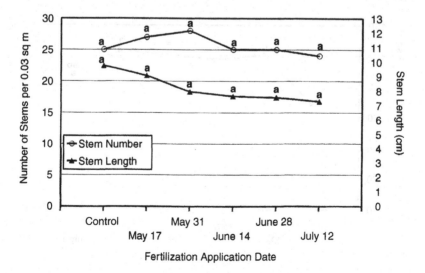

Fertilization Application Date

FIGURE 4. Effect of fertilizer timing on branch number and average length. Four hundred forty-eight kilogram DAP/ha applied on indicated dates. Significance level = 5% for branch number and length.

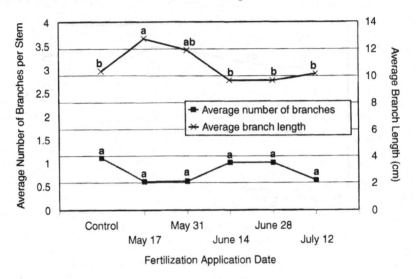

FIGURE 5. Flower buds as influenced by time of DAP application. Four hundred forty-eight kilogram DAP/ha applied on indicated dates. Significance level = 5%.

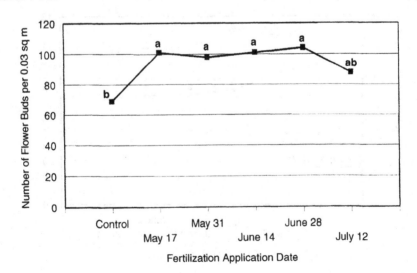

FIGURE 6. Yield as influenced by time of DAP application and stem height at time of application. Four hundred forty-eight kilogram DAP/ha applied on indicated dates. Significance level = 0.01%.

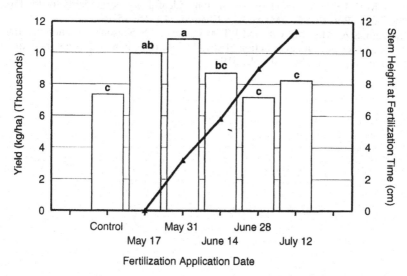

and yield. Fertilizing too late in the prune cycle may not be effective in stimulating growth and flower bud formation because flower bud formation begins at tip dieback. We recommend that fertilizer application on sandy soils should be preemergent or before stem height reaches 4 cm.

## LITERATURE CITED

Conradie, W.J. 1986. Utilisation of nitrogen by the grape-vine as affected by time of application and soil type. South African J. Enol. Viticult. 7:76-83.

Kalra, Y.P. and D.G. Maynard,, 1991. Methods manual for forest soil and plant analysis. p. 116 For. Can., Northwest Reg., North. For. Cen., Edmonton, Alta. Inf. Rep. Nor-X-319.

Smagula, J.M. and W. Litten. 2001. Effects of fertilizer application date on lowbush blueberry. HortScience 36:466.

Throop, P.A. and E.J. Hanson. 1997. Effect of application date on absortion of [15]nitrogen by highbush blueberry. J. Amer. Soc. Hort. Sci. 122:422-426.

Trevett, M.F., P.N. Carpenter, and R.E. Durgin. 1968. A discussion of the effects of mineral nutrient interactions on foliar diagnosis in lowbush blueberries. Maine Agr. Expt. Sta. Bul. 664:1-15.

Trevett, M.F. 1972. A second approximation of leaf analysis standards for lowbush blueberry. Res. Life Sci. Maine Agr. Expt. Sta. 19 (15):15-16.

Weinbaum, S.A., M.L. Merwin, and T.T. Muraoka. 1978. Seasonal variation in nitrate uptake efficiency and distribution of absorbed nitrogen in non-bearing prune trees. J. Amer. Soc. Hort. Sci. 103:516-519.

# Effects of Salt Deposition
# from Salt Water Spray
# on Lowbush Blueberry Shoots

Leonard J. Eaton
Kevin R. Sanderson
Jeff Hoyle

**SUMMARY.** Many commercial lowbush blueberry (*Vaccinium angusti-folium* Ait.) fields border the Bay of Fundy and the Gulf of St. Lawrence. Qualitative producer observations in these areas indicate that salt spray from the marine environment during winter months reduces yield of the lowbush blueberry. To quantitatively examine the effects of ocean spray

Leonard J. Eaton is Research Professor, Department of Environmental Sciences, Nova Scotia Agricultural College, Truro, NS, Canada, B2N 5E3. Kevin R. Sanderson is Research Scientist, Agriculture and Agri-Food Canada, Crops and Livestock Research Centre, 440 University Avenue, Charlottetown, PEI, Canada, C1A 4N6. Jeff Hoyle is Chemistry Professor, Department of Environmental Sciences, Nova Scotia Agricultural College, Truro, NS, Canada, B2N 5E3.

The authors wish to thank producers who allowed them to do experiments in their commercial fields, with special thanks and appreciation to John MacDonald and Gerald Hackett on Prince Edward Island and to Curtis and Stephen Erb and Bragg Lumber Company in Nova Scotia. Gratitude is also expressed to Thomas L. Gallant for his assistance in the preparation of this manuscript. They acknowledge that funding support for this study was provided by Bragg Lumber Company and Agriculture and Agri-Food Canada.

[Haworth co-indexing entry note]: "Effects of Salt Deposition from Salt Water Spray on Lowbush Blueberry Shoots." Eaton, Leonard J., Kevin R. Sanderson, and Jeff Hoyle. Co-published simultaneously in *Small Fruits Review* (Food Products Press, an imprint of The Haworth Press, Inc.) Vol. 3, No. 1/2, 2004, pp. 95-103; and: *Proceedings of the Ninth North American Blueberry Research and Extension Workers Conference* (ed: Charles F. Forney, and Leonard J. Eaton) Food Products Press, an imprint of The Haworth Press, Inc., 2004, pp. 95-103. Single or multiple copies of this article are available for a fee from The Haworth Document Delivery Service [1-800-HAWORTH, 9:00 a.m. - 5:00 p.m. (EST). E-mail address: docdelivery@haworthpress.com].

on the lowbush blueberry, the amount of salt deposited on stems of this species was assessed at several commercial sites in the Canadian provinces of Prince Edward Island and Nova Scotia between 1998 and 2000. Randomly selected areas of commercial fields were protected with 1.8 m × 0.45 m (5.9 ft × 1.48 ft) shelters covered with 4 mil plastic film. Data on growth, yield, and salt deposition on shoots were recorded from both protected and exposed plants. Results varied according to location, weather conditions, and snow cover. Tree line wind protection and snow cover appeared to reduce the severity of salt spray-induced damage to the lowbush blueberry. In general, the exposed plants exhibited more salt deposition (mg $g^{-1}$ dry weight of stems), more dead buds, fewer blossoms and lower yields in comparison to covered specimens. *[Article copies available for a fee from The Haworth Document Delivery Service: 1-800-HAWORTH. E-mail address: <docdelivery@haworthpress.com> Website: <http://www.HaworthPress.com> © 2004 by The Haworth Press, Inc. All rights reserved.]*

**KEYWORDS.** Reproductive growth, winter salt spray exposure, vegetation, salt damage

## INTRODUCTION

The lowbush blueberry (*Vaccinium angustifolium* Ait.) is forced into a biennial production cycle by regular pruning that normally occurs every second year (Kinsman, 1993). Fruit buds develop on 15-20 cm (5.9-7.9 in) upright stems during the first year of the two-year production cycle (Hall et al., 1979), and remain dormant over the winter between the prune and crop years. Flowering, fruit development, and harvest occur during the second year (Eaton, 1988). Lowbush blueberry fruit buds are generally able to withstand the low winter temperatures that are typical of Atlantic Canada (Quamme et al., 1972), but less so when exposed to cold winds during winters that have insufficient snow cover (Cappiello and Dunham, 1994). The lowbush blueberry is susceptible to winter injury through dessication of the fruit buds (Hall et al., 1971). Fruit yields in the Parrsboro area of Nova Scotia, where salt spray periodically drifts off the Bay of Fundy, have been substantially reduced in years following winters with little snow cover (Hall et al., 1979). Kinsman (1993) has documented yield reductions caused by both severe winter temperatures and salt deposition over several winters in Nova Scotia; several producers report similar effects in Nova Scotia (NS) and Prince Edward Island (PEI).

Damage from deicing road salt drifting off NS highways onto nearby blueberry fields has been demonstrated (Eaton et al., 1999; 2001). Winter injury and fruit bud death of roadside shrubs and trees usually results from dessication following exposure to drifting salt sprays during the winter (Smith, 1975; Westing, 1969). Dessication of stems and fruit buds occurs as a result of water loss from cells following deposition of salt (Smith, 1975); effects are more severe when stems are exposed to low temperatures following salt deposition (Leopold and Willing, 1984).

The aim of this study is to assess the effects of sea water spray during the winter on lowbush blueberry shoots at two sites in PEI and at one site in NS.

## MATERIALS AND METHODS

Experimental sites were located in commercial blueberry fields in Norway (47°02'N, 64°02'W) and Big Pond (Hermanville; 46°28'N, 62°16'W), PEI and Fox Point (45°24'N, 64°27'W), NS. The PEI sites are exposed to the Gulf of St. Lawrence, while the NS fields are subjected to salt spray from the Bay of Fundy. Shelters were used at all experimental locations to protect blueberry stems from salt spray deposition. Each shelter was 1.8 m × 0.45 m (5.9 ft × 1.48 ft) in size and covered on three sides and the top with 4 mil white polyethylene film; this apparatus has proven effective as a barrier to salt deposition (Eaton et al., 1999). The long axis of each shelter was placed parallel to the shoreline, with the uncovered side facing away from the shore.

As damage induced from deicing spray is reduced in relation to distance from the salt source (Eaton et al., 1999), experimental trials were established including the element of distance from shoreline. Two fields were used at Big Pond and Norway, PEI; one proximal and one distal relative to the shoreline. Two shelters were located at the seaward (closest to the shore) portion of each field. In Fox Point, only one proximal and distal field were used. The Big Pond experimental shelters were established in 1998, while the Fox Point and Norway sites were set up in 1999. The shelters placed in the proximal field of the Norway site were ~900 m (~2,950 ft) from the shoreline; shelters in the distal field were ~1,000 m (~3,280 ft) from the shoreline. In the proximal field at Big Pond, the shelters were placed ~600 m (1,970 ft) from the shoreline. Two other shelters in Big Pond were placed in the distal field located ~1,250 m (~4,100 ft) further inland from the shore.

The experimental site at Fox Point had an additional objective: to assess winter injury in an exposed field as well as in a field with a tree line wind break. The shelters in the distal field (with the tree line) were placed directly behind a row of pine and spruce trees ~1,200 m (~3,940 ft) from the shore, and 50 m (164 ft) further back in the same field. The shelters placed at the seaward position in the proximal field of the Fox Point site were ~300 m (~984 ft) from the shoreline; shelters at the inland position were ~750 m (~2,460 ft) from the shoreline.

Plant samples were taken from all experimental sites on three occasions (Feb., Mar., and Apr.) during 1999 and 2000–both under (covered) and in close proximity (exposed) to the shelters. In early spring (Apr. or May) of 1999 and 2000, shelters were removed from the fields as to not induce confounding greenhouse effects on the harvested yield data. Prior to shelter removal, stem samples were gathered from within the enclosure, from 10 m (33 ft) line transects close to the apparatus, and from randomly selected areas of the fields to assess salt deposition. Plastic gloves were worn during sample acquisition to prevent salt contamination from human sweat. Stems were washed with chloride free water and the resultant solutions analyzed for chloride using the a chloride specific electrode (Eaton et al., 1999). The concentration of NaCl was expressed as mg $g^{-1}$ of NaCl in the solution. Winter injury was assessed by counting live buds, dead buds, and blossoms on each stem. In addition, the level of damage from winter injury on plant tissue was assessed by slicing the terminal, fourth, and eighth fruit buds (if present) on each stem and searching for signs of necrosis or browning. The level of damage was classified on a scale of 1 to 4 as follows: (1) no damage, (2) slight browning, (3) minimum of one floret brown, and (4) all florets brown. Plant data were compiled and presented as the mean of the three sampling times for salt deposition, and as means of 20 stems within covers and 50 stems outside of covers for bud and blossom numbers. Yield was estimated by harvesting 4 or 5 $m^2$ (43 or 54 $ft^2$) area plots outside of the shelters and by harvesting all the fruit within the shelters in August of the respective years.

## RESULTS

*Big Pond, PEI.* The data obtained from the experimental trials at Big Pond, PEI show a negative relationship between high levels of salt deposition (exposed plants) and the relative health and production of the wild blueberry in terms of buds, blossoms, and yield (Table 1). Yields

TABLE 1. Effects of salt exposure on various vegetative and reproductive characteristics of the lowbush blueberry at site locations in the Atlantic Canadian provinces of Nova Scotia and Prince Edward Island.

| Experiment Site | Field Location Relative to Shore | Position of Trial in Field | Exposure of Plants | Live Buds # | Dead Buds # | Live Buds % | Blossoms # | Yield (t ha$^{-1}$) |
|---|---|---|---|---|---|---|---|---|
| Big Pond, PE 1999 | Proximal[z] | Seaward[y] | Exposed[x] | 2.6 | 8.5 | 23.8 | 11.5 | 4.4 |
| | | | Covered[w] | 8.8 | 0.0 | 99.8 | 54.2 | 13.3 |
| | Distal[v] | Seaward | Exposed | 5.9 | 3.3 | 64.3 | 24.7 | 6.9 |
| | | | Covered | 9.5 | 0.1 | 99.5 | 59.3 | 15.9 |
| Norway, PE 2000 | Proximal | Seaward | Exposed | 3.9 | 7.2 | 35.5 | 21.4 | 3.3 |
| | | | Covered | 5.4 | 2.6 | 67.3 | 31.2 | 13.4 |
| | Distal | Seaward | Exposed | 5.6 | 7.1 | 44.1 | 27.9 | 4.6 |
| | | | Covered | 5.7 | 0.7 | 89.8 | 33.7 | 23.0 |
| Fox Point, NS 2000 | Proximal | Seaward | Exposed | 5.7 | 10.8 | 84.1 | 42.4 | 2.4 |
| | | | Covered | n.d.[t] | n.d. | n.d. | n.d. | n.d. |
| | | Inland[u] | Exposed | 4.1 | 1.1 | 79.3 | 22.0 | 3.7 |
| | | | Covered | 5.2 | 1.2 | 77.4 | 25.9 | 4.1 |
| Fox Point, NS 2000 | Distal | Seaward | Exposed | 3.5 | 0.0 | 99.4 | 16.0 | 3.6 |
| | | | Covered | n.d. | n.d. | n.d. | n.d. | n.d. |
| | | Inland | Exposed | 5.1 | 0.3 | 94.1 | 25.8 | 6.9 |
| | | | Covered | 3.1 | 0.3 | 90.5 | 13.2 | 5.0 |

[z]Fields proximal to the shore were ~300-900 m from waterline; [y]Portion of field closest to waterline; [x]Not protected from salt spray deposition; [w]Protected from salt spray deposition via 4 mil white polyethylene film; [v]Fields distal to the shore were ~1,200 m from waterline; [u]Portion of field furthest from waterline; [t]Data not available.

obtained from the exposed areas of the fields were lower than those within the shelters. The plants located at the seaward edge of both fields showed greater signs of winter injury (data not shown) as levels of dead buds and total salt accumulation on the stems increased, while live bud and blossom numbers decreased. In addition, the exposed stems had much higher levels of salt deposition compared to the covered plants (Table 2). There were greater numbers of buds and blossoms down the length of exposed stems compared to those obtained from more inland fields.

*Norway, PEI.* Data from the experiment in Norway are not as dramatic as at Big Pond, although the values obtained for live buds, dead buds, blossoms, and yields suggest that greater winter injury occurred at the seaward edge of the field and within the exposed areas in comparison to the areas under shelters (Table 1). Winter injury of the three sets of samples collected was observed to be greatest in terminal buds and unprotected stems in comparison to those under shelters (data not shown). Similarly, the amount of salt deposition was higher for the exposed stems in comparison to the plants protected by shelters (Table 2). Stem architecture at this site appeared similar to that at the Big Pond site, in that more buds and blossoms occurred further down the stems in comparison to samples from more inland fields.

*Fox Point, NS.* Although differences in dead buds, live buds, blossoms, and yields between covered and exposed plants did not appear as dramatic as those from the two PEI sites, the trends were similar (Table 1). In both fields, winter injury was greater in fruit buds of exposed plants relative to those which were sheltered (data not shown); greatest winter damage was observed in the field proximal to the shoreline where plants were not sheltered. The amount of salt deposited on stems of exposed plants was higher (Table 2). Salt deposition directly behind the tree line was less than at 50 m (164 ft) further inland. Bud and blossom numbers at this site were similar to those observed at other sites in the area.

## DISCUSSION

Results from the three experimental field sites used in this study indicate that commercial lowbush blueberry fields situated adjacent or close to marine areas with onshore air flows do experience salt spray-induced winter damage to fruiting stems, with subsequently reduced yields. However, the extent of damage appears to depend on a combination of

TABLE 2. Salt deposition (mg g$^{-1}$) per dry weight of lowbush blueberry stems at various sampling times encompassing the three site locations in the Atlantic Canadian provinces of Nova Scotia and Prince Edward Island.

| Experiment Site | Field Location Relative to Shore | Position of Trial in Field | Exposure of Plants | Sampling Time | | | Mean |
| --- | --- | --- | --- | --- | --- | --- | --- |
| | | | | Feb. | Mar. | April | |
| Big Pond, PE 1999 | Proximal[z] | Seaward[y] | Exposed[x] | 38.20 | 31.55 | 49.13 | 39.63 |
| | | | Covered[w] | 1.57 | 3.42 | 0.00 | 1.66 |
| | Distal[v] | Seaward | Exposed | 13.77 | 41.35 | 8.76 | 21.29 |
| | | | Covered | 0.49 | 0.39 | 0.26 | 0.38 |
| Norway, PE 2000 | Proximal | Seaward | Exposed | 0.13 | 0.32 | 0.10 | 0.15 |
| | | | Covered | 0.06 | 0.76 | 0.05 | 0.29 |
| | Distal | Seaward | Exposed | 0.02 | 0.01 | 0.00 | 0.01 |
| | | | Covered | 0.00 | 0.00 | 0.00 | 0.00 |
| Fox Point, NS 2000 | Proximal | Seaward | Exposed | 0.29 | - | 0.00 | 0.14 |
| | | | Covered | 0.06 | - | 0.04 | 0.05 |
| | | Inland[u] | Exposed | 0.07 | - | 0.07 | 0.07 |
| | | | Covered | 0.00 | - | 0.01 | 0.01 |
| Fox Point, NS 2000 | Distal | Seaward | Exposed | 0.00 | - | 0.00 | 0.00 |
| | | | Covered | 0.00 | - | 0.00 | 0.00 |
| | | Inland | Exposed | 0.22 | - | 0.00 | 0.11 |
| | | | Covered | 0.02 | - | 0.00 | 0.01 |

[z]Fields proximal to the shore were ~300-900 m from waterline; [y]Portion of field closest to waterline; [x]Not protected from salt spray deposition; [w]Protected from salt spray deposition via 4 mil white polyethylene film; [v]Fields distal to the shore were ~1,200 m from waterline; [u]Portion of field furthest from waterline.

conditions that varies from location to location. The presence of increased levels of salt on exposed stems and damage to growth at all three sites is consistent with reports of salt spray damage to dormant fruit buds from highway deicing salt (Eaton et al., 1999). However, caution must be taken when interpreting data from studies using shelters because the apparatus also protects plants from exposure to wind and temperature fluctuations (Pritts et al., 1989). We observed that plants within the Big Pond and the Norway fields were unique in that they produced more buds and blossoms down along the stems, rather than mostly at the tops of stems as is characteristic of the *V. angustifolium* species (Zomlefer, 1994). This may be an adaptation to regular exposure to winter salt spray at the Big Pond and Norway sites.

The use of protective covers during this study demonstrates that reducing salt deposition on fruiting stems during the winter prior to harvest can result in more live buds and blossoms per stem, as well as increased yield relative to those from salt stressed plants. Similarly, lowbush blueberry plants sheltered from deicing salt deposition produced greater numbers of live buds, blossoms, and generated greater yields in comparison to plants not protected (Eaton et al., 1999). Our current work has shown that salt deposition on lowbush blueberry stems during the winter can result in dessication and subsequent death of fruit buds. This result has been observed by other researchers for roadside vegetation exposed to deicing salt during the winter months (Westing, 1969; Smith, 1975).

Salt spray damage may be prevented or reduced when there is sufficient snow fall to completely cover the stems during the winter months (Hall et al., 1971;1979; Kinsman, 1993), since snow cover can increase yield by protecting the plants from salt stress. Larger yields from the Big Pond field in 2001 in comparison to 1999 (John MacDonald, pers. commun.) suggest that this did occur; snowfall during 2001 was greater than in 1999. In addition, protecting plants from salt spray with a tree barrier can be effective in reducing the negative effects of salt spray. The data from the Fox Point experiment suggest that a line of trees at the seaward edge of one field did intercept some of the spray drifting off the Bay of Fundy; plausibly due to swirling wind currents or salt deposition on the trees.

## CONCLUSION AND GROWER BENEFITS

The results of this study demonstrate the harmful effects of salt spray on lowbush blueberry reproductive growth. In addition, higher yields were obtained from plants within the shelters, in comparison to those in exposed areas; suggesting considerable winter damage had occurred. It is likely that salt spray damage will remain a problem for commercial blueberry fields located close to coastal waters, particularly along the north shore area of PEI, and to a lesser degree in NS along the Bay of Fundy. As plant status and yield are negatively affected by salt spray, we recommend that producers employ any method at their disposal to reduce salt deposition. Our study also suggests that erecting snow fences to encourage snow accumulation or planting tree lines as a sea breeze wind break can lower the incidence of salt spray-induced damage to the lowbush blueberry.

## LITERATURE CITED

Capiello, P.E. and S.W. Dunham. 1994. Seasonal variation in low-temperature tolerance of *Vaccinium angustifolium* Ait. HortScience 29:302-304.

Eaton, L.J. 1988. Nitrogen cycling in lowbush blueberry stands. PhD Diss., Dalhousie Univ., Halifax, NS.

Eaton, L.J., J. Hoyle, and A. King. 1999. Effects of deicing salt on lowbush blueberry flowering and yield. Can. J. Plant Sci. 79:125-128.

Eaton, L.J., J. Hoyle, and C. Calder. 2001. Salt damage study at Strathlorne, Inverness County, N.S. Final report. Nova Scotia Dept. Transportation and Public Works. Insurance and Risk Mgt. Halifax, NS.

Hall, I.V., F.R. Forsyth, L.E. Aalders, and L.P. Jackson. 1971. Physiology of the lowbush blueberry. Econ. Bot. 26:68-73.

Hall, I.V., L.E. Aalders, N.L. Nickerson, and S.P. Vander Kloet. 1979. The biological flora of Canada. 1. *Vaccinium angustifolium* Ait., sweet lowbush blueberry. Can. Field Nat. 93:415-430.

Kinsman, G.B. 1993. The History of the Lowbush Blueberry Industry in Nova Scotia 1950-1990. Blueberry Producers Assn. of N.S., Springhill, NS.

Leopold, A.C. and R.P. Willing. 1984. Evidence for toxicity effects of salt on membranes. Pp. 67-78 In: R.G. Staples and G.H. Toenniessen (eds.). Salinity tolerance in plants. Strategies for crop improvement. Wiley, New York.

Pritts, M.P., K.A. Worden, and M. Eames-Sheavly. 1989. Rowcover material and time of application and removal affect ripening and yield of strawberries. J. Amer. Soc. Hort. Sci. 114:531-536.

Quamme, H.A., C. Stushnoff, and C.J. Wesiser. 1972. Winter hardiness of several blueberry species and cultivars in Minnesota. HortScience 7:500-502.

Smith, E.M. 1975. Tree stress from salts and herbicides. J. Aboriculture 1:201-205.

Westing, A.H. 1969. Plants and salt in the roadside environment. Phytopathol. 59: 1174-1181.

Zomlefer, W.B. 1994. Guide to flowering plant families. Univ. N.C. Press, Chapel Hill, NC.

# Main and Interactive Effects
# of Vegetative-Year Applications
# of Nitrogen, Phosphorus,
# and Potassium Fertilizers
# on the Wild Blueberry

David Percival
Kevin Sanderson

**SUMMARY.** An experiment examining the main and interactive effects of vegetative (i.e., sprout) year applications of nitrogen, phosphorus, and potassium fertilizers was conducted at two commercial wild

David Percival is Associate Professor, Department of Environmental Sciences, Nova Scotia Agricultural College, P.O. Box 550, Truro, Nova Scotia, B2N 5E3, Canada.

Kevin Sanderson is Research Scientist, Crops and Livestock Research Centre, Agriculture and Agri-Food Canada, 440 University Avenue, Charlottetown, Prince Edward Island, C1A 4N6, Canada.

This research was supported by the Nova Scotia Department of Agriculture and Fisheries and the Prince Edward Island Agricultural Research Investment Fund.

The authors gratefully thank Dr. Tess Astatkie for valuable help with the statistical analysis, and the support of Bragg Lumber Company, the Wild Blueberry Producers Association of Nova Scotia, the Prince Edward Island Blueberry Growers Association, and the Truro AgroMart.

Mention of a product or trade name does not constitute a guarantee or warrantee of the product by the Nova Scotia Agricultural College nor an endorsement over similar products mentioned.

[Haworth co-indexing entry note]: "Main and Interactive Effects of Vegetative-Year Applications of Nitrogen, Phosphorus, and Potassium Fertilizers on the Wild Blueberry." Percival, David, and Kevin Sanderson. Co-published simultaneously in *Small Fruits Review* (Food Products Press, an imprint of The Haworth Press, Inc.) Vol. 3, No. 1/2, 2004, pp. 105-121; and: *Proceedings of the Ninth North American Blueberry Research and Extension Workers Conference* (ed: Charles F. Forney, and Leonard J. Eaton) Food Products Press, an imprint of The Haworth Press, Inc., 2004, pp. 105-121. Single or multiple copies of this article are available for a fee from The Haworth Document Delivery Service [1-800-HAWORTH, 9:00 a.m. - 5:00 p.m. (EST). E-mail address: docdelivery@haworthpress.com].

blueberry fields near Kemptown, Nova Scotia and Mount Vernon, Prince Edward Island during 2000 and 2001. The nitrogen (N), phosphorus (P), and potassium (K) sources consisted of urea (N at 0 to 60 kg·ha$^{-1}$), triple superphosphate ($P_2O_5$ at 0 to 150 kg·ha$^{-1}$), and potash ($K_2O$ at 0 to 60 kg·ha$^{-1}$). A completely randomized rotatable composite design was used with 16 treatments, 4 replications, a plot size of 6 m × 8 m, and 2 m buffers between plots. Overall, leaf tissue N, P, and K levels were increased at both sites with fertilizer applications. Main and interactive effects of the soil applied N, P, and K on stem length, individual stem dry weight, and stem density were also present at both sites. Significant effects of soil-applied N and K, and soil applied K on fruit set were present at the Kemptown and Mount Vernon sites with fruit set increasing up to 20% and 51%, respectively. Harvestable yield varied in response to the N-P-K treatments with no treatment having a significantly higher yield than the non-fertilized treatment at the Kemptown site, and the harvestable yield of the unfertilized treatment being as much as 36% lower than other soil-applied N-P-K treatments at the Mount Vernon site. Therefore, results from this study illustrate the importance of monitoring leaf tissue N, P, and K levels, the ability to alter leaf tissue N, P, and K levels and correct leaf tissue deficiencies, and the beneficial yield component effects of soil-applied N, P, and K when applied to the wild blueberry under nutrient deficient conditions. *[Article copies available for a fee from The Haworth Document Delivery Service: 1-800-HAWORTH. E-mail address: <docdelivery@haworthpress.com> Website: <http://www.HaworthPress. com>* © 2004 by The Haworth Press, Inc. All rights reserved.]*

**KEYWORDS.** *Vaccinium angustifolium* Ait., plant nutrition, yield components, harvestable yield

## INTRODUCTION

The wild blueberry (*Vaccinium angustifolium* Ait.) is an indigenous plant that has developed into an important horticultural commodity in northeastern North America. Although commercially managed, the fields originate when competing vegetation is removed from native plant stands found in forest clearings. Wild blueberries are members of the Ericaceae family and have been characterized as calcifuge (i.e., acid-loving) plants that grow on nutritionally marginal and poorly structured land (Korcak, 1988). Fields are commercially managed on a two-year cycle with the perennial shoot being pruned in alternate years to maximize floral bud initiation, fruit set, yield, and ease of mechanical harvest. Selective herbicides are applied to control competing weeds in the spring of the first year and fertilizers generally containing N, P, and K are also applied.

A tremendous amount of uncertainty presently exists in the nutrient management of wild blueberries. Past nutrient management research has provided valuable information on seasonal growth (Hainstock, 2002), nutrient dynamics (Townsend and Hall, 1970; Trevett et al., 1968), optimum nutrient levels for leaf tissue and subsequent vegetative growth (Korcak, 1988; Korcak, 1989; Penney and MacRae, 2000; Smagula, 1987; Trevett, 1972), the influence of soil pH (Hall et al., 1964), denitrification (Eaton and Patriquin, 1989), nitrification potential (Eaton and Patriquin, 1988), inorganic nitrogen levels (Eaton and Patriquin, 1988; Percival and Privé, 1992), and the impact of inorganic nitrogen formulation (Percival and Privé, 2002; Smagula and Hepler, 1978). In addition, the effects of phosphorus fertilizer applications (Eaton et al. 1997), pruning method (Warman, 1987), organic amendments (Warman, 1987), micronutrient applications including boron (Chen et al., 1998; Perrin, 1999), N-P-K fertilizers (Penney and MacRae, 2000), various timings of N-P-K fertilizer application (Smagula and Hepler, 1978), and methods to improve phosphorus deficiency (Smagula and Dunham, 1995) have been examined. Although the leaf tissue thresholds and the effects of various fertilizer formulations, amendments, and application timings on the wild blueberry are known, the main and interactive effects of N, P, and K remain largely unknown. Hence, the objectives of this study were to examine the main and interactive effects of soil applied N, P, and K on the leaf mineral nutrition, growth, development, and harvestable yield of wild blueberries.

## MATERIALS AND METHODS

The experiment was conducted at two commercial wild blueberry fields in Atlantic Canada during the 2000 and 2001 growing seasons. The commercial wild blueberry fields were situated at Kemptown, Nova Scotia (45°30' N, 63°8' W) and Mount Vernon, Prince Edward Island (46°1' N, 62°45' W). The wild blueberries at the Kemptown and Mount Vernon sites consisted of indigenous and heterogenous phenotypes that were situated on Orthic Podzols belonging to the Cobequid (Webb et al., 1991) and Culloden (MacDougall et al., 1988) soil classifications, respectively.

A three-factor, rotatable, central composite design was used to identify the specific orthogonal treatment combinations required to attain a second order response surface (Cochran and Cox, 1957). Sixteen treatment combinations with five levels (0%, 20%, 50%, 80%, and 100% of full application rate) each of nitrogen, phosphorus, and potassium were used according to design criteria to investigate the main and interactive effects of nitrogen, phosphorus, and potassium (May and Pritts, 1993). A plot size of 6 m × 8 m was used and the N, P, and K application rates consisted of N at 0 to 60 kg·ha$^{-1}$ in the form of urea, $P_2O_5$ at 0 to 150 kg·ha$^{-1}$ in the form of triple-superphosphate, and $K_2O$ at 0 to 70 kg·ha$^{-1}$ in the form of muriate of potash. Fertilizers were applied using a Scott SR2000 rotary fertilizer spreader (Marysville, Ohio). The fertilizer treatments were applied to plants in the vegetative (i.e., "sprout") stage of production on 5 May 2000 and 10 May 2000 for the Kemptown and Mount Vernon sites, respectively. Leaf samples were collected in the vegetative year on 27 and 28 July 2000 at the Mount Vernon and Kemptown sites, respectively. Leaf tissue samples were obtained by randomly collecting leaves from 20 stems per plot, placing the leaves in a 60 EC drying oven until constant dry weight had been achieved, and grinding the leaves in a Wiley mill equipped with a 20 mesh screen. The samples were then sent to the Prince Edward Island Department of Agriculture and Forestry, Soil and Plant Tissue Analytical Laboratory for inorganic nutrient analysis (i.e., N, P, K, Ca, Mg, Fe, Mn, Cu, Zn, and B).

Introduced pollinators (i.e., honeybees) were used at a hive density of approximately four hives·ha$^{-1}$ during the cropping year to ensure adequate pollination. Stem samples for yield component analysis were collected by randomly selecting 20 stems per plots on 19 July 2001 (Kemptown), and 21 July 2001 (Mount Vernon). Measurements of stem length, node number, number of flowering nodes, berry number, length of the fruiting zone and stem dry weight were collected. In addition, stem densities were determined by recording the number of stems in three, randomly selected, 225 cm$^2$ quadrats in each plot. Harvest occurred on 14 August 2001 (Kemptown) and 20 August 2001 (Mount Vernon). Berries were harvested with a forty-tine commercial wild blueberry hand rake from four, randomly selected 1 m$^2$ quadrats in each plot, and harvested berry yield recorded using a digital balance (Mettler PE 6000, Burlington, ON). Average berry weight was determined by collecting a 500 mL composite sample from each plot and weighing 100 randomly selected berries per sample. Analysis of variance was completed using the general linear models (GLM) procedure of SAS (Version 8, SAS Institute, Cary, NC). Models were initially fitted to a second order polynomial function with variables omitted if nonsignificant (i.e., P > |T| > 0.10). The normality of a model was then tested using the Univariate procedure of SAS. Regression coefficients were also obtained using the Proc GLM procedure and means generated using the LSMeans procedure of SAS.

## RESULTS

*Plant nutrition.* Overall, significant main and interactive effects of the soil-applied nitrogen, phosphorus, and potassium fertilizers on leaf tissue N, P, and K were present at both the Kemptown and Mount Vernon sites. With leaf tissue N, main effects of potassium (K*K) and interactive effects of effect of P and K (P*K) were present at the Kemptown site with leaf tissue N levels of the non-fertilized treatment being up to 13% lower than the other treatments (Table 1). At the Mount Vernon site, significant effects of soil-applied N and K were present with leaf tissue N levels increasing up to 50% with applications of K and N (Table 1). Leaf tissue P at the Kemptown site was influenced by main and interactive effects of K and P with leaf tissue P levels increasing up to 81% with applications of K and P (Table 2). At the Mount Vernon site, main and interactive effects of soil-applied N and P were present with leaf tissue P increasing by 23% under conditions of high levels of N, and increasing by 15% under conditions of high levels of P (Table 2). Leaf tissue K was influenced by main and interactive effects of soil applied K and P at the Kemptown site with high levels of P and K increasing leaf tissue K by 15% and 17%, respectively (Table 3). There was no effect of soil-applied K on leaf tissue K at the Mount Vernon site levels with leaf tissue K levels only being influenced by nitrogen and phosphorus applications (Table 3). Leaf tissue K levels at the Mount Vernon site increased by 22% and 24%, respectively, under conditions of high P and moderate N and P application rates, respectively (Table 3).

*Yield components.* Soil-applied N, P, and K affected the yield components at both the Kemptown and Mount Vernon sites. Node number per stem was influenced by soil-applied N, P, and K with N applications increasing node number up to 18% and 29% at the Kemptown and Mount Vernon sites, respectively (data not shown). Stem length was also affected by applications of P and K at the Kemptown site and N and K at the Mount Vernon sites. The fertilized treatments had stem lengths that were up to 26% and 20% greater than the unfertilized treatment at

TABLE 1. Sprout year applications of nitrogen, phosphorus, and potassium affect leaf tissue nitrogen levels (%) of wild blueberries during the 2000 growing season at two commercial fields in Atlantic Canada.

| | | Kemptown, Nova Scotia | | | | |
|---|---|---|---|---|---|---|
| | Nitrogen | 0 | 0.40 | 1.0 | 1.6 | 2.0 |
| Phosphorus | Potassium | | | | | |
| 0 | 0 | 1.68 | | | | |
| | 1.0 | | | 1.77 | | |
| 0.40 | 0.40 | | | 1.70 | 1.74 | |
| | 1.6 | | | 1.89 | 1.86 | |
| 1.0 | 0 | | | 1.90 | | |
| | 1.0 | 1.74 | | 1.78 | | 1.76 |
| | 2.0 | | | 1.86 | | |
| 1.6 | 0.40 | | | 1.82 | 1.83 | |
| | 1.6 | | | 1.76 | 1.75 | |
| 2.0 | 1.0 | | | 1.80 | | |
| Sig. effects (P = 0.05) | | K*K, P*K (R² = 0.581) | | | | |
| SEM = 0.0403 | | | | | | |
| | | Mount Vernon, Prince Edward Island | | | | |
| 0 | 0 | 0.809 | | | | |
| | 1.0 | | | 0.914 | | |
| 0.40 | 0.40 | | | 0.769 | 0.992 | |
| | 1.6 | | | 0.798 | 0.906 | |
| 1.0 | 0 | | | 0.891 | | |
| | 1.0 | 0.938 | | 0.948 | | 1.21 |
| | 2.0 | | | 0.907 | | |
| 1.6 | 0.40 | | | 0.717 | 0.987 | |
| | 1.6 | | | 0.737 | 0.998 | |
| 2.0 | 1.0 | | | 0.903 | | |
| Sig. effects (P = 0.05) | | N, K (R² = 0.741) | | | | |
| SEM = 0.0208 | | | | | | |

Values in white represent those within the standard $2^3$ factorial design.
Shaded areas include 6 star points, a center point (1, 1, 1), and a control (0, 0, 0: no soil-applied fertilizer treatment).

the Kemptown and Mount Vernon sites, respectively (Table 4). Significant effects of soil-applied N, P, and K on individual upright stem dry weight were also observed with stem dry weights of the fertilized treatments being as much as 36% and 60% greater than the unfertilized treatment at the Kemptown and Mount Vernon sites, respectively (Table 5). Stem density was affected by soil-applied N, P, and K, with P applications reducing stem density up to 19% at the Kemptown site, and appli-

TABLE 2. Sprout year applications of nitrogen, phosphorus, and potassium affect the leaf tissue phosphorus levels (%) of wild blueberries during the 2000 growing season at two commercial fields in Atlantic Canada.

| | | Kemptown, Nova Scotia | | | | |
|---|---|---|---|---|---|---|
| | Nitrogen | 0 | 0.40 | 1.0 | 1.6 | 2.0 |
| Phosphorus | Potassium | | | | | |
| 0 | 0 | 0.130 | | | | |
| | 1.0 | | | 0.0911 | | |
| 0.40 | 0.40 | | 0.141 | | 0.140 | |
| | 1.6 | | 0.156 | | 0.149 | |
| 1.0 | 0 | | | 0.202 | | |
| | 1.0 | 0.145 | | 0.215 | | 0.145 |
| | 2.0 | | | 0.235 | | |
| 1.6 | 0.40 | | 0.193 | | 0.189 | |
| | 1.6 | | 0.198 | | 0.191 | |
| 2.0 | 1.0 | | | 0.198 | | |
| Sig. effects (P = 0.05) SEM = 0.00407 | | P, P*K, K*K *(R$^2$ = 0.511)* | | | | |
| | | Mount Vernon, Prince Edward Island | | | | |
| 0 | 0 | 0.105 | | | | |
| | 1.0 | | | 0.0914 | | |
| 0.40 | 0.40 | | 0.124 | | 0.108 | |
| | 1.6 | | 0.0991 | | 0.0853 | |
| 1.0 | 0 | | | 0.128 | | |
| | 1.0 | 0.0938 | | 0.138 | | 0.130 |
| | 2.0 | | | 0.131 | | |
| 1.6 | 0.40 | | 0.104 | | 0.0889 | |
| | 1.6 | | 0.09425 | | 0.0998 | |
| 2.0 | 1.0 | | | 0.121 | | |
| Sig. effects (P = 0.05) SEM = 0.00568 | | P, P*P, N*P *(R$^2$ = 0.706)* | | | | |

Values in white represent those within the standard $2^3$ factorial design.
Shaded areas include 6 star points, a center point (1, 1, 1), and a control (0, 0, 0: no soil-applied fertilizer treatment).

cations of N and K causing an increase in stem density by as much as 28% at the Mount Vernon site (Table 6).

The soil-applied N, P, and K affected reproductive yield components at both the Kemptown and Highland Village sites. Main and interactive effects of N and K on fruit were present at Kemptown with the unfertilized treatment having up to 20% fewer fruit per stem than treatments with applications of N and K (Table 7). Significant effects of the soil-

TABLE 3. Sprout year applications of nitrogen, phosphorus, and potassium affect the leaf tissue potassium levels (%) of wild blueberries during the 2000 growing season at two commercial fields in Atlantic Canada.

| | | Kemptown, Nova Scotia | | | | |
|---|---|---|---|---|---|---|
| | Nitrogen | 0 | 0.40 | 1.0 | 1.6 | 2.0 |
| Phosphorus | Potassium | | | | | |
| 0 | 0 | 0.461 | | | | |
| | 1.0 | | | 0.541 | | |
| 0.40 | 0.40 | | | 0.523 | 0.536 | |
| | 1.6 | | | 0.595 | 0.489 | |
| 1.0 | 0 | | | 0.553 | | |
| | 1.0 | | 0.493 | 0.503 | | 0.533 |
| | 2.0 | | | 0.541 | | |
| 1.6 | 0.40 | | | 0.505 | 0.526 | |
| | 1.6 | | | 0.579 | 0.585 | |
| 2.0 | 1.0 | | | 0.531 | | |
| Sig. effects (P = 0.05) SEM = 0.0229 | | K, P*K $(R^2 = 0.592)$ | | | | |
| | | Mount Vernon, Prince Edward Island | | | | |
| 0 | 0 | 0.403 | | | | |
| | 1.0 | | | 0.451 | | |
| 0.40 | 0.40 | | | 0.439 | 0.491 | |
| | 1.6 | | | 0.477 | 0.479 | |
| 1.0 | 0 | | | 0.484 | | |
| | 1.0 | | 0.463 | 0.501 | | 0.493 |
| | 2.0 | | | 0.504 | | |
| 1.6 | 0.40 | | | 0.495 | 0.496 | |
| | 1.6 | | | 0.506 | 0.475 | |
| 2.0 | 1.0 | | | 0.489 | | |
| Sig. effects (P = 0.05) SEM = 0.0275 | | P, N*P $(R^2 = 0.401)$ | | | | |

Values in white represent those within the standard $2^3$ factorial design.
Shaded areas include 6 star points, a center point (1, 1, 1), and a control (0, 0, 0: no soil-applied fertilizer treatment).

applied K on fruit set were also present at the Mount Vernon site with K applications increasing fruit set by as much as 51% and maximum fruit set occurring at K levels of $K_2O$ at 48 kg·ha$^{-1}$ (Table 7).

Harvestable yield was also impacted by soil-applied N, P, and K, with significant interactive effects of N and P existing at the Kemptown site, and interactive effects of soil-applied N, P, and K existing at the Mount Vernon site (Table 8). Upon examining the differences be-

TABLE 4. Sprout year applications of nitrogen, phosphorus, and potassium affect stem length (cm) of wild blueberries during the 2001 growing season at two commercial fields in Atlantic Canada.

| | | Kemptown, Nova Scotia. | | | | |
|---|---|---|---|---|---|---|
| | Nitrogen | 0 | 0.40 | 1.0 | 1.6 | 2.0 |
| Phosphorus | Potassium | | | | | |
| 0 | 0 | 13.6 | | | | |
| | 1.0 | | | 16.3 | | |
| 0.40 | 0.40 | | 15.1 | | 15.0 | |
| | 1.6 | | 15.5 | | 15.9 | |
| 1.0 | 0 | | | 16.3 | | |
| | 1.0 | 16.3 | | 16.3 | | |
| | 2.0 | | | 16.3 | | 16.3 |
| 1.6 | 0.40 | | 15.2 | | 15.5 | |
| | 1.6 | | 17.1 | | 15.4 | |
| 2.0 | 1.0 | | | 16.3 | | |
| Sig. effects (P = 0.05) SEM = 0.504 | | N*P, N*K, P*K $(R^2 = 0.614)$ | | | | |
| | | Mount Vernon, Prince Edward Island | | | | |
| 0 | 0 | 14.7 | | | | |
| | 1.0 | | | 17.6 | | |
| 0.40 | 0.40 | | 16.8 | | 16.9 | |
| | 1.6 | | 15.4 | | 15.8 | |
| 1.0 | 0 | | | 16.0 | | |
| | 1.0 | 15.9 | | 17.1 | | |
| | 2.0 | | | 15.8 | | 16.3 |
| 1.6 | 0.40 | | 16.5 | | 16.8 | |
| | 1.6 | | 16.9 | | 16.9 | |
| 2.0 | 1.0 | | | 17.1 | | |
| Sig. effects (P = 0.05) SEM = 0.445 | | N, N*N, K*K $(R^2 = 0.561)$ | | | | |

Values in white represent those within the standard $2^3$ factorial design.
Shaded areas include 6 star points, a center point (1, 1, 1), and a control (0, 0, 0: no soil-applied fertilizer treatment).

tween the unfertilized treatment and the other soil applied N, P, and K treatments, there was no treatment with a significantly higher harvestable yield at the Kemptown site (Table 8). However, applications of N increased harvestable yield up to 31% at the Mount Vernon site (Table 8).

TABLE 5. Sprout year applications of nitrogen, phosphorus, and potassium affect individual upright stem weight (g) of wild blueberries during the 2001 growing season at two commercial fields in Atlantic Canada.

| | | Kemptown, Nova Scotia | | | | |
|---|---|---|---|---|---|---|
| | Nitrogen | 0 | 0.40 | 1.0 | 1.6 | 2.0 |
| Phosphorus | Potassium | | | | | |
| 0 | 0 | 0.760 | | | | |
| | 1.0 | | | 0.901 | | |
| 0.40 | 0.40 | | 0.851 | | 1.03 | |
| | 1.6 | | 0.948 | | 0.931 | |
| 1.0 | 0 | | | 0.977 | | |
| | 1.0 | 0.931 | | 0.963 | | 1.03 |
| | 2.0 | | | 0.958 | | |
| 1.6 | 0.40 | | 0.871 | | 0.936 | |
| | 1.6 | | 0.976 | | 0.972 | |
| 2.0 | 1.0 | | | 0.952 | | |
| Sig. effects (P = 0.05) SEM = 0.0865 | | N, N*P ($R^2 = 0.707$) | | | | |
| | | Mount Vernon, Prince Edward Island | | | | |
| 0 | 0 | 0.663 | | | | |
| | 1.0 | | | 0.995 | | |
| 0.40 | 0.40 | | 0.845 | | 0.971 | |
| | 1.6 | | 0.871 | | 0.973 | |
| 1.0 | 0 | | | 0.792 | | |
| | 1.0 | 0.870 | | 0.952 | | 0.961 |
| | 2.0 | | | 0.771 | | |
| 1.6 | 0.40 | | 0.810 | | 0.963 | |
| | 1.6 | | 0.863 | | 0.981 | |
| 2.0 | 1.0 | | | 1.06 | | |
| Sig. effects (P = 0.05) SEM = 0.0848 | | N, K*K ($R^2 = 0.762$) | | | | |

Values in white represent those within the standard $2^3$ factorial design.
Shaded areas include 6 star points, a center point (1, 1, 1), and a control (0, 0, 0: no soil-applied fertilizer treatment).

## DISCUSSION

*Plant nutrition.* In addition to the inherent genotypic variability challenges present in native wild blueberry stands, difficulties in nutrition studies have been encountered as a result of blueberries being a calcifuge plant with relatively low nutrient requirements (Korcak, 1988) and the absence of a suitable soil analysis technique to accurately determine the

TABLE 6. Sprout year applications of nitrogen, phosphorus, and potassium affect wild blueberry stem density (stems per m$^2$) during at the conclusion of the 2001 growing season at two commercial fields in Atlantic Canada.

| Phosphorus | Potassium | Nitrogen 0 | 0.40 | 1.0 | 1.6 | 2.0 |
|---|---|---|---|---|---|---|
| Kemptown, Nova Scotia |
| 0 | 0 | 2039 | | | | |
| | 1.0 | | | 2016 | | |
| 0.40 | 0.40 | | 2182 | | 2217 | |
| | 1.6 | | 2001 | | 2011 | |
| 1.0 | 0 | | | 1879 | | |
| | 1.0 | 1868 | | 1939 | | 1888 |
| | 2.0 | | | 1798 | | |
| 1.6 | 0.40 | | 1801 | | 1698 | |
| | 1.6 | | 1771 | | 1687 | |
| 2.0 | 1.0 | | | 1717 | | |
| Sig. effects (P = 0.05) SEM = 188.5 | | P (R$^2$ = 0.691) | | | | |
| Mount Vernon, Prince Edward Island |
| 0 | 0 | 1606 | | | | |
| | 1.0 | | | 1950 | | |
| 0.40 | 0.40 | | 1867 | | 2057 | |
| | 1.6 | | 1985 | | 1989 | |
| 1.0 | 0 | | | 1877 | | |
| | 1.0 | 2050 | | 1896 | | 2004 |
| | 2.0 | | | 2021 | | |
| 1.6 | 0.40 | | 1883 | | 1986 | |
| | 1.6 | | 2005 | | 1779 | |
| 2.0 | 1.0 | | | 1933 | | |
| Sig. effects (P = 0.05) SEM = 216.8 | | N*K (R$^2$ = 0.534) | | | | |

Values in white represent those within the standard 2$^3$ factorial design.
Shaded areas include 6 star points, a center point (1, 1, 1), and a control (0, 0, 0: no soil-applied fertilizer treatment).

availability of various macronutrients including phosphorus (Ring, 2001). Additional factors hindering wild blueberry nutrition studies can be attributed to 75% to 85% of the total dry matter of the wild blueberry existing in the form of roots and rhizomes which provide a large nutrient reservoir (i.e., N at 7,500 to 8,000 kg·ha$^{-1}$) (Hainstock, 2002), and the direct, indirect, and interactive effects of inorganic nutrients on plant nutrition, growth, development, and yield (May and Pritts, 1993). Given

TABLE 7. Sprout year applications of nitrogen, phosphorus, and potassium affect the fruit set (fruit per stem) of wild blueberries during the 2001 growing season at two commercial fields in Atlantic Canada.

| | | Kemptown, Nova Scotia | | | | |
|---|---|---|---|---|---|---|
| | Nitrogen | 0 | 0.40 | 1.0 | 1.6 | 2.0 |
| Phosphorus | Potassium | | | | | |
| 0 | 0 | 16.9 | | | | |
| | 1.0 | | | 17.8 | | |
| 0.40 | 0.40 | | 18.3 | | 19.4 | |
| | 1.6 | | 20.0 | | 19.5 | |
| 1.0 | 0 | | | 18.2 | | |
| | 1.0 | 18.8 | | 18.9 | | 19.0 |
| | 2.0 | | | 19.9 | | |
| 1.6 | 0.40 | | 17.8 | | 18.9 | |
| | 1.6 | | 20.2 | | 19.4 | |
| 2.0 | 1.0 | | | 18.9 | | |
| Sig. effects (P = 0.05) SEM = 1.24 | | K, N*K ($R^2$ = 0.413) | | | | |
| | | Mount Vernon, Prince Edward Island | | | | |
| 0 | 0 | 15.5 | | | | |
| | 1.0 | | | 23.0 | | |
| 0.40 | 0.40 | | 18.8 | | 19.7 | |
| | 1.6 | | 19.4 | | 20.0 | |
| 1.0 | 0 | | | 15.5 | | |
| | 1.0 | 22.9 | | 23.4 | | 22.4 |
| | 2.0 | | | 15.5 | | |
| 1.6 | 0.40 | | 18.3 | | 20.3 | |
| | 1.6 | | 20.9 | | 21.0 | |
| 2.0 | 1.0 | | | 19.8 | | |
| Sig. effects (P = 0.05) SEM = 1.67 | | K*K ($R^2$ = 0.523) | | | | |

Values in white represent those within the standard $2^3$ factorial design.
Shaded areas include 6 star points, a center point (1, 1, 1), and a control (0, 0, 0: no fertilizer treatment).

these challenges, it is understandable why conflicting results with soil-applied nutrient analysis have been obtained, and the continued need for long-term nutrient management studies. Despite these challenges and the drought conditions experienced at the Kemptown site during the 2001 season, beneficial effects of the soil-applied N, P, and K treatments on leaf tissue N, P, and K were present at both locations in this study.

TABLE 8. Sprout year applications of nitrogen, phosphorus, and potassium affect the harvestable yield (berries harvested per $m^2$) of wild blueberries in 2001 at two commercial fields in Atlantic Canada.

| | | Kemptown, Nova Scotia | | | | |
|---|---|---|---|---|---|---|
| | Nitrogen | 0 | 0.40 | 1.0 | 1.6 | 2.0 |
| Phosphorus | Potassium | | | | | |
| 0 | 0 | 645 | | | | |
| | 1.0 | | | 622 | | |
| 0.40 | 0.40 | | 610 | | 560 | |
| | 1.6 | | 647 | | 631 | |
| 1.0 | 0 | | | 567 | | |
| | 1.0 | 587 | | 680 | | 537 |
| | 2.0 | | | 587 | | |
| 1.6 | 0.40 | | 548 | | 604 | |
| | 1.6 | | 678 | | 632 | |
| 2.0 | 1.0 | | | 608 | | |
| Sig. effects (P = 0.05) SEM = 72.0 | | N*P $(R^2 = 0.331)$ | | | | |
| | | Mount Vernon, Prince Edward Island | | | | |
| 0 | 0 | 634 | | | | |
| | 1.0 | | | 737 | | |
| 0.40 | 0.40 | | 669 | | 694 | |
| | 1.6 | | 809 | | 859 | |
| 1.0 | 0 | | | 707 | | |
| | 1.0 | 827 | | 789 | | 777 |
| | 2.0 | | | 719 | | |
| 1.6 | 0.40 | | 802 | | 829 | |
| | 1.6 | | 661 | | 682 | |
| 2.0 | 1.0 | | | 773 | | |
| Sig. effects (P = 0.05) SEM = 61.8 | | N*K, K*P $(R^2 = 0.784)$ | | | | |

Values in white represent those within the standard $2^3$ factorial design.
Shaded areas include 6 star points, a center point (1, 1, 1), and a control (0, 0, 0: no fertilizer treatment).

With leaf tissue N levels, the complimentary effect of the soil-applied potassium on leaf tissue N may have been due to the beneficial impact of K on photosynthesis (Marschner, 1997) and protein synthesis (Wyn-Jones and Pollard, 1979). The main and interactive effects of soil-applied N and P on leaf tissue P observed at the Mount Vernon site were effective in raising the leaf tissue P levels above the minimum threshold of 0.125 (Trevett, 1972) (Table 2). These results are similar in magnitude to those reported by Penney and MacRae (2000), and more pronounced than the single factor response observed by Eaton et al. (1997). The mechanisms associated with the interactive effects of soil-applied N and K on leaf tissue P levels may have been due to increases in carbohydrate and biomass production and the subsequent facilitative effects of increased carbohydrate status on the ericoid mycorrhizal association of the wild blueberry (Jeliazkova and Percival, 2002). Extensive (i.e., 69% to 72%) colonization of the wild blueberry root system has been reported (Jeliazkova and Percival, 2002), and although the importance of ericoid mycorrhiza has been mainly attributed to their role in N assimilation, this association may supply P to the plant as well as nutrients from organic sources that are normally unavailable to host roots (Goulart et al., 1993; Read and Stribley, 1973).

   *Yield components.* Main and interactive effects of soil-applied N, P, and K were observed on an individual stem and unit area (i.e., stems per m²) basis in this study with individual stem length, individual stem dry weight, stem density, and the number of set fruit per stem being significantly affected by the soil-applied N, P, and K treatments (Tables 4, 5, and 6). Of particular interest from this study was the importance of K to the yield potential, with the soil-applied K influencing stem density at the Mount Vernon site and the number of set fruit at both the Kemptown and Mount Vernon sites (Table 7). These results suggest that K availability and uptake may exert more of an effect on fruit set in the wild blueberry than other calcifuges (Korcak, 1988), and may warrant further attention in future studies.

   *Harvestable yield.* Ultimately, significant main and interactive effects of the soil-applied N, P, and K treatments on harvestable yield (i.e., berries harvested per m²) were present at both sites. Despite beneficial effects of the soil-applied N and K on fruit set, no beneficial effects of the N, P, and K treatments on harvestable yield were present at the Kemptown site. This result may have been caused by a combination of the leaf tissue levels of N, P, and K being within the nutrient sufficiency thresholds established by Trevett (1972), and also the confounding of treatment effects by environmental factors. Extensive drought conditions were encountered during the 2001 growing season at the Kemptown site with soil moisture levels of < 8% ($m^3$ $H_2O$ per $m^3$ soil) for most of June, July, and August (data not reported), resulting in the possible negating of treatment effects due to excessively unharvestable berry size or premature fruit loss due to berry shatter. However, at the Mount Vernon site, beneficial effects of the soil-applied N, P, and K treatments on harvestable yield were present with harvestable yield increasing as much as 31% compared to the unfertilized treatment (Table 8). This may have been due to the leaf tissue N, P, and K levels of the unfertilized treatment being less than the 1.60%, 0.125%, and 0.40% thresholds for leaf tissue N, P, and K, respectively, established by Trevett (1972). The leaf tissue N, P, and K levels increased to levels in which leaf tissue P and K deficiencies were no longer present (Tables 2 and 3). Therefore, results from this investigation confirm the importance of examining the direct and indirect and main and interactive effects of soil-applied N, P, and K on yield, and subsequently, may provide insight into possible mechanisms contributing the a lack of response observed in previous studies. Overall, a more pronounced N effect may have been observed in this study if a different N formulation such as ammonium sulfate had been used. Excessive N losses have been observed in previous broadcast (i.e., surface applied and non-incorporated) fertility studies with urea as a result of lack of rainfall and subsequent volatilization losses of N (Demeyer et al., 1995).

## CONCLUSIONS

Results from this study have provided insight into the main and interactive effects of soil applied N, P, and K on the plant nutrition, yield components, and harvestable yield of wild blueberries. Despite large levels of inherent phenotypic variability associated with the native wild blueberry stands, main and interactive effects of N, P, and K were present on all of the plant nutrition and yield component variables examined. In addition, results from this study clearly indicated that leaf tissue nutrient levels were increased with applications of soil-applied N, P, and K. The benefits to blueberry producers that can be obtained from this study include insight into the importance of using leaf tissue nutrient analysis as part of a regular nutrient management program, an improved understanding of the main and interactive effects of N, P, and K on the growth and development of the wild blueberry, and the beneficial effects of N, P, and K fertilizers on harvestable yield under nutrient deficient conditions. However, before commercial recommendations are made, further replication is required.

## LITERATURE CITED

Chen, Y., J. Smagula, W. Litten, and S. Dunham. 1998. Effect of boron and calcium foliar sprays on pollen germination and development, fruit set, seed development, and berry yield of lowbush blueberry. J. Amer. Soc. Hort. Sci. 123:524-531.

Cochran, W.C. and G.M. Cox. 1957. Experimental Designs. (2nd ed.). J. Wiley and Sons, New York. pp. 346-353.

Demeyer, P., G. Hofman, and O. VanCleemput. 1995. Fitting ammonia volatilization dynamics with a logistic equation. Soil Science Soc. Amer. J. 59:261-265.

Eaton, L.J. and D.G. Patriquin. 1989. Denitrification in lowbush blueberry soils. Can. J. Soil Sci. 69:303-312.

Eaton, L.J. and D.G. Patriquin. 1988. Inorganic nitrogen levels and nitrification potential in lowbush blueberry soils. Can. J. Soil Sci. 68:63-75.

Eaton, L.J., K.R. Sanderson, and G.W. Stratton. 1997. Fertilizer phosphorus in lowbush blueberries: Effects and fate. Acta Hort. 446:477-486.

Goulart, B.L., M.L. Schroeder, K. Demchak, J.P. Lynch, J.R. Clark, R.L. Darnell, and W.F. Wilcox. 1993. Blueberry mycorrhizae: Current knowledge and future directions. Acta Hort. 346:230-239.

Hainstock, L.J. 2002. Seasonal phytochemistry, growth dynamics and carbon allocation of the wild blueberry (*Vaccinium angustifolium* Ait.). MS Diss., Dalhousie University, Halifax, NS.

Hall, I.V., L.E. Aalders, and L.R. Townsend. 1964. The effects of soil pH on the mineral composition and growth of the lowbush blueberry. Can. J. Plant Sci. 44:433-438.

Jeliazkova, E. and D. Percival. 2003. The influence of drought stress on the ericoid mycorrhizal association in wild blueberry (*Vaccinium angustiflium* Ait.) roots. Can. J. Plant Sci. *In-press.*

Korcak, R.F. 1988. Nutrition of blueberries and other calcifuges. Hort. Rev. 10:183-227.

Korcak. R.F. 1989. Variation in nutrient requirements of blueberries and other calcifuges. HortScience 24:573-578.

MacDougall, J.I., C. Veer, and F. Wilson. 1988. Soils of Prince Edward Island: Prince Edward Island Soil Survey. Agr. Can. L.R.R.C. Contrib. No. 84-85.

Marschner, H. 1997. Mineral Nutrition of Higher Plants. 2nd Ed. Academic Press, London.

May, G.M. and M.P. Pritts. 1993. Phosphorus, zinc and boron influence yield components of 'Earliglow' strawberry. J. Amer. Soc. Hort. Sci. 118:43-49.

Penney, B.G. and K.B. McRae. 2000. Herbicidal weed control and crop-year NPK fertilization improves lowbush blueberry (*Vaccinium angustifolium* Ait.) production. Can. J. Plant Sci. 80:351-361.

Percival, D. and J.P. Privé. 2002. Nitrogen formulation and application date influence plant nutrition, growth, development, and yield of wild blueberry. Acta Hort. 574:347-355.

Perrin, G.D. 1999. Main and interactive effects of boron on lowbush blueberry nutrition, growth, development, and yield. MS Diss., Dalhousie University, Halifax, NS.

Read, D.J. and D.P. Stribley. 1973. Effect of mycorrhizal infection on nitrogen and phosphorus nutrition of ericaceous plants. Nature 244:81-82.

Ring, R. 2001. A comparison of five extraction methods for determining available phosphorus in Nova Scotia blueberry soils. MS Diss., Dalhousie University, Halifax, NS.

Smagula, J.M. 1987. Lowbush blueberry nutrition series. N-P-K. Univ. Maine Coop. Ext. Serv. Factsheet No. 223.

Smagula, J.M. and S. Dunham. 1995. Diammonium phosphate corrects phosphorus deficiency in lowbush blueberry. J. Small Fruit Viticult. 3:183-191.

Smagula, J.M. and P.R. Hepler. 1978. Comparison of urea and sulfur-coated urea as nitrogen source for lowbush blueberries growing on a colton gravelly sand loam. J. Amer. Soc. Hort. Sci. 103:818-820.

Townsend, L.R. and I.V. Hall. 1970. Trends in nutrient levels of lowbush blueberry leaves during four consecutive years of sampling. Naturaliste Can. 97:461-466.

Trevett, M.F. 1970. Soil tests in lowbush blueberry fields. Res. Life Sci. (Spring): 21-23.

Trevett, M.F. 1972. A second approximation of leaf analysis standards for lowbush blueberry. Research in the Life Sciences. Maine Agr. Sta. Bul. 19:15-16.

Trevett, M.F., P.N. Carpenter, and R.E. Durgin. 1968. Seasonal trend and interrelation of mineral nutrients in lowbush blueberry leaves. Maine Agr. Exp. Sta. Bul. 665.

Warman, P.R., 1987. The effects of pruning and fertilizers on the lowbush blueberry. Plant and Soil 101:67-72.

Webb, K.T., R.L. Thompson, G.J. Beke, and J.L. Nowland. 1991. Soils of Colchester County, Nova Scotia. Agr. Can. L.R.R.C. Contrib. No. 19.

Wyn-Jones, R.G. and A. Pollard. 1983. Proteins, enzymes and inorganic ions, Vol. 15b, pp. 528-562. In: A. Läuchli and R.L. Bieleski, (eds.). Encyclopedia of Plant Physiology, new series. Springer-Verlag, Berlin.

# Efficient Mowing
# for Pruning Wild Blueberry Fields

Leonard J. Eaton
Robbie W. Glen
J. Doug Wyllie

**SUMMARY.** Commercial wild blueberry (*Vaccinium angustifolium* Ait.) fields were mowed in spring and autumn at low (2.5-5 cm; 1-2 inches), medium (5-7.5 cm; 2-3 inches), and high (> 7.5 cm; > 3 inches) heights with a flail mower and also with a rotary mower (> 7.5 cm; > 3 inches), in order to determine optimal heights for mowing. Initial stem lengths reflected differences in mowing heights at both sites, but there were no differences in plant heights at the end of the pruning year growth, or in the spring of the crop year. There were no differences in buds per stem or in fresh fruit yields among the treatments at the Adams field, or among the flail mowed plots at the Murray Siding field. Yields in rotary mowed plots were lower than yields in all other plots at the

---

Leonard J. Eaton is Research Professor, Department of Environmental Services, Nova Scotia Agricultural College, Truro, NS B2N 5E3, Canada.

Robbie W. Glen is Field Foreman, C.L. Stonehouse Enterprises, Ltd., Debert, NS B0M 1G0, Canada.

J. Doug Wyllie is Manager, C.L. Stonehouse Enterprises, Ltd., Debert, NS B0M 1G0, Canada.

Special thanks to the Stonehouse Enterprises Ltd. crew for assistance in setting up these experiments and for doing the mowing, and to summer research assistants over two years for their assistance in sampling and measuring. Finally, the authors appreciate the interest and financial support provided by Bragg Lumber Company for this study.

[Haworth co-indexing entry note]: "Efficient Mowing for Pruning Wild Blueberry Fields." Eaton, Leonard J., Robbie W. Glen, and J. Doug Wyllie. Co-published simultaneously in *Small Fruits Review* (Food Products Press, an imprint of The Haworth Press, Inc.) Vol. 3, No. 1/2, 2004, pp. 123-131; and: *Proceedings of the Ninth North American Blueberry Research and Extension Workers Conference* (ed: Charles F. Forney, and Leonard J. Eaton) Food Products Press, an imprint of The Haworth Press, Inc., 2004, pp. 123-131. Single or multiple copies of this article are available for a fee from The Haworth Document Delivery Service [1-800-HAWORTH, 9:00 a.m. - 5:00 p.m. (EST). E-mail address: docdelivery@haworthpress.com].

Murray Siding field, and also stems were more branched than were stems in the other treatment plots. These results suggest that producers can mow their fields at higher heights without impact on plant growth and production, as long as they use the flail mower. Mowing at greater heights results in less damage to equipment, plants and soil, and is more economical than the low heights of mowing presently recommended for the industry. *[Article copies available for a fee from The Haworth Document Delivery Service: 1-800-HAWORTH. E-mail address: <docdelivery@haworthpress. com> Website: <http://www.HaworthPress.com> © 2004 by The Haworth Press, Inc. All rights reserved.]*

**KEYWORDS.** *Vaccinium angustifolium*, pruning, flail mower, rotary mower, production, yields

## INTRODUCTION

The wild blueberry (*Vaccinium angustifolium* Ait.) is a perennial calcifuge shrub native to much of northeastern North America. It is described by Vander Kloet (1978) as a genetically diverse species composed of three distinct polymorphic forms that interbreed freely, resulting in considerable variation in form and growth habit (Barker et al., 1964; Hepler and Yarborough, 1991; Trevett, 1962). Plants initially establish from seedlings and spread by extensive rhizome systems (Trevett, 1956), eventually forming large clones that often intermingle with neighbouring clones (Barker et al., 1964). The rhizome system allows the blueberry to withstand destruction of stems and leaves by pruning (Hall et al., 1979). The entire aerial portion is replaced by vigorous new shoots (Kender and Eggert, 1966), which grow vegetatively the first season after pruning, and produce blossoms and fruit in the second or crop year (Blatt et al., 1989). Wild blueberries typically grow on Orthic Humo-Ferric Podsols (Canada Soil Survey Committee, 1978) which are acid and infertile (Trevett, 1962). Ecological dominance of the blueberry in commercial production is maintained through pruning and use of selective herbicides (Jensen, 1986; Yarborough et al., 1986).

To sustain maximum production, wild blueberry fields are regularly pruned (Chiasson and Argall, 1995), either by burning (Trevett, 1959) or mowing (Hanson et al., 1982; Ismail and Yarborough,1981), in order to force the blueberry into a two year production cycle (Ismail and Yarborough, 1981). Burning, with a Woolery burner or with straw, is more expensive than the alternate method, mowing with a flail mower, or with a rotary mower (Hanson et al., 1982; McIsaac, 1999). Both methods provide equal pruning, and yields are similar where mowers follow the field contours and do not damage roots and rhizomes by knocking off the tops of hummocks (Ismail and Yarborough, 1979; 1984). Ismail et al. (1981) and DeGomez (1988) recommend that blueberry plants be pruned as close to the ground as possible, or to a height of 1 cm (1/2 inch). This practice has been followed by a number of managers and producers, but extensive damage to soils and plants has been noted, especially where fields have not been leveled. Producers have, therefore, varied mowing heights to accommodate uneven field terrain, and to allow for pruning with less damage to the soil surface. At the same time, however, there are concerns among producers that higher mowing heights might result in reduced yields, especially in areas where fruit buds are not removed by pruning.

The objectives of this study were: (1) to assess the effects of several mowing heights on blueberry plant growth and reproduction and (2) to determine if plants can be successfully pruned to heights greater than the recommended height, allowing faster mowing and reduced production costs.

## *MATERIALS AND METHODS*

The experiments were initiated on the Adams field, East Mines (45° 26' N, 63° 32' W) in April 1998, and on the Murray Siding field, Truro (45° 22' N, 63° 12' W) in October 1998. Four mowing treatments were used, with treatments (1) to (3) with the flail mower (three gang mower, mowing width 2.4 m (8 feet), Doug Bragg Enterprises Ltd., Collingwood, NS), and treatment (4) with a rotary mower (John Deer 2 m rotary mower, 2.1 m (7 feet) mowing width). The treatments were: (1) Low: 2.5-5 cm (1-2 inches); (2) Medium: 5-7.5 cm (2-3 inches); (3) High: > 7.5 cm (> 3 inches); (4) Rotary Mow: > 7.5 cm (> 3 inches). Treatments were replicated in five random blocks, with plot size 4 m × 50 m, except for the bush mow at the Adams field which was a 25 m × 50 m single plot. Speeds recorded for the treatments at the Murray Siding field were: (1) Low: second gear turtle, 2.41-3.22 km h⁻¹ (1.5-2.0 mph); (2) Medium: second gear rabbit, 3.22-4.02 km h⁻¹ (2.0-2.5 mph); (3) High: third gear turtle, 4.02-4.83 km h⁻¹ (2.5-3.0 mph); (4) Rotary mow: third gear turtle, 4.83-5.63 km h⁻¹ (3.0-3.5 mph). Actual speeds under field conditions varied somewhat with changes in plant stand and surface variations in field contours. The surface at the Adams field was very uniform, whereas that at the Murray Siding field was somewhat rough. The mowing treatments were completed by an experienced tractor operator.

Plant growth and production were assessed in each field over a two year production cycle. This cycle was 1998 (prune) and 1999 (crop) for the Adams field which was pruned in April 1998, and 1999 (prune) and 2000 (crop) for the Murray Siding field which was pruned in October 1998 and did not begin vegetative growth until the following spring.

Following pruning, initial plant heights were estimated in the field by measuring 40 randomly selected stems in each plot with a ruler. Two 50-stem samples from each plot, taken along a 10 m line transect were taken from each plot and assessed for stem length and dry weights in the prune year, and for stem length, branches, buds, and dry weight in the crop year (Eaton, 1994). In the Adams field, numbers of branches per stem were obtained by counting branches on plants in the field; those in the Murray Siding field were obtained from plant samples taken in the spring of the crop year. Fresh fruit weights were obtained in August 1999 and August 2000. Data were analyzed using the ANOVA program of Sigmastat® (Jandel, 1995).

## RESULTS AND DISCUSSION

Initial plant heights (stem lengths) differed among the treatments at each site and also between the two sites (Table 1). Measurements were taken directly following spring mowing at the Adams field and early in the year following fall mowing at the Murray Siding field, but prior to observable plant growth. The longer initial stem lengths (height of cut stems) at the Murray Siding site are possibly due to the more uneven soil surface at that site than at the Adams field. At both sites, there were significant differences in stem lengths early in the growing season, but these differences had disappeared by the August sampling dates (Table 1). The mean stem lengths in all treatment plots were similar at the end of the season regardless of the initial cut length. The plants that were pruned in the fall grew more rapidly during the early part of the season; however, those pruned in the spring had caught up by August (Table 1). The differences in plant height between the two fields most likely reflect the effects of different growing seasons in the two fields, as well as possible differences in nutrient levels. Earlier researchers (i.e., Hanson et al., 1982; Ismail and Yarborough, 1981; 1984) reported similar results for final stem lengths in their studies, but did not follow growth throughout the prune year.

TABLE 1. Mean stem length (cm) after pruning and during the prune year at two sites. Adams field was pruned April 1998 and Murray Siding was pruned October 1998.

| Treatment | Average stem length (cm) | | | |
|---|---|---|---|---|
| | Initial | June | July | August |
| a) Adams: | | | | |
| Low (2.5-5 cm) | 1.43 d[z] | 4.70 a | 13.89 a | 15.04 a |
| Medium (5-7.5 cm) | 3.10 c | 4.97 a | 13.05 ab | 15.84 a |
| High (> 7.5 cm) | 5.10 b | 4.75 a | 12.54 ab | 14.40 a |
| Rotary Mow (>7.5 cm) | 8.38 a | 4.51 a | 12.00 b | 15.32 a |
| | | | | |
| b) Murray Siding: | | | | |
| Low (2.5-5 cm) | 3.47 d[z] | 8.81 c | 13.07 a | 13.12 a |
| Medium (5-7.5 cm) | 5.01 c | 8.88 c | 12.56 a | 13.77 a |
| High (> 7.5 cm) | 6.11 b | 11.40 b | 13.34 a | 12.32 a |
| Rotary Mow (> 7.5 cm) | 9.84 a | 14.33 a | 11.76 a | 12.78 a |

[z] Values in columns followed by the same letter do not differ significantly at $P \leq 0.05$ (Tukey means separation test).

Stem length, branches per stem, buds per stem and fruit yields (kg ha$^{-1}$) at the Adams field were similar in all treatment plots, but dry weights of stems from the lowest mowing height were greater than those from the single rotary mowed plot (Table 2). Similarly, there were no differences among treatments for stem length, dry weights and buds per stem at the Murray Siding field, but there were more branches on stems from the bush mowed plots than in other plots (Table 2). In the Adams field, yields were similar in all treatment plots, even though the yields from the rotary mowed plot appeared to be considerably lower than those from the other plots. By contrast, yields in the rotary mowed plots at Murray Siding were significantly lower than those in other treatment plots at that site. The rougher terrain at the Murray Siding field resulted in longer initial stem lengths for all treatments, and may have contributed to reduced vegetative and reproductive growth in the rotary mowed plots relative to the flail mowed plots.

There are advantages to high mowing heights, including effects on soil, plants, and equipment, as well as economic benefits. With high mowing, there are fewer incidents of the mower blades hitting hum-

TABLE 2. Crop year plant data from two pruning experiments. Adams field was pruned April 1998 and Murray Siding was pruned October 1998. Branches per stem data were collected from Adams field in the prune year; those from Murray Siding were obtained in the crop year.

| Treatment | Length (cm) /stem | Branches /stem | Dry Wt. (g) /50 stems | Buds /stem | Yields kg ha$^{-1}$ |
|---|---|---|---|---|---|
| a) Adams: | | | | | |
| Low (2.5-5 cm) | 16.04 a[z] | 3.23 a | 10.57 b | 5.02 a | 5919 a |
| Medium (5-7.5 cm) | 22.84 a | 3.69 a | 11.99 ab | 5.54 a | 5958 a |
| High (>7.5 cm) | 16.70 a | 2.68 a | 11.22 ab | 5.68 a | 6259 a |
| Rotary Mow (>7.5 cm) | 19.78 a | 3.76 a | 13.58 a | 6.52 a | 4163 a |
| b) Murray Siding: | | | | | |
| Low (2.5-5 cm) | 13.76 a[z] | 0.030 b | 7.91 a | 3.89 a | 5749 a |
| Medium (5-7.5 cm) | 13.01 a | 0.032 b | 7.52 a | 3.26 a | 6611 a |
| High (> 7.5 cm) | 12.67 a | 0.140 ab | 7.23 a | 3.43 a | 5249 a |
| Rotary Mow (> 7.5 cm) | 13.33 a | 0.384 a | 7.59 a | 3.07 a | 3684 b |

[z] Values in columns followed by the same letter do not differ significantly at $P \leq 0.05$ (Tukey means separation test).

mocks or rocks within the blueberry field, or with the blades destroying exposed rhizomes at the edges of clones. Reduced damage to plants and soil will prevent or reduce soil loss due to exposure and subsequent erosion, and will contribute to longer health of the system. There are also economic advantages, including reduced replacement and repair costs for cutting blades and other parts of the mowers. Finally, operating costs can be reduced by mowing at faster ground speeds and covering larger areas of the fields (Table 3). The data in Table 3 were calculated using a range of ground speeds, flail mower and rotary mower widths of 2.2 and 2.1 m (7.9 and 6.9 ft), respectively, and a single hourly rate of $55.00 per hour for the tractor and operator. The costs of mowing listed in Table 3 do not account for variations in plant stand, size of the field, roughness, and other factors that influence the efficiency of the mowing. Hanson et al. (1982) use an efficiency rating of 0.7 in their calculations to account for overlap, turning time, etc., and we have included that factor in our calculations. From the cost values given in Table 3, we conclude that mowing at higher heights will reduce management costs related to pruning. For example, mowing at the highest height, 7.5 cm

TABLE 3. Costs related to mowing wild blueberries at different heights. Cost of operator and tractor = $55.00/hr [(km ha$^{-1}$ × 1609.3 = m hr$^{-1}$ (mph × 5280 = ft/hr); 1 ha = 10000 m$^2$ (1 acre = 43560 sq. ft); flail mower width = 2.2 m (8 ft); rotary mower width = 2.1 m (7 ft)]. Final values for ha (acres) mowed per hr are multiplied by 0.7 to reflect an approximate efficiency ratio suggested by Hanson et al. (1982).

**a) Costs per hectare**

| Mowing Height | Speed (km ha$^{-1}$) | Ha mowed per h | $ Cost per ha |
|---|---|---|---|
| Low (2.5-5 cm) | 2.41-3.22 | 0.41-0.55 | 133.24-99.93 |
| Medium (5-7.5 cm) | 3.22-4.02 | 0.55-0.68 | 99.93-80.41 |
| High (> 7.5 cm) | 4.02-4.83 | 0.68-0.83 | 80.41-65.66 |
| Rotary mow (> 7.5 cm) | 4.83-5.63 | 0.83-1.03 | 65.66-53.30 |

**b) Costs per acre**

| Mowing Height | Speed (mph) | Acres mowed per h | $ Cost per acre |
|---|---|---|---|
| Low (2.5-5 cm) | 1.5-2.0 | 1.02-1.36 | 53.92-40.44 |
| Medium (5-7.5 cm) | 2.0-2.5 | 1.36-1.69 | 40.44-32.54 |
| High (> 7.5 cm) | 2.5-3.0 | 1.69-2.04 | 32.54-26.57 |
| Rotary mow (> 7.5 cm) | 3.0-3.5 | 2.18-2.55 | 25.23-21.57 |

(3 inch), allows pruning to be completed at savings of approximately $34.00 to $53.00 per hectare ($14.00 to $21.00 per acre) when compared to the costs of mowing at the lowest height.

A possible negative effect of high mowing may be the large amount of branching noted in Table 2. More branching may interfere with the efficient harvest of the crop. This possibility appears to be more certain with rotary mowing than with flail mowing.

## CONCLUSIONS AND GROWER BENEFITS

The results of these studies suggest that all heights of mowing result in similar final plant height in the prune year, and in similar plant growth, buds, and yields in crop years, as long as the pruning is by the flail mower. No differences were noted between spring or autumn pruning. Mowing at high heights reduces management costs, without reducing yields, and also reduces potential for plant and environmental damage within blueberry fields. Rotary mowing is not recommended as a general procedure, as it results in more branching and decreased yields relative to the flail mowing heights examined in this study. We would recommend, therefore, that flail mowers be set at the medium or high heights for the pruning operation.

The results of this study will enable producers and managers of commercial wild blueberry stands to reduce costs of pruning by mowing with flail mowers at higher heights than previously recommended. In addition to the direct savings, further savings will be realized by reducing equipment damage, as well as reducing plant and soil damage to the plants.

## LITERATURE CITED

Barker, W.G., I.V. Hall, L.E. Aalders, and G.W. Wood. 1964. The lowbush blueberry in eastern Canada. Econ. Bot. 18:357-365.

Blatt, C.R., L.R. Crozier, I.V. Hall, K.I.N. Jensen, W.T.A. Neilson, P.E. Hildebrand, N.L. Nickerson, R.K. Prange, P.D. Lidster, and J.D. Sibley. 1989. Lowbush Blueberry Production. Agriculture Canada Publication 1477/E.

Canada Soil Survey Committee. 1978. The Canadian system of soil classification. Can. Dept. Agric. Publ. 1646. Supply and Services Canada, Ottawa, Ont.

Chaisson, G. and J. Argall. 1995. Pruning wild blueberry fields. Wild blueberry factsheet A.5.0. Wild blueberry production guide, New Brunswick Dept. of Agr. and Rural Dev.

DeGomez, T. 1988. Pruning lowbush blueberry fields. Wild Blueberry Fact Sheet. Fact Sheet No. 229, Univ. of Maine Co-Operative Ext., Orano, Maine.

Eaton, L.J. 1994. Long term effects of herbicides and fertilizers on lowbush blueberry growth and production. Can. J. Plant Sci. 74: 341-345.

Hall, I.V., L.E. Aalders, N.L. Nickerson, and S.P. Vander Kloet. 1979. The biological flora of Canada. I. *Vaccinium angustifolium* Ait., sweet lowbush blueberry. Can. Field Nat. 93:415-430.

Hanson, E.J., A.A. Ismail, and H.B. Metzger.1982. A cost analysis of pruning procedures in lowbush blueberry production. Life Sci. Agri. Expt. Sta. Univ. Maine Orono Bul. 780.

Hepler, P.R. and D.E. Yarborough. 1991. Natural variability in yield of lowbush blueberries. HortScience 26:245-246.

Ismail, A.A. and D.E. Yarborough. 1979. Pruning lowbush blueberries–a review and update. Proc. 4th North Amer. Blueberry Res. Workers Conf. Univ. Ark., Fayetteville, Ark., Oct. 16-18, 1979. pp. 87-95.

Ismail, A.A., J.M. Smagula, and D.E. Yarborough. 1981. Influence of pruning method, fertilizer and Terbacil on the growth and yield of the lowbush blueberry. Can. J. Plant Sci. 61:61-71.

Ismail, A.A. and D.E. Yarborough.1981. A comparison between flail mowing and burning for pruning lowbush blueberries. HortScience 16: 318-319.

Ismail, A.A. and D.E. Yarborough 1984. Flail mowing for pruning lowbush blueberries. Pp. 158-167 In: T.E. Crocker and P. Lyrene (eds.) Proc. of the Fifth North Amer. Blueberry Res. Workers Conf., Feb 1-3 1984, Gainesville, FL.

Jandel Corporation 1995. Sigmastat® Stastical Software User's Manual.

Jensen, K.I.N. 1986. Response of lowbush blueberry to weed control with atrazine and hexazinone. HortScience 21: 1143-1144.

Kender, W.J., and F. Eggert 1966. Several soil management practices influencing the growth and rhizome development of the lowbush blueberry. Can. J. Plant Sci. 46:141-149.

McIsaac, D.W. 1999. Wild blueberry production and marketing in Nova Scotia–A situation report 1999. Lowbush Blueberry Fact Sheet. N.S. Depart. Agri. Mktg. Truro, N.S.

Trevett, M.F. 1956. Observations on the decline and rehabilitation of lowbush blueberry fields. Maine Agr. Exp. Sta., Orono, Maine. Misc. Publ. 626.

Trevett, M.F. 1959. Growth studies of the lowbush blueberry. Bull. 581. Maine Agri. Expt. Sta., Univ. Maine Orono.

Trevett, M.F. 1962. Nutrition and growth of the lowbush blueberry. Maine Agr. Exp. Sta. Bull. 605.

Vander Kloet, S.P. 1978. Systematics, distribution and nomenclature of the polymorphic *Vaccinium angustifolium* Ait. Rhodora 80: 358-376.

Yarborough, D.E., J.J. Hanchar, S.P. Skinner, and A.A. Ismail. 1986. Weed response, yield and economics of hexazinone and nitrogen use in lowbush blueberry production. Weed Sci. 34: 723-729.

# Effect of Soil Calcium Applications on Blueberry Yield and Quality

Eric J. Hanson
Steven F. Berkheimer

**SUMMARY.** The response of mature highbush blueberries (*Vaccinium corymbosum* L. cv 'Jersey') to soil applied Ca was studied for five years. The study site was a mature field in southwest Michigan with a relatively low soil pH (4.2) and Ca content (85 ppm). The treatments were: (1) nontreated control; (2) 1,100 kg/ha calcitic limestone; and (3) 550 kg/ha gypsum (calcium sulfate). Lime and gypsum were spread in a 5 ft wide band under the plants in May of 1996, 1997, 1999, and 2000. Lime, and to a lesser extent gypsum, increased soil pH and Ca levels, but had inconsistent affects on Ca levels in leaves and fruit. No treatment affected berry yield or size (1999-2001), firmness (1996-2001) or fruit rot incidence (2001). *[Article copies available for a fee from The Haworth Document Delivery Service: 1-800-HAWORTH. E-mail address: <docdelivery@haworthpress.com> Website: <http://www.HaworthPress.com> © 2004 by The Haworth Press, Inc. All rights reserved.]*

**KEYWORDS.** Fruit firmness, anthracnose, *Alternaria*

Eric J. Hanson is Professor, and Steven F. Berkheimer is Research Technician, Department of Horticulture, Michigan State University, East Lansing, MI 48824-1325.

The authors acknowledge the Michigan Agricultural Experiment Station and MBG-Marketing, Grand Junction, Michigan, for support of this research.

[Haworth co-indexing entry note]: "Effect of Soil Calcium Applications on Blueberry Yield and Quality." Hanson, Eric J., and Steven F. Berkheimer. Co-published simultaneously in *Small Fruits Review* (Food Products Press, an imprint of The Haworth Press, Inc.) Vol. 3, No. 1/2, 2004, pp. 133-139; and: *Proceedings of the Ninth North American Blueberry Research and Extension Workers Conference* (ed: Charles F. Forney, and Leonard J. Eaton) Food Products Press, an imprint of The Haworth Press, Inc., 2004, pp. 133-139. Single or multiple copies of this article are available for a fee from The Haworth Document Delivery Service [1-800-HAWORTH, 9:00 a.m. - 5:00 p.m. (EST). E-mail address: docdelivery@haworthpress.com].

*133*

## INTRODUCTION

Highbush blueberries have low Ca requirements relative to other temperate fruit crops. Healthy bushes typically contain 0.3% to 0.8% Ca in leaf tissue (Eck, 1988) compared to 1% to 3% in temperate tree crops (Shear and Faust, 1980). Korsak (1988) referred to blueberries as calcifuge or lime avoiding plants because they require acidic soils that typically contain low Ca levels. Optimum Ca levels in blueberry leaves are not well understood since deficiencies in field plants have not been reported. Eck (1988) reviewed available information at the time and proposed that leaf levels below 0.2% may indicate a deficiency. Although extremely acidic soils (pH < 4.0), which are typically accompanied by low Ca levels, can cause reduced vigor, dieback, or plant death (Harmer, 1944; Merrill, 1944; Spiers, 1984), this is rare in Michigan fields. Injury associated with extreme acidity may result from toxic accumulation of aluminum or manganese, or deficiencies of nutrients such as calcium, magnesium, or potassium.

Calcium nutrition has recognized effects on the quality of various fruits, even when the plants are adequately supplied with Ca (Bangerth, 1979; Poovaiah et al., 1988; Shear, 1975). Foliar Ca applications have increased the firmness of strawberries (*Fragaria* ×*ananassa* Duch.) (Cheour et al., 1990; Eaves et al., 1962) and raspberries (*Rubus ideas* L.) (Eaves et al., 1972), but not blueberries (Hanson, 1995). However, highbush blueberry firmness was increased by dipping harvested berries in Ca solutions (Hanson et al., 1993). This treatment left surface residues and was not practical for fresh market berries, but the results implied that other methods of increasing fruit Ca might increase firmness. The goal of the current study was to test whether Ca applied to the soil as lime or gypsum might enhance the Ca status of bushes and improve berry firmness.

## MATERIALS AND METHODS

The study site, a 20 year-old 'Jersey' field in Grand Junction, Michigan, was selected because pre-treatment soil analyses (means of 4 observations) indicated a relatively low pH (4.2) and Ca level (85 ppm). The soil was a Pipestone-Kingsville complex (sandy mixed mesic Typic Endoquad, mixed, mesic Mollic Psammaquents). The treatments were: (1) non-treated control, (2) 1,120 lb/acre calcitic limestone (96% Ca-carbonate) and (3) 560 lb/acre gypsum (calcium sulfate). Lime and gypsum were applied in May of 1996, 1997,1999, and 2000 to a 1.5 m wide band under the plants. Plots were single rows, 14 bushes long, and treatments were replicated 5 times in a randomized complete block design.

Soil samples were collected in April or May of 1997, 1998, 1999, and 2000 (0-15 cm depth) and in May 2001 (0-15 cm and 16-30 cm), and analyzed for Bray-Kurtz-1 extractable P, ammonium acetate extractable Ca, K, and Mg, and pH (1:1 soil/water slurry). Composite samples of leaves (20 per plot) were collected each August from 1996 to 2001 for analysis of all nutrients (minus N) by DC plasma emission spectrophotometry. Fruit from the first picking of each year were freeze-dried and similarly analyzed.

Fruit samples were picked by hand just prior to the first commercial harvest between 1996 and 1999, packed in 0.5 L perforated clamshell containers, and placed in plastic bags in a 2°C cooler. Samples of 10 berries were removed from each container after storage periods of 2, 7, and 14 d (1996), 1, 9, and 16 d (1997), 7 and 21 d (1998), or 0, 21, 35, 42, and 49 d (1999), allowed to warm to room temperature, and tested for firmness with a FirmTech1 instrument (Bioworks, Stillwater, OK). In 2000 and 2001, samples were picked just prior to the first and second commercial harvest, and stored in a similar manner for 7 or 14 d before firmness measurements. These fruit were also sorted to determine the percentage of berries exhibiting rot due to *Alternaria*, *Colletotrichum*, or other fungi. Berry yields were measured in 1999, 2000, and 2001 by harvesting the plots twice with a BEI Inc. (South Haven, MI) over-the-row harvester after fruit samples for firmness were removed.

## RESULTS AND DISCUSSION

Lime, and to a lesser extent, gypsum increased pH (Figure 1) and Ca concentrations (Figure 2) in the top 15 cm of soil between 1998 and 2001. Treatments did not affect soil P or K concentrations in any year (means across all years 71 mg/kg P, 27 mg/kg K). In 2001, soil Mg levels were significantly higher in plots receiving lime (77 mg/kg) than in plots receiving gypsum (40 mg/kg) or controls (37 mg/kg), but levels were not affected in any other years. Subsoil (15-30 cm depths) collected in 2001 showed no effects of treatments on pH (overall means 4.5) or nutrient levels (means: 42, 14, 79, and 18 mg/kg of P, K, Ca, and Mg, respectively).

Leaf Ca levels were affected by treatments only later in the study (Figure 3). Lime resulted in higher leaf Ca levels than controls in 1999

FIGURE 1. Effect of lime and gypsum applications on soil pH. Letters indicated significant differences between treatment means (P ≤ 0.05).

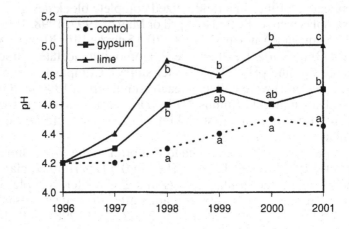

FIGURE 2. Effect of lime and gypsum applications on soil Ca concentrations. Letters indicated significant differences between treatment means (P ≤ 0.05).

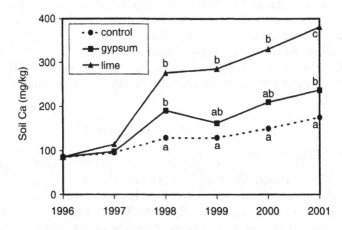

and 2001, whereas gypsum increased leaf Ca levels above controls only in 2000. Treatments affected leaf levels of other nutrients inconsistently if at all (data not shown). Compared to controls, lime significantly increased leaf P levels in 1996 and 1997, but decreased levels in 1998. Lime also increased leaf Zn in 1997 and decreased leaf Mn in 1998, compared to controls. Treatments did not affect leaf K (mean across

FIGURE 3. Effect of lime and gypsum applications on leaf Ca concentrations. Letters indicated significant differences between treatment means (P ≤ 0.05).

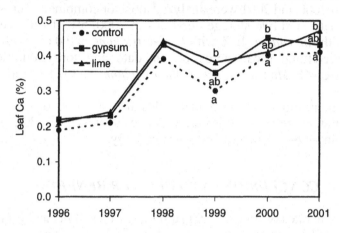

years and treatments: 0.34%), Mg (0.15%), Cu (5.7 ppm), Fe (140 ppm), and Al (94 ppm).

Lime applications increased fruit Ca levels compared to controls only in 1998 (0.070% vs. 0.032% Ca, dry weight basis). Gypsum resulted in higher fruit Ca levels than controls only in 1999 (0.034% vs. 0.030% Ca). Cummings and Lilly (1980) also observed no effect of a single lime application on Ca concentrations in blueberry fruit. Fruit Ca concentrations averaged 0.039% across years and treatments, which is comparable to levels previously reported in blueberry fruit (Cummings and Lilly, 1980; Hanson, 1995). Levels of other elements in fruit tissues were not affected by treatments (overall means: 0.061% P, 0.31% K, 0.031% Mg, 6.4 ppm B, 3.5 ppm Cu, 19 ppm Fe, 3.6 ppm Mn, 6.4 ppm Zn, 16 ppm Al).

Treatments did not affect berry firmness. Over this six-year study, 25 separate comparisons of firmness were made using berries from the first and second pickings and held in cold storage for varying lengths of time. In no instance was there a significant effect of treatments on firmness. Firmness data from 2000 and 2001 were combined for analysis because the protocol each year was identical (fruit from 1st and 2nd harvest evaluated after 1 or 2 weeks storage). No treatment effects were observed when the data were combined or analyzed separately. Firmness averaged 167 g/mm deformation over these two years. Overall, firmness declined with storage duration.

The percentage of berries exhibiting rot associated with *Alternaria*, *Colletotrichum* or other fungi were not affected by treatments when data from 2000 and 2001 were analyzed alone or combined. Across two picking dates and two storage durations in 2000, 5% of fruit were infected with *Alternaria*, 10% with *Colletotrichum*, and 0.6% with other unidentified fungi. Means across picking dates and storage durations in 2001 were 4% *Alternaria*, 7% *Colletotrichum*, and 0.1% with other fungi.

Treatments did not affect annual yields in 1999, 2000, or 2001, or total cumulative yield over this period. Yields averaged 3.1, 3.8, and 2.4 kg/bush in 1999, 2000, and 2001, respectively.

## CONCLUSIONS AND GROWER BENEFITS

Moderate rates of lime and gypsum increased soil pH and Ca levels, and resulted in modest increases in leaf Ca levels after three years. These treatments had minimal effects on fruit Ca levels and no effect on berry yields or quality. Results suggest that moderate applications of lime and gypsum, which do not increase soil pH beyond the recommended range for blueberries are not likely to benefit blueberries on native blueberry soils.

## LITERATURE CITED

Bangerth, F. 1979. Calcium-related physiological disorders of plants. Ann. Rev. Phytopath. 17:97-122.

Cheour, F., C. Willemot, J. Arul, Y. Desjardins, J. Makhlouf, P.M. Charest, and A. Gosselin. 1990. Foliar application of calcium chloride delays postharvest ripening of strawberry. J. Amer. Soc. Hort. Sci. 115:789-792.

Cummings, G.A. and J.P. Lilly. 1980. Influence of fertilizer and lime rates on nutrient concentrations in highbush blueberry fruit. HortScience 15:752-754.

Eaves, C.A. and J.S. Leefe. 1962. Note on the influence of foliar sprays of calcium on the firmness of strawberries. Can. J. Plant Sci. 42:746-747.

Eaves, C.A., C.L. Lockhart, R. Stark, and D.L. Craig. 1972. Influence of preharvest sprays of calcium salts and wax on fruit quality of red raspberry. J. Amer. Soc. Hort. Sci. 97:706-707.

Eck, P. 1988. Blueberry Science. Pp. 106-109. Rutgers Univ. Press. New Brunswick, NJ.

Hanson, E.J. 1995. Preharvest calcium sprays do not improve highbush blueberry (*Vaccinium corymbosum* L.) quality. HortScience 30:977-978.

Hanson, E.J., J.L. Beggs, and R.M. Beaudry. 1993. Applying calcium chloride postharvest to improve highbush blueberry firmness. HortScience 28:1033-1034.

Harmer, P.M. 1944. The effect of varying the reaction of organic soil on the growth and productivity of the domesticated blueberry. Soil Sci. Soc. Proc. 9:133-141.

Korsak, R.F. 1988. Nutrition of blueberry and other calcifuges. Hort. Rev. 10:183-227.

Merril, T.A. 1944. Effects of soil treatments on the growth of the highbush blueberry. J. Agr. Res. 69:9-20.

Poovaiah, B.W., G.M. Glenn, and A.S.N. Reddy. 1988. Calcium and fruit softening: Physiology and biochemistry. Hort. Rev. 10:107-143.

Shear, C.B. 1975. Calcium-related disorders of fruits and vegetables. HortScience 10:361-365.

Shear, C.B. and M. Faust. 1980. Nutritional ranges in deciduous tree fruits and nuts. Hort. Rev. 2:142-163.

Spiers, J.M. 1984. Influence of lime and sulfur soil additions on growth, yield, and leaf nutrient content of rabbiteye blueberry. J. Amer. Soc. Hort. Sci. 109:559-562.

# Effect of In-Row Spacing and Early Cropping on Yield and Dry Weight Partitioning of Three Highbush Blueberry Cultivars the First Two Years After Planting

Bernadine Strik
Gil Buller

**SUMMARY.** The effect of early cropping (no blossom removal the first two years) and in-row spacing at 0.45 m and 1.2 m (1.5 ft and 4 ft) are being studied in 'Duke', 'Bluecrop', and 'Elliott' blueberries (*Vaccinium corymbosum* L.) planted in October 1999. No yield was produced on the non-cropped plants in 2000 and 2001. In the early-cropped treatments, yield at 0.45 m was about three times that at 1.2 m in all cultivars in 2000 and 2001. 'Duke' and 'Elliott' produced the highest yield in 2000. In 2001, yield increased 8 to 16 fold at the 0.45 m spacing, depending on

Bernadine Strik is Professor and Gil Buller is Research Assistant, Department of Horticulture and the North Willamette Research and Extension Center, Oregon State University, 4017 ALS, Corvallis, OR 97331-7304 (E-mail: strikb@science.oregonstate. edu).

The authors appreciate the financial support of the Oregon Blueberry Commission and the Northwest Center for Small Fruits Research and the assistance from Fall Creek Farm and Nursery with plant costs.

[Haworth co-indexing entry note]: "Effect of In-Row Spacing and Early Cropping on Yield and Dry Weight Partitioning of Three Highbush Blueberry Cultivars the First Two Years After Planting." Strik, Bernadine, and Gil Buller. Co-published simultaneously in *Small Fruits Review* (Food Products Press, an imprint of The Haworth Press, Inc.) Vol. 3, No. 1/2, 2004, pp. 141-147; and: *Proceedings of the Ninth North American Blueberry Research and Extension Workers Conference* (ed: Charles F. Forney, and Leonard J. Eaton) Food Products Press, an imprint of The Haworth Press, Inc., 2004, pp. 141-147. Single or multiple copies of this article are available for a fee from The Haworth Document Delivery Service [1-800-HAWORTH, 9:00 a.m. - 5:00 p.m. (EST). E-mail address: docdelivery@haworthpress.com].

cultivar. Pruning weight per plant was affected by cultivar, in-row spacing, and early cropping. In winter 2000/01, after one year of early cropping, there was no treatment effect on the percentage of fruit buds per lateral. However, in winter 2001/02, early-cropped plants had a lower percentage of fruit buds in 'Bluecrop' and 'Duke' than plants that were not cropped early. Plants spaced at 0.45 m also had a lower percentage of fruit buds than those at 1.2 m in 'Duke' and 'Elliott'. Total plant dry weight in winter 2001/02 was affected by cultivar and early-cropping. Early cropping reduced plant size in all cultivars. *[Article copies available for a fee from The Haworth Document Delivery Service: 1-800-HAWORTH. E-mail address: <docdelivery@haworthpress.com> Website: <http://www. HaworthPress.com> © 2004 by The Haworth Press, Inc. All rights reserved.]*

**KEYWORDS.** Pruning, growth, berry size, planting

## *INTRODUCTION*

Blueberry production has been increasing steadily in Oregon and Washington with approximately 80 to 120 ha (200 to 300 acres) being planted per year, on average, over the last ten years. Growers have been following recommendations (Pritts and Hancock, 1992; Strik et al., 1993) in removing blossom buds the first two years after planting. This standard procedure is thought to be necessary to promote good root and vegetative growth. However, preliminary findings in *Vaccinium corymbosum* L. 'Bluecrop' showed that cropping plants the first two years, actually increased production in years three through eight 18% compared to plants that were not cropped early (Strik, unpublished). It is not known how other cultivars would respond to early cropping. If growers were able to crop plants early (in years one and two), then they would not only derive some income from the fruit, but would save an estimated $125-$250/ha ($50-$100/acre) in not having to prune off the blossom buds.

Most of the mature acreage in Oregon and Washington is spaced at 1.2 m (4 ft) in the row with 3 m (10 ft) between rows. However, growers are tending now to establish plantings at higher density, particularly at 0.76 to 0.91m (30 inches to 3 ft) in the row based, in part, on research done by Strik and Buller (2002). Moore et al. (1993), in a five-year spacing study with 'Bluecrop' and 'Blueray', found no differences between cultivars and that yield was highest at the 0.6 m (2 ft) spacing. Plants were not cropped early in his study. Strik and Buller (2002) found that cumulative yield of 'Bluecrop' from years three through seven was 104% higher at a 0.45 m (1.5 ft) than at 1.2 m (4 ft) spacing. Early cropping in combination with higher planting densities could mean faster economic returns for highbush blueberry growers.

The objectives of this study were to determine the effect of early cropping and in-row spacing on yield and dry-weight partitioning of 'Bluecrop', 'Duke', and 'Elliott'.

## MATERIALS AND METHODS

The research planting was established at the North Willamette Research and Extension Center (NWREC) in Aurora, Oregon in October 1999. The planting site was fumigated with methyl bromide/chloropicrin with sawdust and fertilizer incorporated and raised beds formed prior to planting two-year-old container stock. The treatments were: cultivar (Duke, Bluecrop, Elliott); in-row spacing (0.45 m, 1.2 m [1.5 ft, 4 ft]); and early cropping (with or without blossom bud removal). In the no early crop treatments, blossom buds were pruned off the plants in October 1999 and in February 2001 in addition to standard pruning of young plants (Strik et al., 1990) to have no crop in 2000 and 2001. The early-cropped plants were not pruned in winter 1999/00 and 2000/01 other than to remove any diseased wood or very low growth. There were five replicates of each treatment combination arranged in a randomized complete block design for a total of 60 plots. Each plot was 6 m (20 ft) long (with 13 or 5 plants per plot at the 0.45 or 1.2 m spacing, respectively). The planting was flanked by guard rows.

Data collected included yield and picking time per harvest per plot, average berry weight (25 berries per harvest), pruning weight, cane number and age distribution, and percent fruit bud set (2000/01 and 2001/02). In February 2002, one plant per plot was destructively harvested and divided into its parts and dry weights obtained.

The results presented here are for the first two years of harvests (early crop treatments only) and the dry weight data in 2002.

## RESULTS AND DISCUSSION

Total yield, berry weight, and hand picking efficiency in 2000 and 2001 were significantly affected by cultivar and in-row spacing. There was no cultivar by spacing interaction. However, data are presented by cultivar here for ease of interpretation. 'Duke' and 'Elliott' had the highest yield in year one, whereas 'Bluecrop' and 'Elliott' had the highest yield in year two (Table 1).

Yield at the 0.45 m spacing was about three times the yield of the 1.2 m spaced plots in all cultivars the year after planting, 2000, and in 2001 (Table 1). Although yield at the traditional 1.2 m spacing only averaged 403 g/plot (580 lb/acre) the first year after planting (2000), yield increased to an average of 4 kg/plot (1,915 lb/acre) in 2001 (Table 1). Yield for plants spaced at 45cm in the row averaged 12.1 kg/plot (5,851 lb/acre) in 2001 (Table 1). Yield of 'Bluecrop' in this study in year one was similar to what we observed in our preliminary early-cropping study (Strik, unpublished). However, yield in this study was 2.5 to 6.5 times higher in year two than in our earlier study. This may have been a result of our earlier planting being on flat ground in heavy soil whereas this planting was established on raised beds and plants were observed to be more vigorous in this study.

Berry weight was significantly higher at the closer in-row spacing in both years. In-row spacing had inconsistent effects on berry weight of 'Bluecrop' in an earlier study done by Strik and Buller (2002). 'Elliott' tended to have the lowest picking efficiency (Table 1).

TABLE 1. The effect of in-row spacing and cultivar on total yield, berry weight, and picking efficiency of early-cropped plants in 2000 and 2001. Plants were established in October 1999.

| Treatment | Total Yield (g/plot) | | Berry Weight (g) | | Picking Efficiency (g/min) | |
|---|---|---|---|---|---|---|
| | 2000 | 2001 | 2000 | 2001 | 2000 | 2001 |
| **Duke** | | | | | | |
| 0.45 m | 1243 | 9729 | 1.16 | 1.66 | 73 | 111 |
| 1.2 m | 431 | 3097 | 1.14 | 1.58 | 74 | 100 |
| **Bluecrop** | | | | | | |
| 0.45 m | 852 | 13587 | 2.08 | 1.56 | 130 | 81 |
| 1.2 m | 305 | 4103 | 1.94 | 1.38 | 113 | 68 |
| **Elliott** | | | | | | |
| 0.45 m | 1192 | 13075 | 1.44 | 0.98 | 85 | 50 |
| 1.2 m | 474 | 4774 | 1.30 | 0.96 | 78 | 39 |
| **Significance[z]** | | | | | | |
| Cultivar | *** | *** | *** | *** | *** | *** |
| Spacing | *** | *** | * | * | NS | ** |
| Cult. × Space | NS | NS | NS | NS | NS | NS |

[z] NS, *, **, *** = non-significant or significant at P < 0.05, 0.01 or 0.001, respectively

There was no effect of in-row spacing on the number of one-, two- or three-year-old canes per plant (data not shown). Pruning (crop or no crop) affected the number of three-year-old canes per plant, as only the early-cropped plants had older canes (data not shown).

In winter 2000/01, after one year of early cropping, there was no effect of cultivar, spacing, or early cropping on the percentage of fruit buds per lateral (data not shown). However, in winter 2002, plants that were cropped early had a lower percentage of fruit buds in 'Bluecrop' and 'Duke' (Table 2). Plants spaced at 1.2 m in the row had a higher percentage of fruit buds in 'Duke' and 'Elliott'. 'Duke' had the highest percentage of fruit buds, averaging 63.5% compared to 58.1% and 45.8% in 'Elliott' and 'Bluecrop', respectively (Table 2).

Early cropping had no significant effect on pruning weight per plant in winter 2002 (Table 3). Plants at 1.2 m had a higher pruning weight than those at 0.45 m, particularly in 'Elliott' (Table 3).

Total plant dry weight, including roots (after pruning in 2002) was significantly affected by cultivar ($P < 0.001$) and early cropping ($P < 0.001$), but not in-row spacing. Plants that were not cropped early had significantly larger root systems and more one-, two-, and three-year-old wood and crowns than plants that were cropped early (Figure 1). 'Bluecrop' plants were smaller than those of 'Duke' and 'Elliott' in the early cropped and no crop systems (Figure 1). Roots accounted for 31% to 58% of the total plant dry weight depending on cultivar and early cropping.

## CONCLUSIONS

Young plants spaced at high density (0.45 m; 1.5 ft) compared to the more traditional spacing of 1.2 m (4 ft) produced about three times the

TABLE 2. Effect of cultivar, plant spacing, and early cropping on percent fruit buds (%) on one-year-old wood in winter 2002. Plants were established in October 1999.

| Plant Spacing | Bluecrop | | Duke | | Elliott | |
|---|---|---|---|---|---|---|
| | No crop | Early crop | No crop | Early crop | No crop | Early crop |
| 0.45 m | 50.1 | 41.2 | 63.8 | 55.6 | 52.6 | 55.0 |
| 1.2 m | 51.6 | 40.3 | 69.2 | 65.5 | 61.7 | 62.9 |
| Significance[z] | Cultivar: ***; Spacing: **; Early Cropping: *; Cultivar × Spacing: NS; Cultivar × Pruning: *; Spacing × Pruning: NS; Cultivar × Spacing × Pruning: NS | | | | | |

[z] NS, *, **, *** = non-significant or significant at $P < 0.05$, 0.01 or 0.001, respectively

TABLE 3. Effect of cultivar, plant spacing and early cropping on pruning weight per plant (g) in winter 2002. Plants were established in October 1999.

| Plant Spacing | Bluecrop | | Duke | | Elliott | |
|---|---|---|---|---|---|---|
| | No crop | Early crop | No crop | Early crop | No crop | Early crop |
| 0.45 m | 130.2 | 146.8 | 157.2 | 120.5 | 156.2 | 169.9 |
| 1.2 m | 142.5 | 151.6 | 175.3 | 161.6 | 245.5 | 256.5 |
| Significance[z] | Cultivar: ***; Spacing: ***; Early Cropping: NS; Cultivar × Spacing: **; Cultivar × Pruning: NS; Spacing × Pruning: NS; Cultivar × Spacing × Pruning: NS | | | | | |

[z] NS, *, **, *** = non-significant or significant at $P < 0.05$, 0.01 or 0.001, respectively

FIGURE 1. The effect of cultivar and early-cropping (the first two years after planting) on plant dry weight (roots, one-year-old wood, two-year-old wood, and crown + three-year-old and older wood) in February 2002. Averaged over in-row plant spacing.

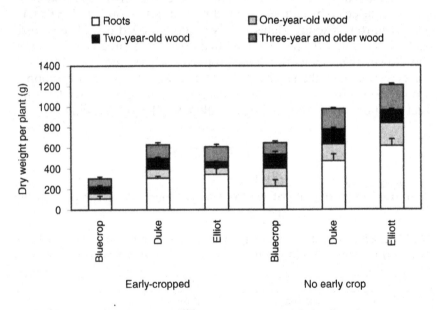

yield in this study for all cultivars. Our results on the effect of in-row spacing on yield are similar to those reported by Strik and Buller (2002) in 'Bluecrop'. Early cropping produced economical yields, particularly in year two (averaging 5,851 lb/acre). However, early cropping seemed to "stress" plants as evidenced by a reduced total plant dry weight after year two. In particular, the weight of the root system was reduced by 42% by early cropping, averaged over all cultivars and in-row spacing. The impact of this on yield of all treatments will be measured in year three as this study continues.

## GROWER BENEFITS

Early cropping, particularly at a high planting density, could provide a significant source of income for growers. For example, in 'Duke', cumulative yield in years one and two was 5,792 kg/ha (5,265 lb/acre); at $1.10/kg this would be a gross income of $6,370/ha. However, early cropping was shown to reduce plant size including the amount of roots and one-year-old fruiting wood. At this point it is not known what impact this will have on yield in year three when we will compare early-cropped plants to those that had blossoms removed.

## LITERATURE CITED

Moore, J.N., M.V. Brown, and B.P. Bordelon. 1993. Yield and fruit size of 'Bluecrop' and 'Blueray' highbush blueberries at three plant spacings. HortScience 28:1162-1163.

Pritts, M.P. and J.F. Hancock (eds.) 1992. Highbush blueberry production guide, NRAES-55, Ithaca, NY.

Strik, B., D. Brazelton, and R. Penhallegon. 1990. Grower's guide to pruning highbush blueberries. Oregon State University Extension Service Video, VTP002, Corvallis, OR.

Strik, B., C. Brun, M. Ahmedullah, A. Antonelli, L. Askham, D. Barney, P. Bristow, G. Fisher, J. Hart, D. Havens, R. Ingham, D. Kaufman, R. Penhallegon, J. Pscheidt, B. Scheer, C. Shanks, and R. William. 1993. Highbush blueberry production. Oregon State University Extension Service Publication, PNW215, Corvallis, OR.

Strik, B. and G. Buller. 2002. Improving yield and machine harvest efficiency of 'Bluecrop' through high-density planting and trellising. Acta Hort. 574:227-231.

# Effects of Nitrogen Sources on Growth and Leaf Nutrient Concentrations of 'Tifblue' Rabbiteye Blueberry Under Water Culture

Takato Tamada

**SUMMARY.** Growth responses and concentrations of mineral elements in leaves of 'Tifblue' rabbiteye blueberry were studied using water culture. In experiment 1, combinations of three N sources [$(NH_4)_2SO_4$, $NH_4NO_3$, $NaNO_3$] and four pH levels (average pH 3.5, 4.5, 5.5, 6.5) were evaluated. In general, top growth and root growth were superior with $(NH_4)_2SO_4$ and $NH_4NO_3$, and inferior with $NaNO_3$. Growth was best in the N source at pH combinations of $NH_4NO_3$ at pH 3.5-4.0, $NH_4NO_3$ at pH 4.1-5.0, and $(NH_4)_2SO_4$ at pH 4.1-5.0, and poorest with $NaNO_3$ at pH 6.1-7.0. However, growth with $NaNO_3$ at pH 3.5-4.0 was the same as with $(NH_4)_2SO_4$ at pH 6.1-7.0. In experiment 2, ten N sources, including their different elemental components, were evaluated. The top and root growth were superior with the $NH_4$-N forms [$NH_4Cl$, $NH_4H_2PO_4$, $(NH_4)_2SO_4$], and inferior with $NO_3$-N forms [$Ca(NO_3)_2 \cdot 4H_2O$, $KNO_3$]. Growth was intermediate with $Mg(NO_3)_2 \cdot 6H_2O$, $NaNO_3$, $NH_4NO_3$, or a mixture of $(NH_4)_2SO_4$ and $NaNO_3$. The growth with $NH_2CONH_2$ (urea) was similar to that with $NH_4Cl$. In ex-

Takato Tamada is Vice-President and Secretary-General, Japan Blueberry Association, 1104 Itoopia-Hamarikyu, 1-6-1 Kaigan, Minato-ku, Tokyo 106-0022, Japan.

[Haworth co-indexing entry note]: "Effects of Nitrogen Sources on Growth and Leaf Nutrient Concentrations of 'Tifblue' Rabbiteye Blueberry Under Water Culture." Tamada, Takato. Co-published simultaneously in *Small Fruits Review* (Food Products Press, an imprint of The Haworth Press, Inc.) Vol. 3, No. 1/2, 2004, pp. 149-158; and: *Proceedings of the Ninth North American Blueberry Research and Extension Workers Conference* (ed: Charles F. Forney, and Leonard J. Eaton) Food Products Press, an imprint of The Haworth Press, Inc., 2004, pp. 149-158. Single or multiple copies of this article are available for a fee from The Haworth Document Delivery Service [1-800-HAWORTH, 9:00 a.m. - 5:00 p.m. (EST). E-mail address: docdelivery@haworthpress.com].

Digital Object Identifer: 10.1300/J301v03n01_15

periment 3, three N sources [$(NH_4)_2SO_4$, $NH_4NO_3$, $NaNO_3$] were replaced with one of the alternative sources half way (at July 16) through the growing period. In general, the top and root growth were superior with $(NH_4)_2SO_4$ and $NH_4NO_3$ applied during the first half of the growing period (from April 18 to July 16), whereas, growth with $NaNO_3$ was inferior. Applying $(NH_4)_2SO_4$ during the later half of the growing period (from July 17 to October 16) increased growth, while $NaNO_3$ applied at this time decreased growth. A comparison of leaf mineral concentrations in the three experiments suggests N was highest when $(NH_4)_2SO_4$ was the N source and lowest when $NaNO_3$ was the N source. An imbalance between cations was observed. It is concluded that fertilizers containing the $NH_4$-N form are essential for optimum growth of 'Tifblue' rabbiteye blueberry. *[Article copies available for a fee from The Haworth Document Delivery Service: 1-800-HAWORTH. E-mail address: <docdelivery@haworthpress. com> Website: <http://www.HaworthPress.com> © 2004 by The Haworth Press, Inc. All rights reserved.]*

**KEYWORDS.** *Vaccinium ashei*, plant nutrition, N sources, Japan

## INTRODUCTION

Many studies of mineral nutrition in blueberries have been well documented by Ballinger (1966), Cain and Eck (1966), Eck (1988), Gough (1994), Korcak (1988, 1989), and Tamada (1997a). These studies can be divided into three categories; the examination of physiological characteristics of blueberry plants; the establishment of fertilizer recommendations; and the diagnosis of nutrient condition. It is recognized that blueberries are acid loving plants, prefer ammonium nitrogen, and have lower inorganic leaf nutrient compositions than other fruit species. However, fertilizer requirements differ with climate, soil conditions, and varieties.

The major problems with blueberry culture in Japan are that most of the blueberry growing regions are relatively new and climate and soil conditions differ from location to location. Also, the method for diagnosis of nutrient condition is not well established. Tamada (1989, 1993, 1997b, 2002a, 2002b) and Tamada et al. (1994a, 1994b, 1997) studied the effect of N sources on the growth, seasonal changes in leaf and fruit mineral concentrations, and symptoms of essential element deficiency and excess, to aid in the assessment of nutrient conditions of blueberries in Japan.

Three water culture experiments were conducted to obtain baseline date for the N nutrition and for the establishment of the method for diagnosis of nutrient condition. In experiment 1, the influence of three N sources and four pH level on growth were investigated. In experiment 2, ten N sources, including their different elemental components, were tested on growth. In experiment 3, replacement with one of the three alternative N sources half way through the growing period was evaluated on growth of 'Tifblue' rabbiteye blueberry.

## MATERIALS AND METHODS

One-year-old rooted cuttings of 'Tifblue' rabbiteye blueberry were planted in 1/5000 Wagner pots (one plant per pot), and grown under water culture in the glasshouse (an unheated glasshouse) from early April to early October 1999. The N sources evaluated in experiment 1 and 3 were $(NH_4)_2SO_4$, $NH_4NO_3$, and $NaNO_3$. Ten N sources, including their different elements components, were evaluated in experiment 2. The basic nutrient solution composition and treatment details are presented in Table 1. Six plants received each treatment. The dry weights of leaves, new and old shoots, and roots were measured, and the mineral elements of leaves were assayed.

## RESULTS AND DISCUSSION

*Effects of three N sources and four pH levels on growth.* Combinations of three N sources [$(NH_4)_2SO_4$, $NH_4NO_3$, $NaNO_3$] and four pH levels (average pH 3.5, 4.5, 5.5, 6.5) were evaluated in experiment 1. Whole plant, top and root growth were superior with $(NH_4)_2SO_4$ and $NH_4NO_3$, and inferior with $NaNO_3$ (Figure 1). These results were similar to the many reports on the N nutrition of blueberry (Cain, 1961; Peterson et al., 1987; Spiers, 1978; Townsend, 1967). However, the top growth with $NaNO_3$ at pH 3.5-4.0 was similar to that with $(NH_4)_2SO_4$ at pH 6.1-7.0.

Top and root growth was best in the N treatment combinations of $NH_4NO_3$ at pH 3.5-4.0, $NH_4NO_3$ at pH 4.1-5.0 and $(NH_4)_2SO_4$ at pH 4.1-5.0. In these treatments, new shoots elongated vigorously, leaves were large with healthy green color, and roots had many rootlets with light brown color. However, in both of $(NH_4)_2SO_4$ and $NH_4NO_3$ treatments, the growth had a tendency to decrease at pH higher than 5.0.

TABLE 1. Components of the basic nutrient solution common to the three experiments.

| Element | Concentration (ppm) | Salts Used | Notes |
|---|---|---|---|
| N | 56 | $(NH_4)SO_4$ | 1. Other microelements were applied. |
| N | 56 | $NH_4NO_3$ | |
| N | 56 | $NaNO_3$ | 2. Stock solution was diluted with tap water. |
| P | 30 | $NaH_2PO_4 \cdot 4H_2O$ | |
| K | 40 | $K_2SO_4$ | 3. Nutrient solution was changed at four week intervals. |
| Ca | 40 | $CaCl_2 \cdot 7H_2O$ | |
| Mg | 24 | $MgSO_4 \cdot 6H_2O$ | 4. Solutions were intermittently aerated 12 hours per day by air compressor. |
| Fe | 0.05 | Fe-EDTA | |

*N source differed in experiment 2.

Remark: • pH of solution was adjusted with $H_2SO_4$ and NaOH three times (Mon., Wed., and Fri.) per week.
• In experiment 1, pH was adjusted to average pH 3.5, 4.5, 5.5, and 6.5.
• In experiment 2 and 3, pH was adjusted to pH 4.1-5.0.

Top and root growth was poorest with $NaNO_3$ at pH 6.1-7.0. Growth tended to decrease in the $NaNO_3$ treatment when pH was raised over 4.0, and chlorosis symptoms appeared on young leaves at pH 5.1-6.0 and pH 6.1-7.0. Also, the rootlets were few and dark brown in color, and with raised pH level root color became nearly black.

It was very difficult to maintain the pH level of solutions. The pH level of $(NH_4)_2SO_4$ sources had a tendency to decrease with time; the $NaNO_3$ sources had a tendency to increase with time; but in the $NH_4NO_3$ treatment, the changes in solution pH were small.

N concentration in leaves of the $(NH_4)_2SO_4$ and $NH_4NO_3$ were higher than that of the $NaNO_3$ (Figure 2). Mineral concentration in leaves differed among N sources. In the $(NH_4)_2SO_4$ treatment, N, K, Ca, Mg concentration in leaves decreased as pH levels increased from 3.5-4.0 to 6.1-7.0. In the $NH_4NO_3$ treatments, N and Ca concentration in leaves changed little with pH level, but K concentration in leaves increased with raising pH level. In the $NaNO_3$ treatments, N concentrations in leaves were highest at the lower and highest pH level. K concentration was highest at pH 5.1-6.0 and lowest at pH 6.1-7.0. P concentration increased and Ca decreased with raising pH level.

FIGURE 1. Effects of N source and pH level on top and root growth of 'Tifblue' rabbiteye blueberry under water culture (experiment 1).

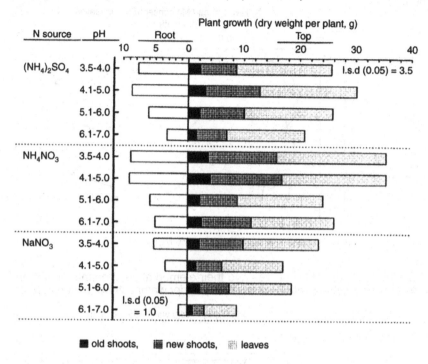

■ old shoots,　▦ new shoots,　▒ leaves

The results obtained in experiment 1 suggest that for best growth of 'Tifblue' rabbiteye blueberry, the N source should be $(NH_4)_2SO_4$ at a pH level of 4.1-5.0.

*Effects of ten N fertilizer including their different elemental components on growth.* Ten N sources, including their different elemental components, were evaluated in experiment 2. Very few studies have been reported on growth response to this many N sources (Tamada, 1993). In general, top and root growth were superior with the $NH_4$-N forms [$NH_4Cl$, $NH_4H_2PO_4$, $(NH_4)_2SO_4$], and inferior with $NO_3$-N forms [$Ca(NO_3)_2.4H_2O$, $KNO_3$, $Mg(NO_3)_2.6H_2O$, $NaNO_3$] (Figure 3). Growth with $NH_4NO_3$ or a mixture of $(NH_4)_2SO_4$ and $NaNO_3$ was intermediate. However, the top growth of $Mg(NO_3)_2·6H_2O$ was similar to that of $NH_4Cl$, $NH_4H_2PO_4$, $NH_4NO_3$ or mixture of $(NH_4)_2SO_4$ and $NaNO_3$. Growth with $NH_2CONH_2$ (urea) was almost the same as that with the $NH_4Cl$.

FIGURE 2. Effects of N source and pH level on the concentration of mineral elements in leaves of 'Tifblue' rabbiteye blueberry under water culture (experiment 1).

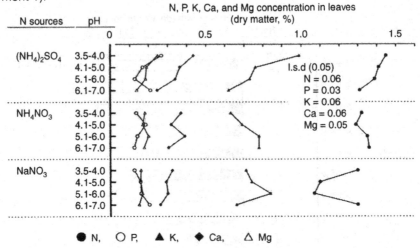

N, P, K, Ca, and Mg concentration in leaves (dry matter, %)

l.s.d (0.05)
N = 0.06
P = 0.03
K = 0.06
Ca = 0.06
Mg = 0.05

● N,   ○ P,   ▲ K,   ◆ Ca,   △ Mg

FIGURE 3. Effects of N sources and their elemental components on top and root growth of 'Tifblue' rabbiteye blueberry under water culture (experiment 2).

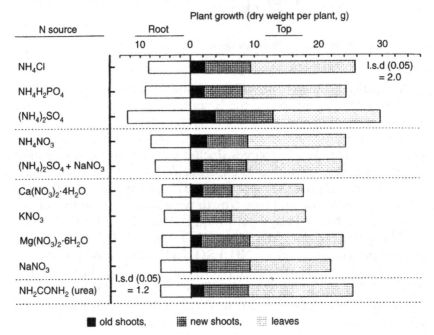

Plant growth (dry weight per plant, g)

l.s.d (0.05) = 2.0

l.s.d (0.05) = 1.2

■ old shoots,   ▦ new shoots,   ░ leaves

FIGURE 4. Effects of N sources and their elemental components on concentration of mineral elements in leaves of 'Tifblue' rabbiteye blueberry under water culture (experiment 2).

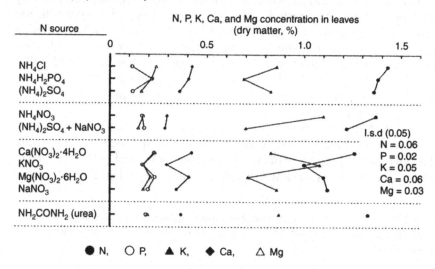

The appearance of leaves and roots was affected by N form. In the $NH_4$-N form, leaves were large with healthy green color and rootlets were abundant with light brown in color. For the $NO_3$-N forms, however, leaf color was pale green and the rootlets were changed from dark brown color to black brown color.

Plants receiving $NH_4$-N form had higher leaf N concentrations than for those receiving the $NO_3$-N form, but P concentration was higher with the $NO_3$-N form (Figure 4). N and P concentrations in leaves of $NH_4NO_3$ were intermediate.

Differences in other leaf mineral concentrations also occurred with different N sources. With the $NH_4$-N form, when $NH_4H_2PO_4$ was applied, P concentration in leaves increased, and K concentration decreased. With the $NO_3$-N form, when $Ca(NO_3)_2.4H_2O$ was applied, Ca concentration in leaves increased, while K concentration decreased. However, when $KNO_3$ was applied, K concentration in leaves increased and conversely, N, P, Ca and Mg concentrations decreased. When $Mg(NO_3)_2 \cdot 6H_2O$ was applied, Mg, P, and Ca concentrations in leaves became higher, and K concentration decreased.

These results suggest that the different elemental components of N sources had little effects on the top and root growth of 'Tifblue' rabbiteye blueberry.

*Effects of replacing a solution with an alternative N source halfway through the growing period.* Three N sources [$(NH_4)_2SO_4$, $NH_4NO_3$, $NaNO_3$] were replaced with one of the alternative sources halfway through the growing period (July 16) in experiment 3. In general, top and root growth were superior with $(NH_4)_2SO_4$ and $NH_4NO_3$ applied during the first half of the growing period (from April 18 to July 16), and growth with $NaNO_3$ was inferior (Figure 5). Applying $(NH_4)_2SO_4$ during the second half of the growing period (from July 17 to Oct. 16) increased the top and root growth while $NaNO_3$ applied at this time decreased the growth.

When $(NH_4)_2SO_4$ was applied during the second half of growing period, leaves generally had a healthy green color and the roots had a light brown color. Application of $NaNO_3$ during the second half of the growing period, however, resulted in a slightly pale green leaf coloration and

FIGURE 5. Effects of N source replacement half way through the growing period on top and root growth of 'Tifblue' rabbiteye blueberry under water culture (experiment 3).

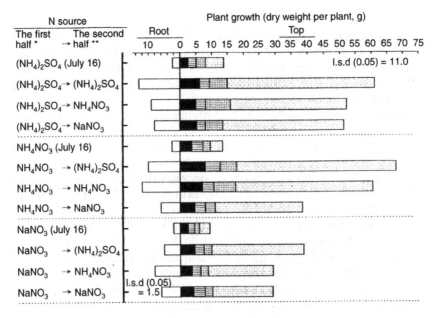

* from April 18 to July 16,    ** from July 17 to October 16
■ old shoots,    ☰ spring shoots and leaves,    ▦ summer shoots and leaves,
▦ long shoots over the 30 cm length and leaves

a dark brown root coloration. These results should lead to further study of root morphological changes caused by different N forms. Using $(NH_4)_2SO_4$ in the second half of growing period resulted in higher N, Ca, and Mg concentrations in leaves than using $NaNO_3$ (Figure 6). In the $NH_4NO_3$ treatment, N and K concentrations in leaves was intermediate.

The poor top and root growth that resulted from using $NaNO_3$ in the first half of the growing period was improved by using $(NH_4)_2SO_4$ during the second half of the growing period.

## *CONCLUSIONS*

The results obtained in the three experiments showed that the top and root growth with $NH_4$-N and $NH_4NO_3$ forms were superior to that with $NO_3$-N form. A comparison of leaf mineral concentrations suggests N was highest when $NH_4$-N form was the N source and lowest when $NO_3$-N was the N source. The tendency to an imbalance between cations with $NO_3$-N form was larger than that with $NH_4$-N form.

It is concluded that fertilizers containing the $NH_4$-N form are essential for optimum growth of 'Tifblue' rabbiteye blueberry.

FIGURE 6. Effects of N source replacement half way through the growing period on concentration of mineral elements in leaves of 'Tifblue' rabbiteye blueberry under water culture (experiment 3).

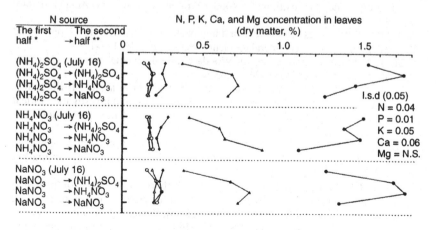

* from April 18 to July 16,   ** from July 17 to October 16
● N,   ○ P,   ▲ K,   ◆ Ca,   △ Mg

# LITERATURE CITED

Allinger, W. E. 1966. Soil management, nutrition, and fertilizer practices. Pp. 132-178. In: Blueberry culture. P. Eck and N. F. Childers (eds.). Rutgers University, New Brunswick, NJ.

Cain, J. C. 1961. A comparison of ammonium and nitrate nitrogen for blueberries. Proc. Amer. Soc. Hort. Sci. 59: 161-167.

Cain, J. C. and P. Eck. 1966. Blueberry and cranberry. Pp. 101-129. In: Fruit nutrition. N. F. Childers (ed.). Horticultural Publications, New Brunswick, NJ.

Eck, P. 1988. Blueberry science. Rutgers University Press. Pp. 91-159. New Brunswick, NJ.

Gough, R. E. 1994. The highbush blueberry and its management. Pp. 109-136. Food Products Press. NY.

Korcak, R. F. 1988. Nutrition of blueberry and other calcifuges. Pp. 183-227. In: Horticultural reviews. J. Janick (ed.). Vol. 10. Timber press, Portland, OR.

Korcak, R. F. 1989. Variation in nutrient requirements of blueberries and other calcifuges. HortScience. 24: 573-578.

Peterson, L. A., E. J. Stang, and M. N. Dana. 1988. Blueberry response to $NH_4$-N and $NO_3$ N. J. Amer. Soc. Hort. Sci. 112: 612-616.

Spiers, J. M. 1978. Effects of pH level and nitrogen sources on elemental leaf content of 'Tifblue' rabbiteye blueberry. J. Amer. Soc. Hort. Sci. 103: 705-708.

Tamada, T. 1989. Nutrient deficiencies of rabbiteye and highbush blueberries. Acta Hort. 241: 132-138.

Tamada, T. 1993. Effects of the nitrogen source on the growth of rabbiteye blueberry under soil culture. Acta Hort. 346: 207-213.

Tamada, T. 1997a. Fundamentals of blueberry production [18]. Agri. and Hort. Vol. 72(8)-72(12) (In Japanese).

Tamada, T. 1997b. Effect of manganese, copper, zinc and aluminum application rate on the growth and composition of 'Woodard' rabbiteye blueberry. Acta Hort. 446: 497-506.

Tamada, T. 2002a. Stages of rabbiteye and highbush blueberry fruit development and the associated changes in mineral elements. Acta Hort. 574: 127-129.

Tamada, T. 2002b. Growth response of 'Woodard' rabbiteye blueberry to higher rates of macronutrients. Acta Hort. 574: 355-361.

Tamada, T., S. Nishikawa, Y. Yamakoshi, and K. Iida. 1994a. Seasonal distribution of mineral elements concentration in the leaf of blueberries, and investigation of leaf analysis of rabbiteye blueberry orchard at some places in Chiba prefecture. Bull. Chiba Agri. College 7: 33-45 (In Japanese).

Tamada, T., T. Ishigami, H. Saito, and T. Miyama. 1994b. Macronutrient deficiency in rabbiteye blueberry. Bull. Chiba Agri. College 7: 47-58 (In Japanese with English summary).

Tamada, T., H. Ooki, K. Noji, T. Kurauchi, and T. Kobori. 1997. Effects of the higher applied rate of nitrogen, phosphorus, potassium, calcium magnesium, iron and manganese on the growth and mineral elements in the leaf of 'Tifblue' rabbiteye blueberry. Bull. Chiba Agri. College 8: 13-26 (In Japanese with English summary).

Townsend, L. R. 1967. Effect of ammonium N and nitrate N separately and in combination on growth of the highbush blueberry. Can. J. Plant Sci. 47: 555-462.

# Stem and Leaf Diseases
# and Their Effects on Yield
# in Maine Lowbush Blueberry Fields

Seanna L. Annis
Constance S. Stubbs

**SUMMARY.** In 2001, six lowbush blueberry (*Vaccinium angustifolium* Ait.) fields in crop production were examined for the effects of stem and leaf diseases on yield. Symptomatic and control stems were tagged during bloom and the flowers per stem were counted. In late July, the berries were counted. Fungi commonly found on stems and leaves included *Alternaria, Aureobasidium, Cladosporium, Colletotrichum,* and *Gloeosporium.* For one field, stems with disease at their base showed a significant reduction in yield compared to healthy stems. Fungicides being evaluated for control of *Monilinia* blight had no significant effect on either stem disease or leaf spot incidence. *[Article copies available for a fee from The Haworth Document Delivery Service: 1-800-HAWORTH. E-mail address: <docdelivery@haworthpress.com> Website: <http://www.HaworthPress. com> © 2004 by The Haworth Press, Inc. All rights reserved.]*

Seanna L. Annis is Assistant Professor of Mycology and Constance S. Stubbs is Senior Research Scientist, Department of Biological Sciences, Deering Hall, University of Maine, Orono, ME 04469 USA.

Address correspondence to: Seanna L. Annis (E-mail: sannis@maine.edu).

The authors would like to gratefully acknowledge the support of the Wild Blueberry Commission of Maine and the cooperating growers. Maine Agricultural and Forest Experimental Station external publication # 2624.

[Haworth co-indexing entry note]: "Stem and Leaf Diseases and Their Effects on Yield in Maine Lowbush Blueberry Fields." Annis, Seanna L., and Constance S. Stubbs. Co-published simultaneously in *Small Fruits Review* (Food Products Press, an imprint of The Haworth Press, Inc.) Vol. 3, No. 1/2, 2004, pp. 159-167; and: *Proceedings of the Ninth North American Blueberry Research and Extension Workers Conference* (ed: Charles F. Forney, and Leonard J. Eaton) Food Products Press, an imprint of The Haworth Press, Inc., 2004, pp. 159-167. Single or multiple copies of this article are available for a fee from The Haworth Document Delivery Service [1-800-HAWORTH, 9:00 a.m. - 5:00 p.m. (EST). E-mail address: docdelivery@haworthpress.com].

**KEYWORDS.** Fungicides, leaf spot, lowbush blueberry, stem blight, *Vaccinium angustifolium*, yield

## INTRODUCTION

The Maine blueberry industry has 24,000 ha of lowbush blueberries yielding 35 million kg per year (5 year average) (Yarborough, 2002). Fields predominantly consist of wild clonal plants of *Vaccinium angusti-folium* Ait. and *V. myrtilloides* Michx. that are occasionally interspersed with other species. Blueberry fields are thus comprised of a patchwork of genetically diverse clones of several species that are extensively managed to produce a commercial crop.

Fields of lowbush blueberries in Maine are managed on a two year cycle to produce increased uniformity in the blueberry stand and in-creased yields from the plants. After harvest, blueberry fields are severely pruned followed by vegetative growth for one year and production of berries in the following year. In the last few decades, changes in man-agement practices have increased yields, but these changes may affect the long-term health of the plants and their susceptibility to disease.

Mummy berry disease caused by *Monilinia vaccinii-corymbosi* (Reade) Honey is the major disease of lowbush blueberry (Lambert, 1990). Widespread use of fungicides to control mummy berry disease has usually decreased the impact of this disease on yield, but this prac-tice may change the populations of other fungal pathogens in the fields. Other diseases of lowbush blueberry have been reported (Caruso and Ramsdell, 1995), but only a few have been well studied and very few have had their effect on yield determined. An example, red leaf disease of lowbush blueberry, caused by *Exobasidium vaccinii* (Fuckel) Wor., significantly decreased the number of flowers and fruit on infected stems compared to healthy stems (Hildebrand et al., 2000). Hildebrand et al. (2000) also found that the loss of yield was proportional to the inci-dence of red leaf in fields. Many other fungal diseases are known from studies on highbush (*V. corymbosum* L.) or rabbiteye blueberries (*V. ashei* Reade) (Caruso and Ramsdell, 1995). However, these diseases may have different development and severity on lowbush blueberries than on other species of blueberry.

Maine lowbush blueberry growers have reported increased symp-toms of stem dieback and leaf spot with early leaf drop in their fields and have expressed concern about the impact of disease on production. Extensive surveys of lowbush blueberry fields in Maine have demonstrated a higher incidence of leaf spot and stem diseases in bearing fields than in non-bearing fields from 1999 to 2001 (Annis and Stubbs, 2002). The objectives of the present research were to determine the impact of stem and leaf spot diseases on the yield of lowbush blueberry plants in 2001 and the effect, if any, of fungicides applied to control mummy berry disease on other stem and leaf spot diseases.

## MATERIALS AND METHODS

The study sites were six lowbush blueberry fields, which were bearing (in their fruit producing year) near the towns of Deblois (fields D1, D2, D3), Hope (field H1), Sedgwick (field S1) and Stockton Springs (field SS1) in Maine. To examine the effect of stem and leaf spot diseases on yield, 20 randomly selected equally spaced 0.25 $m^2$ plots were established along a W transect in six bearing fields in May 2002, prior to full bloom. The fields were a mixture of organic (with no pesticides applied) and managed fields (fungicide sprays for mummy berry applied during leaf bud break).

Two additional bearing fields, D4 and CF1, in Deblois and Columbia Falls, respectively, were used to examine the effects of fungicides, which were being evaluated for the control of mummy berry disease, on the incidence of stem and leaf spot diseases and their subsequent effect on yield. Each field had four blocks and each block had eight 2 m × 10 m plots; a control treatment and seven different fungicide treatments (8 treatments, 4 replications per treatment per field). The fungicide plots were sprayed three times (3 May, 6 and 11 June) with the same type and quantity of fungicide (Table 1) with a backpack $CO_2$ sprayer (see Yarborough, 2001 for spray application details). Within each plot within each block, a 0.25 $m^2$ plot was established to examine stem and leaf spot diseases.

In all eight fields studied, stems with symptoms of stem disease at the tip, middle, or bottom of the stem and healthy stems were tagged during bloom (25 May-6 June). A maximum of 30 stems per health condition was tagged per field. All flowers on the tagged stems were counted and recorded.

In late July fruits were harvested and counted on the tagged stems and the percentage of flowers that produced berries was determined. All tagged stems were collected from these plots, stored in plastic bags, and refrigerated until the location and symptoms of disease on the stems were later diagnosed in the laboratory.

TABLE 1. Incidence of blighted stems and stems with leaf spots per 0.25 m$^2$ plot and percentage yield (flowers that produced fruit) in fungicide treatment plots for two lowbush blueberry fields.

| Treatment[z] | Avg. # of diseased stems per plot | Avg. % of stems with leaf spots per plot | Avg. % yield |
|---|---|---|---|
| **Columbia Falls** | | | |
| **Field 1** Control | 7.0 | 3.2 | 44.3 |
| Azoxystrobin (1.14 L/ha) | 2.5 | 1.2 | 48.2 |
| Azoxystrobin (1.28 L/ha) | 5.2 | 34.2 | 57.4 |
| Propiconazole (0.29 L/ha) | 3.5 | 20.0 | 58.3 |
| Propiconazole (0.44 L/ha) | 2.8 | 8.2 | 39.8 |
| Propiconazole + Azoxystrobin[y] | 2.0 | 26.5 | 62.4 |
| Pyraclostrobin + Boscalid (2:1) (1.1 kg/ha) | 3.5 | 150 | 34.6 |
| Pyraclostrobin + Boscalid (2:1) (1.74 kg/ha) | 3.2 | 4.2 | 57.6 |
| **Deblois Field 4** | | | |
| Control | 3.5 | 0.2 | 74.6 |
| Azoxystrobin (1.14 L/ha) | 4.0 | 1.8 | 73.2 |
| Azoxystrobin (1.28 L/ha) | 4.0 | 27.5 | 64.4 |
| Propiconazole (0.29 L/ha) | 3.2 | 18.8 | 77.1 |
| Propiconazole (0.44 L/ha) | 3.5 | 1.5 | 69.8 |
| Propiconazole + Azoxystrobin[y] | 3.5 | 4.2 | 54.2 |
| Pyraclostrobin + Boscalid (2:1) (1.1 kg/ha) | 5.2 | 14.8 | 71.2 |
| Pyraclostrobin + Boscalid (2:1) (1.74 kg/ha) | 2.0 | 0.2 | 62.2 |

[z] The treatments had no significant effects on the avg. # of diseased stems per plot, the avg. % of stems with leaf spot per plot, and the avg. % yield.
[y] Propiconazole applied at 0.29 L/ha and Azoxystrobin applied at 1.14 L/ha.

The incidence of leaf spot disease was determined by visually estimating the percentage of stems showing leaf spot symptoms for each plot. In 0.25 m$^2$ plots adjacent to the tagged stem plots, all diseased stems within the plot were counted and rated for location of disease on the stems. The total number of stems within a plot was determined for four of the 0.25 m$^2$ plots per field in order to estimate stem density.

From each field, the tagged stems and leaves showing symptoms of leaf spot were surface sterilized in 10% bleach for 3 min, rinsed three times in sterile water, and plated on malt yeast extract agar and/or water agar. Fungi growing out of the stems and leaves were identified using standard keys.

Data analyses were performed using the SAS program (SAS Institute Inc., Cary, NC). The data were not normally distributed and thus were analyzed using non-parametric statistics. Significant effects were determined using the Kruskal-Wallis test ($P < 0.05$). Dunn's test was used for multiple comparisons of ranked data (Neave and Worthington, 1988).

## *RESULTS AND DISCUSSION*

The six bearing fields surveyed in 2001 significantly varied in the severity of stem diseases as measured by the average number of diseased stems per plot (Figure 1). Fields D1 and H1 had higher average numbers of diseased stems than the other fields (Figure 1). There were no significant differences in stem density among the fields. Each field has its own unique patchwork of genetically different blueberry clones (Hepler and Yarborough, 1991) that may vary in susceptibility to disease. Hepler and Yarborough (1991) reported high variability in yield among different lowbush blueberry clones.

The fields significantly differed in the incidence of leaf spot diseases (Figure 2). The level of incidence of leaf spot in the fields did not correspond to the incidence of stem disease found in the fields (Figures 1-2). For example, field S1 had a high level of leaf spot (62%) but only a moderate level of stem disease. This suggests that susceptibility of lowbush blueberry clones to leaf spot diseases may be unrelated to susceptibility to stem diseases.

The six fields also significantly differed in their yields as measured

FIGURE 1. The average number of diseased stems per 0.25 $m^2$ plot (20 plots per field) in the six lowbush blueberry fields sampled for stem and leaf spot diseases. Vertical lines indicate one standard error of the mean. Bars with the same letters indicate nonsignificance (Dunn's test, $P \leq 0.05$).

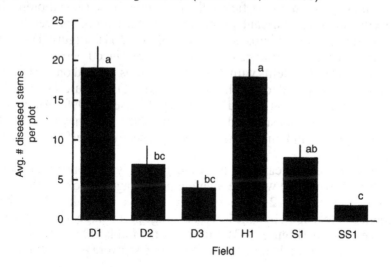

FIGURE 2. The average percentage of stems with leaf spots per 0.25 m² plot (20 plots per field) in six lowbush blueberry fields sampled for stem and leaf spot diseases. Vertical lines indicate one standard error of the mean. Bars with the same letters indicate nonsignificance (Dunn's test, P ≤ 0.05).

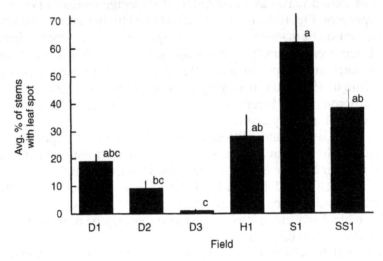

by the average percentage of flowers that produced berries (Figure 3). The severity of stem disease in the fields did not correspond to the average percentage of flowers that produced berries in each field. Disease on different parts of the stem may have different effects on berry production. For five out of six fields, the location of disease symptoms on the stems did not significantly correspond to the average percentage of flowers that produced berries (for example field H1, Figure 4). However, the average percentage of flowers that produced berries was significantly lower for stems with disease symptoms located on the bottom of the stems compared to healthy stems for field D1 (Figure 4). Field D1 had extensive infection by mummy berry disease in 1999 (Annis, unpublished results) and suffered drought in 2000 which may have weakened the plants and made them more prone to the effects of stem infection.

The incidence of leaf spot per field generally demonstrated an inverse correspondence with the average percentage of flowers that produced berries (Figures 2-3). For example, field S1, which had the highest incidence of leaf spot, had the lowest average percentage of flowers that produced berries (Figures 2-3). Similarly, the fields D2 and D3 had the lowest incidence of leaf spot and the highest average percentage of

FIGURE 3. The average percentage of flowers that produced blueberries on tagged stems in 0.25 m² plots (20 plots per field) in six lowbush blueberry fields sampled for stem and leaf spot diseases. Vertical lines indicate one standard error of the mean. Bars with the same letters indicate nonsignificance (Dunn's test, P ≤ 0.05).

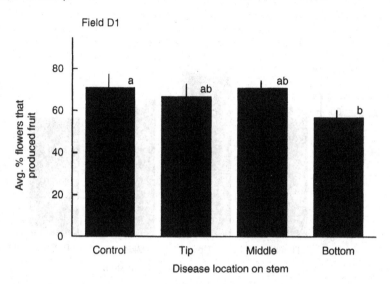

Field D1

flowers that produced berries. High levels of leaf spot may decrease the photosynthetic capacity of the leaves and thereby affect fruit production in the current year. Ojiambo et al. (2002) found that the level of *Septoria* leaf spot was positively related to defoliation in southern highbush blueberry, and this may also be occurring with leaf spot diseases in lowbush blueberry. *Septoria* leaf spot may also have long-term effects on blueberry plants. Ojiambo et al. (2002) found that a high incidence of *Septoria* leaf spot and defoliation in southern highbush blueberries in the fall decreased the flower bud set in the winter and decreased fruit production in the following spring.

Many lowbush blueberry growers apply two applications of a protectant fungicide from bud break until mid-bloom to control primary leaf infection of mummy berry disease. Different amounts or combinations of fungicides were tested for their ability to protect against mummy berry disease (Yarborough, 2001). We examined the effects of these fungicides on other stem and leaf diseases in two fields, CF1 and D4 (Table 1). None of the fungicide treatments had a significant effect on the incidence of stem disease or leaf spot disease or on the average per-

FIGURE 4. The average percentage of flowers that produced blueberries on tagged stems diagnosed as healthy, or with disease symptoms at the tip, middle or bottom of the stem in 0.25 m² plots (20 plots per field) in Field D1 and Field H1. Vertical lines indicate one standard error of the mean. Bars with the same letters indicate nonsignificance (Dunn's test, P ≤ 0.05).

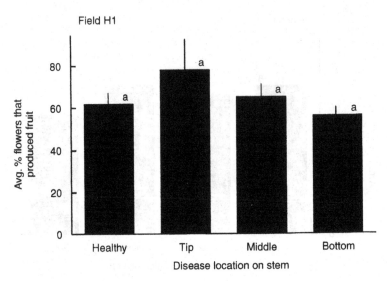

centage of flowers that produced berries in the two fields examined. Disease incidence was low in these fields and may have contributed to the inability to detect any effect of the fungicides. A significantly lower incidence of stem disease was found in Maine blueberry fields in 2001 compared to 1999 and 2000 (Annis and Stubbs, 2002), which may be due in part to the warm dry spring followed by a drought in August that blueberry growing areas in Maine experienced in 2001. Other possible reasons for not detecting any effect of the fungicides are that the pathogens causing stem and leaf diseases do not infect the developing leaf buds or these pathogens are unaffected by the fungicides tested. Other modes of infection that pathogens may be using include wounds caused by pruning damage, breakage and/or insects. Our observations of insect damage near lesions on stems and of lesions occurring where branches join to the stems support the involvement of wounds in disease infection.

Fungi commonly identified from tagged diseased stems in 2001 were *Alternaria, Aureobasidium, Cladosporium, Colletotrichium, Epicoccum, Gloeosporium, Oidiodendron, Penicillium, Phoma,* and *Phomopsis.*

Common fungi identified from leaves with leaf spot were *Alternaria, Aureobasidium,* and *Botrytis.* Numerous other genera of fungi were less commonly identified from infected stem and leaf tissue. Some fungi are reported to cause both stem and leaf diseases of blueberry (Caruso and Ramsdell, 1995), and some overlap in genera was found between the fungi identified from diseased stems and leaves. Work is in progress to determine which of these commonly identified fungi are causing disease in blueberry.

## CONCLUSIONS AND GROWER BENEFITS

The fields varied significantly in the incidence of stem and leaf spot diseases and in yield. For one field, a significant reduction in fruit production occurred in stems with disease at their base compared to healthy stems. Fungicides tested for effectiveness for control of mummy berry disease had no significant effect upon stem and leaf spot diseases and yield.

## LITERATURE CITED

Annis, S.L. and C.S. Stubbs. 2002. Stem and leaf diseases in Maine wild blueberry fields, (1999-2001). P. 7, In: Program abstracts 2002 Annu. Mtg. Wild Blueberry Res. Ext. Workers. Univ. Maine Coop. Ext., Orono, Maine.

Caruso, F.L. and D.C. Ramsdell. 1995. Compendium of blueberry and cranberry diseases. APS Press, St. Paul, MN.

Hepler, P.R. and D.E. Yarborough. 1991. Natural variability in yield of lowbush blueberries. HortScience. 26:245-246.

Hildebrand, P.D., N.L. Nickerson, K.B. McRae, and X. Lu. 2000. Incidence and impact of red leaf disease caused by *Exobasidium vaccinii* in lowbush blueberry fields in Nova Scotia. Can. J. Plant Path. 22:364-367.

Lambert, D.H. 1990. Effects of pruning method on the incidence of mummy berry and other lowbush blueberry diseases. Plant Dis. 74:199-201.

Neave, H.R. and P.L. Worthington. 1988. Distribution-free tests. Unwin Hyman, London.

Ojiambo, P.S., H. Scherm, and P. Brannen. 2002. Septoria leaf spot intensity, defoliation, and yield loss relationships in Southern blueberry (Abstr.) Phytopathology. 92:1025.

Yarborough, D.E. 2001. Evaluation of fungicide efficacy in wild blueberry fields. P. 125, In: Annu. Wild Blueberry CSREES Prog. Rep. Univ. of Maine, Orono, Maine.

Yarborough, D.E. 2002. Wild blueberry newsletter. August 2002. Univ. Maine Coop. Ext., Orono, Maine.

# Row Covers to Delay or Advance Maturity in Highbush Blueberry

Peter Hicklenton
Charles Forney
Carolyn Domytrak

**SUMMARY.** A wide variety of hardy highbush blueberry (*Vaccinium corymbosum* L.) cultivars are suitable for cultivation in parts of North-Eastern Canada, and especially Nova Scotia's Annapolis Valley (latitude 45 EN). The object of this research was to investigate the potential to expand late market opportunities for exported fruit from northern areas by using a combination of controlled atmosphere (CA) storage and in-field row covers to delay fruit maturity in two cultivars ('Bluegold' and 'Brigitta'), and to advance maturity in the cultivar 'Elliott', that normally matures too late for full harvest at this latitude. Covering 'Bluegold' and 'Brigitta' with 50% shading from time of fruit set to harvest

Peter Hicklenton and Charles Forney are Research Scientists; Carolyn Domytrak is Research Technician, Atlantic Food and Horticulture Research Centre, Agriculture and Agri-Food Canada, 32 Main Street, Kentville, Nova Scotia, Canada, B4N 1J5.

Address correspondence to: Peter Hicklenton (E-mail: HicklentonP@agr.gc.ca).

The authors thank Nova-Agri Associates Ltd. for their cooperation with this research. The technical support of Michael Jordan in maintaining controlled atmosphere chambers, and Alex MacDonald, Julia Reekie and Andrea Braun with field work is also gratefully acknowledged. Funding for the project was provided by the Matching Investment Initiative Program of Agriculture and Agri-Food Canada.

[Haworth co-indexing entry note]: "Row Covers to Delay or Advance Maturity in Highbush Blueberry." Hicklenton, Peter, Charles Forney, and Carolyn Domytrak. Co-published simultaneously in *Small Fruits Review* (Food Products Press, an imprint of The Haworth Press, Inc.) Vol. 3, No. 1/2, 2004, pp. 169-181; and: *Proceedings of the Ninth North American Blueberry Research and Extension Workers Conference* (ed: Charles F. Forney, and Leonard J. Eaton) Food Products Press, an imprint of The Haworth Press, Inc., 2004, pp. 169-181. Single or multiple copies of this article are available for a fee from The Haworth Document Delivery Service [1-800-HAWORTH, 9:00 a.m. - 5:00 p.m. (EST). E-mail address: docdelivery@haworthpress.com].

http://www.haworthpress.com/web/SFR
Digital Object Identifer: 10.1300/J301v03n01_17

delayed full harvest by about 2.5 weeks in the first year. Yield decreases were recorded in the 2nd year of covering suggesting that covering with this density may diminish long-term yield potential. Fruit quality following after 6 weeks of CA storage ($10\%$ $CO_2$, $16\%$ $O_2$, $0°C$) was not affected by shading and after 42 days of storage fruit showed very little decay. Decay increased substantially after 63 days of storage. A removable row cover (6 mil polyethylene) advanced maturity in 'Elliott' in the first year by between 10 and 14 days, but this advancement was not repeated in the second season. In the first year, covering the crop only until petal drop was nearly as effective as covering throughout the season. Yield from covered plants was increased about 25% in the first year as compared with the controls, but this increase was not realized in the second year. 'Elliott' fruit from all treatments stored successfully for up to 42 days, and up to 63 days in control plants; fruit from covered plants showed a 30% decline in marketability when subjected to an additional 7 days in air at $7°C$ following 63 days in CA. *[Article copies available for a fee from The Haworth Document Delivery Service: 1-800-HAWORTH. E-mail address: <docdelivery@haworthpress.com> Website: <http://www.HaworthPress. com>]*

**KEYWORDS.** Shading, maturity manipulation, controlled-atmosphere storage, *Vaccinium corymbosum* L., 'Elliott', 'Bluegold', 'Brigitta'

## *INTRODUCTION*

In the early part of the 21st Century most berry fruit is available year round in markets throughout the world. Wholesale prices, however, may vary greatly depending on season, production factors, and market conditions. In North America peak production of highbush blueberries occurs in July and August, but prices on world markets tend to peak later, in October and early November before southern hemisphere fruit becomes widely available (Figure 1). This suggests that there are opportunities for northern growers to use a combination of in-field maturity control and postharvest storage techniques to extend the marketing period for their berries into the mid-Fall to capitalize on high market prices. A number of strategies are needed in order to maintain availability of fruit in late October and early November. First, cultivars that produce mature fruit relatively late in the season must be selected. Then, it may be necessary to manipulate fruit maturity in the field, and finally storage techniques must be devised to maintain fruit quality until the target marketing period. While a range of cultivars are available to pro-

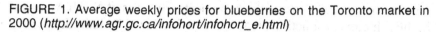

FIGURE 1. Average weekly prices for blueberries on the Toronto market in 2000 (*http://www.agr.gc.ca/infohort/infohort_e.html*)

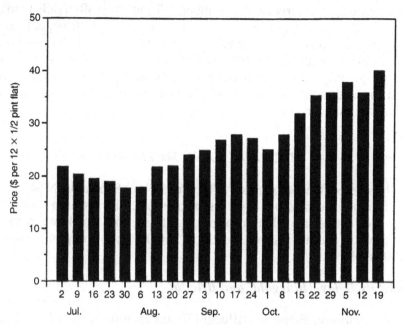

duce fruit from early to late season, there is little information on in-field techniques to control maturity. Plastic tunnel culture has been used to provide early fruit from southern highbush blueberries (Sampson and Spiers, 2002). Bal et al. (1993) have reported the use of row-covers to delay maturity by up to 3 weeks in The Netherlands. Shutak et al. (1956) reported that reducing the light intensity incident on developing fruit delayed ripening, suggesting that shading might be used to extend the harvest season. Still, the concept of using row covers on selected cultivars in combination with post-harvest storage methods to extend the marketing season has not been reported.

In this study we have investigated the potential for in-field manipulation of fruit maturity in one late-maturing cultivar 'Elliott', and two mid-season cultivars 'Bluegold' and 'Brigitta'. We have also evaluated the effects of maturity manipulation on post-harvest fruit quality, and investigated techniques to successfully store fruit for up to 63 d. While growers in long-season districts favor 'Elliott' for late-season marketing and stability of quality in storage, it has proven difficult to reliably produce fruit before frost in the more northerly parts of North America. In the principal production areas in Nova Scotia (45°N latitude), where frost may occur as early as 9 September, 'Elliott' fruit often fails to mature. On the other hand mid-season cultivars such as 'Bluegold' and 'Brigitta' generally mature too early to capitalize on mid-Fall marketing opportunities. Accordingly our research has focused on two strategies: the use of polyethylene row covers to advance maturity and provide frost protection for 'Elliott', and shading to delay fruit maturity in 'Bluegold' and Brigitta.

## MATERIALS AND METHODS

The site for all experiments was Medford, Nova Scotia (45°N 64.5°W) at a farm owned by Nova-Agri Associates Ltd. The research was conducted on established plants at a spacing of 1 m, with 2 m between adjacent rows. Experiments were conducted on the same group of plants in 2000 and 2001.

### Maturity Delay–'Bluegold' and Brigitta

*Plant culture.* Plants of 'Bluegold' and Brigitta were selected for study in spring 2000 from various existing plantings and formed two experimental blocks. Plants in block 1 had been planted 4 years before, while those in block 2 had been in place for 3 years. In each block, structures measuring 52 m long × 14.5 m wide × 3 m high were erected and covered with a neutral density woven material providing 50% shade. The fabric extended from the sides of the structure to ensure complete shading of the plants. Shading material was applied following fruit set on 30 June and removed in early September. Rows adjacent to the shade structures provided un-shaded control plants. Ten shaded, and ten un-shaded plants in each block were randomly selected for study. In 2001, the ten shaded plants were subdivided into two groups of 5 in each block; one group was re-shaded, while the other was returned to full light conditions for the entire season.

*Environmental monitoring.* Air temperature and photosynthetically active radiation (400-700 nm; PAR) were measured using shielded thermistors (Campbell Scientific Ltd. (CSI) Edmonton, Alta. Canada) and quantum sensors (Apogee Instruments Ltd., Logan, Utah) in shaded and control environments. All data were recorded using a CSI CR10-A datalogger. Temperatures were recorded at 60 s intervals and averaged over each hour.

*Fruit harvest.* Harvest began when 30% of the fruit in the first-maturing group were fully blue. All ripe fruit from all environments were harvested by hand on the same day. Harvests continued at 2 to 3 d intervals.

*Experimental design and statistical analysis.* Data were analyzed according to the randomized complete block arrangement of shade treatments. Plants within treatments were considered sub-samples and all data were converted to a log-scale prior to analysis. Year to year data were analyzed in a mixed model to assess the effects of change of treatment (e.g., shade in 2000 to full sun in 2001) on fruit yield.

### Maturity Acceleration–Elliott

*Plant culture.* In 2000, a row of 30, 4-year old 'Elliott' plants was covered with a 28.3 m long × 3.1 m wide × 2.5 m high hoop structure covered on 4 May with 6-mil greenhouse-grade polyethylene. Only a single row of 'Elliott' plants was available at the Medford farm. Twenty plants were selected at random and assigned to one of two treatment groups: Covered throughout the season, or covered until petal drop on 20 June (10 per group). The same plants were re-covered and subjected to the same treatments in 2001. Another 10 plants in the same row beyond the hoop structure were designated as controls.

Environmental monitoring and harvest procedures were identical to those described for 'Bluegold' and 'Brigitta'. The small area of the farm planted to 'Elliott' precluded the use of experimental blocks. Individual plants within the single hoop structure were considered replicates for analysis of yield data.

### Fruit Quality Evaluation and Post-Harvest Treatments

Three commercial hand harvests were carried out in each year and in each environment using fruit on plants not designated for cumulative yield analysis (see above). Pre-cooled fruit were packaged by hand in 125 g plastic clamshells (11 cm × 5 cm top; 8.5 cm × 7 cm bottom; 3 cm depth) at 2°C. Immediately following packaging, the fruit were placed in cardboard flats and cooled to 0°C. They were then transferred to 320 L stainless steel controlled atmosphere (CA) storage chambers in a cold room at 0°C. Atmospheres in all chambers were maintained at 10% $CO_2$ and 16% $O_2$ using an Oxystat 2002 Controlled Atmosphere System (David Bishop Instruments, Heathfield, E. Sussex, UK).

Immediately following each harvest, 3 clamshells of fruit were evaluated for initial quality. After 21, 42, or 63 d of CA storage, 6 clamshells for each field treatment were removed from the storage chambers and warmed to room temperature. Three clamshells were opened immediately and the fruit evaluated for marketability. Any fruit showing shrivel, split, or decay were considered unmarketable. The remaining 3 clamshells were held for an additional 7 days at 7°C and were then evaluated.

## RESULTS AND DISCUSSION

*Yield–'Brigitta' and 'Bluegold'.* The harvest season for 'Brigitta' and 'Bluegold' began in early August in both 2000 and 2001. During 2000 both cultivars showed a strong response to the 50% shade treatments. By 5 August for example nearly 20% of fruit had been harvested from 'Bluegold' control plants, but only 4% from those under shade (Figure 2). Similarly cumulative yield in 'Brigitta' reached 40% on 19 August when shaded plants had yielded only 11% of their seasonal total (Figure 3). The time taken to reach 80% of full seasonal yield differed by about 16 d in both cultivars. Since commercial harvests are conducted two or three times on blueberry plants, such a delay could be used effectively to extend the marketing season; harvesting would commence, and finish later. Total cumulative yield in 'Bluegold' (16 September) was unaffected by shade treatments in 2000 (Table 1) whereas 'Brigitta' showed a 24% reduction in yield at the end of the season (Table 2).

During the second growing season, the delay in maturity obtained by covering plants in both seasons (Figures 2 and 3) was less than that observed in 2000. Interpolating between the end-of-week cumulative yields showed only about a 6-day delay to reach 80% of full-season yield in both 'Bluegold' and 'Brigitta'. The most serious effect of the second season shading, however, was a drastic reduction in yield in both cultivars (Tables 1 and 2). At the end of the harvest period 'Bluegold' and 'Brigitta' plants subjected to shading in both seasons showed yield reductions of 28% and 73%, respectively, whereas for those that were shaded in 2000 and grown in full sun in 2001, the equivalent reductions were 37% and 58%, respectively. Floral differentiation in many blueberry cultivars occurs between 60 and 90 d after full bloom (Gough et al., 1978) and may be impaired by reduced light intensity (Hall and Ludwig, 1961). Thus while reduced light intensity clearly delays fruit maturation, the commercial application of the technique must be considered in relation to the potential yield loss.

FIGURE 2. Cumulative yield of blueberry 'Bluegold' in 2000 and 2001 expresssed as percentage of maximum yield within each treatment. Open bar: Control plants (full sun); Closed bar: 50% shade in 2000 and 2001.

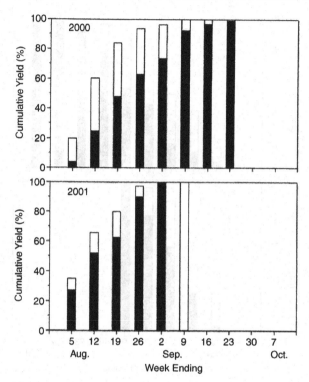

*Yield–'Elliott'.* During the 2000 season fruit maturity was advanced in 'Elliott' plants grown under plastic row covers. Harvest began on 21 August; by 7 September nearly 90% of the fruit from the full season cover treatment had been picked (Figure 4). Partial season cover (for 37 days before and during bloom period) resulted in harvest of about 80% of full season yield by 7 September, when control plants had yielded only 63% of their total. Yield was increased by about 25% in both partial and full-season cover treatments (Table 3).

The 2001 season failed to produce similar results. While fruit maturity was slightly advanced during the first two weeks of harvest (weeks ending 24 August and 31 August), the advancement was not sustained and time to reach 80% maturity (between 14 September and 21 September) was similar in all treatments. Yields were also reduced in both the

FIGURE 3. Cumulative yield of blueberry 'Brigitta' in 2000 and 2001 expressed as percentage of maximum yield within each treatment. Open bar: Control plants (full sun); Closed bar: 50% shade in 2000 and 2001.

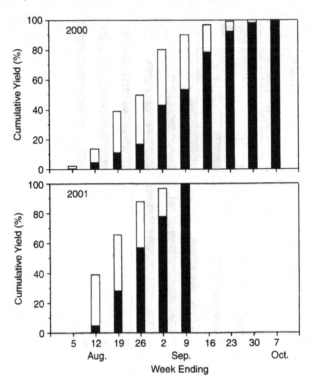

TABLE 1. Yield of blueberry 'Bluegold' from 50% shade and control (full sun) treatments in 2000 and 2001. Means are shown ± standard error.

| | \multicolumn{6}{Week ending} | | | | | |
|---|---|---|---|---|---|---|
| | 7 Aug. | | 27 Aug. | | 16 Sep. | |
| Treatment | 2000 yield (g) | 2001 yield (g) | 2000 yield (g) | 2001 yield (g) | 2000 yield (g) | 2001 yield (g) |
| Control | 735 ± 23[z] | 2300 ± 105 | 2830 ± 120 | 3850 ± 145 | 3020 ± 137 | 3900 ± 153 |
| Shade 00 only | 330 ± 18 | 1730 ± 57 | 2010 ± 89 | 2305 ± 95 | 3040 ± 97 | 2450 ± 104 |
| Shade 00 and 01 | - | 854 ± 23 | - | 2506 ± 87 | - | 2820 ± 107 |

[z]mean ± SE

TABLE 2. Yield of blueberry 'Brigitta' from 50% shade and control (full sun) treatments in 2000 and 2001.

| | Week ending | | | | | |
|---|---|---|---|---|---|---|
| | 7 Aug. | | 27 Aug. | | 23 Sep. | |
| Treatment | 2000 yield (g) | 2001 yield (g) | 2000 yield (g) | 2001 yield (g) | 2000 yield (g) | 2001 yield (g) |
| Control | 130 ± 13[z] | 200 ± 12 | 1100 ± 83 | 1050 ± 54 | 1590 ± 84 | 1200 ± 27 |
| Shade 00 only | 25 ± 5 | 185 ± 8 | 410 ± 21 | 500 ± 23 | 1210 ± 73 | 505 ± 18 |
| Shade 00 and 01 | - | 10 ± 2 | - | 225 ± 19 | - | 325 ± 13 |

[z]mean ± SE

FIGURE 4. Cumulative yield of blueberry 'Elliott' in 2000 and 2001 expresssed as percentage of maximum yield within each treatment.

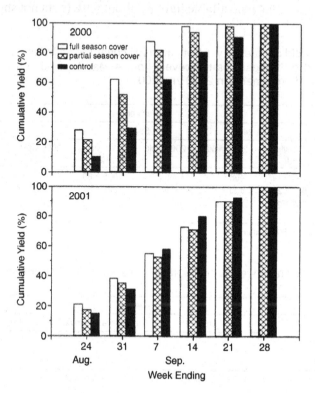

partial and full-season cover plants (Table 3). Fruit from the covered plants was smaller than the controls in both seasons (Table 4), perhaps pointing to lower pollination and fertilization. While parthenocarpy is common in highbush blueberry (MacKenzie, 1997) and can yield fruit of commercial size, fruit size and harvested yield is reduced when flowers are not effectively pollinated. Covering plants during the bloom period may reduce visits by both managed pollinators (honey bees from hives placed in the field but outside the plastic tunnel) and native bees. The introduction of bumble bees or other pollinators into tunnels (Sampson and Spiers, 2002) may improve pollination and result in larger berries and higher yields. The difference in cumulative yield distribution between the 2000 and 2001 seasons was unexpected. Generally warmer climatic conditions prevailed during 2001 although the differences in temperature under the tunnels were minor (maximum temperatures exceeded 35°C about 25% of the time during both seasons). Still, Bal et al. (1997) have pointed out that excessive heat can delay fruit ripening in highbush blueberry although the effect has not been quantified. The plastic cover reduced available light by about 33% (data not shown) and

TABLE 3. Yield of blueberry 'Elliott' from full cover (polyethylene cover from 4 May to end of harvest), partial cover (from 4 May to 10 June) and control (non-covered) treatments in 2000 and 2001.

| | Week ending | | | | | |
|---|---|---|---|---|---|---|
| | Aug. 24 | | Sep. 14 | | Sep. 28 | |
| Treatment | 2000 yield (g) | 2001 yield (g) | 2000 yield (g) | 2001 yield (g) | 2000 yield (g) | 2001 yield (g) |
| Control | 150 ± 14[z] | 408 ± 16 | 1825 ± 134 | 1905 ± 89 | 2205 ± 145 | 2380 ± 156 |
| Full Cover | 680 ± 23 | 400 ± 21 | 2570 ± 128 | 1025 ± 64 | 2753 ± 146 | 1310 ± 55 |
| Partial Cover | 653 ± 20 | 385 ± 18 | 2558 ± 130 | 1010 ± 45 | 2750 ± 135 | 1330 ± 46 |

[z]Means are shown ± Standard error.

TABLE 4. Average weight of 'Elliott' fruit in 2000 and 2001 growing season.

| | Average fruit weight (g) | | |
|---|---|---|---|
| | Control | Partial season cover | Full season cover |
| 2000 season | 2.0a[z] | 1.63b | 1.58b |
| 2001 season | 1.56a | 1.33b | 1.29b |

[z]Means in row designated with the same letter are not significantly different by LSD test at 5% probability level.

the combined effects of repeated shading and high temperatures require further investigation. Further work is also needed on the effects of row covers on pollinators, including studies on the introduction of managed pollinators (e.g., bumblebees, *Bombus impatiens*) under tunnels.

*Post-harvest storage.* The percentage of marketable fruit remaining after various storage periods was little affected by in-field covering treatments in either 'Bluegold' or 'Brigitta'. 'Brigitta' fruit stored well in CA for 42 d and showed only a minor decrease in quality when held for a further 7 d at 7°C in air (Table 5). 'Bluegold' fruit also stored well for 42 d, but quality and marketability decreased quickly when fruit was removed from CA (Table 6). 'Elliott' showed the best response to storage. In-field treatments had no significant effect on percentage of marketable fruit after 42 d of holding in CA, or even when fruit was subjected to a further 7 d at 7°C in air (Table 7). Even after 63 d fruit from both control and covered plants showed very high quality (little split, decay, or shrivel) immediately upon removal from CA. Following the additional 7 d in air, however, marketability declined in fruit from covered plants, while 84% of control fruit remained fully marketable.

TABLE 5. Marketable yield of blueberry 'Brigitta' fruit on removal and after a week in air at 7°C following various periods in controlled atmosphere storage (0°C, 10% $CO_2$, and 16% $O_2$) for the 2000 season.

| Storage duration (days) | Marketable yield (%) | |
|---|---|---|
| | Removal | 7 days in air |
| 21 | 98 | 92 |
| 42 | 92 | 86 |
| 63 | 83 | 69 |

TABLE 6. Marketable yield of blueberry 'Bluegold' fruit on removal and after a week in air at 7°C following various periods in controlled atmosphere storage (0°C, 10% $CO_2$, and 16% $O_2$) for the 2000 season.

| Storage duration (days) | Marketable yield (%) | |
|---|---|---|
| | Removal | 7 days in air |
| 21 | 97 | 89 |
| 42 | 86 | 60 |
| 63 | 68 | 40 |

TABLE 7. Marketable yield of blueberry 'Elliott' fruit on removal and after a week in air at 7°C following various periods in controlled atmosphere storage (0°C, 10% $CO_2$, and 16% $O_2$) for the 2000 season.

| | Marketable yield (%) | | | |
| --- | --- | --- | --- | --- |
| | Control | | Full season cover | |
| Storage duration (days) | Removal | 7 days in air | Removal | 7 days in air |
| 21 | 98 | 95 | 98 | 96 |
| 42 | 94 | 91 | 96 | 92 |
| 63 | 95 | 84 | 96 | 70 |

## CONCLUSIONS AND GROWER BENEFITS

The effects of row covers on maturity of blueberry fruit are inconsistent across seasons. While fruit ripening can be substantially delayed by 50% shading during the first year in 'Brigitta' and 'Bluegold', the effects are much smaller in the second season, and are accompanied by a significant reduction in yield. Future studies must focus on the effects of lower shade densities and timing of covering to modify the plant response. Advancing maturity of 'Elliott' fruit by enclosing plants in a plastic tunnel either throughout the season or only during bloom may be feasible, but requires more study on pollination effectiveness and the influence of high temperatures and reduced light intensity on fruit maturation. The quality and percentage of marketable fruit following CA storage (10% $CO_2$, 16% $O_2$, 0°C) is little affected by in-field covering, but is dependent on cultivar. Longevity of fruit in storage is greatest in 'Elliott', intermediate in 'Brigitta' and least in 'Bluegold'. In general, market season extension is best achieved by implementing CA storage of fruit, although in-field maturity control techniques show some promise.

## LITERATURE CITED

Bal, J.J.M., J.M. Wijsmuller, M.A. Jansen, and J. Dijkstra. 1993. Experiments with blueberries in the Netherlands. Acta Hort. 336: 272-276.

Bal, J.J.M. 1997. Blueberry culture in greenhouses, tunnels and under raincovers. Acta Hort. 446: 327-331.

Gough, R.E., V.G. Shutak, and R.L. Hauke. 1978. Growth and development of highbush blueberry. II. Reproductive growth, histological studies. J. Amer. Soc. Hort. Sci. 103:476-479.

Hall, I.V. and R. A. Ludwig. 1961. The effects of photoperiod, temperature and light intensity on the growth of the lowbush blueberry. Can. J. Bot. 39:1733-1739.

MacKenzie, K.E. 1997. Pollination requirements of three highbush blueberry (*Vaccinium corymbosum* L.) cultivars. J. Amer. Soc. Hort. Sci. 122:891-896.

Sampson, B.J. and J.M. Spiers. 2002. Evaluating bumblebees as pollinators of 'Misty' southern highbush blueberry growing inside plastic tunnels. Acta Hort. 574: 53-59.

Shutak, V.G., R. Hindle, and E.P. Christopher. 1956. Factors associated with ripening of highbush blueberry fruits. Proc. Amer. Soc. Hort. Sci. 68:178-183.

# Rabbiteye Blueberry Field Trials with the Growth Regulator CPPU

D. Scott NeSmith

H. Marcus Adair

**SUMMARY.** The growth regulator CPPU [N-(2-chloro-4-pyridyl)-N'-phenylurea] has recently been shown to increase both fruit set and berry size in rabbiteye blueberries (*Vaccinium ashei* Reade) in Experiment Station research plots. In order to validate these findings, several field experiments were conducted during 2001 and 2002 on commercial farms across south Georgia with the rabbiteye cultivars 'Climax' and 'Tifblue'. The standard CPPU concentration used was 10 mg/L applied at 10 to 18 days after 50% bloom. In 2001, most CPPU applications were with a back-pack sprayer, but in 2002, applications were made using commercial air-blast sprayers. In general, results from both years showed a positive benefit of CPPU with respect to fruit set, especially for the cultivar 'Climax'. Depending on location and year, fruit set was increased by as much as three-fold for 'Climax'. Fruit set increases were also observed for 'Tifblue', but the effect was less pronounced. With respect to berry size, CPPU generally increased 'Climax' size by 12 to 22% in both years. Berry size increases were observed for 'Tifblue', although, responses were not as consistent as for 'Climax'. A slight delay

---

D. Scott NeSmith is Professor, University of Georgia, Department of Horticulture, Griffin Campus, 1109 Experiment Street, Griffin, GA 30223-1797.

H. Marcus Adair is Field Technical Representative, Valent BioSciences Corporation, 214 Cross Breeze Drive, Memphis, TN 38018-2910.

[Haworth co-indexing entry note]: "Rabbiteye Blueberry Field Trials with the Growth Regulator CPPU." NeSmith, D. Scoot, and H. Marcus Adair. Co-published simultaneously in *Small Fruits Review* (Food Products Press, an imprint of The Haworth Press, Inc.) Vol. 3, No. 1/2, 2004, pp. 183-191; and: *Proceedings of the Ninth North American Blueberry Research and Extension Workers Conference* (ed: Charles F. Forney, and Leonard J. Eaton) Food Products Press, an imprint of The Haworth Press, Inc., 2004, pp. 183-191. Single or multiple copies of this article are available for a fee from The Haworth Document Delivery Service [1-800-HAWORTH, 9:00 a.m. - 5:00 p.m. (EST). E-mail address: docdelivery@haworthpress.com].

Digital Object Identifer: 10.1300/J301v03n01_18

in fruit ripening sometimes occurred for CPPU treated plants. These results suggest that CPPU may be useful for increasing yield of rabbiteye blueberries by increasing fruit set and/or berry size. *[Article copies available for a fee from The Haworth Document Delivery Service: 1-800-HAWORTH. E-mail address: <docdelivery@haworthpress.com> Website: <http://www. HaworthPress.com> © 2004 by The Haworth Press, Inc. All rights reserved.]*

**KEYWORDS.** *Vaccinium ashei* Reade, fruit set, berry size, blueberry production, yield

## INTRODUCTION

Research efforts in Georgia over the past ten years with the growth regulator gibberellic acid ($GA_3$) have overcome some of the fruit set problems with rabbiteye blueberry and have led to significant yield increases (NeSmith et al., 1995, 1999; NeSmith and Krewer, 1992, 1997a, 1997b, 1999). Even though research has shown positive benefits from using $GA_3$ in many instances, there are still some problems with small, late ripening fruit when using the growth regulator (NeSmith and Krewer, 1999). The cytokinin compound N-(2-chloro-4-pyridyl)-N'-phenylurea (CPPU) has shown some positive results in increasing fruit size and fruit set in a number of fruit crops including table grapes, kiwifruit, apples, table olives, and Japanese persimmon (Antognozzi et al., 1993a, and 1993b; Greene, 1989, 1993; Looney, 1993; Reynolds et al., 1992; Sugiyama and Yamaki, 1995). Recently, research at The University of Georgia has shown that CPPU may also be beneficial in rabbiteye blueberry production (NeSmith, 2002). The objective of this research was to test the usage of CPPU in a number of commercial rabbiteye blueberry fields during the 2001 and 2002 growing seasons for potential benefits of increased fruit set and berry size.

## MATERIALS AND METHODS

These field trials were part of an Experimental Use Permit project (EUP #71409). The cultivars Tifblue and Climax represent the two most widely grown rabbiteye blueberry cultivars in the state of Georgia, therefore, these were selected for the CPPU trials. Basic treatments at each site consisted of controls (no CPPU) and 10 mg/L of CPPU applied with a non-ionic surfactant. CPPU was applied using a back-pack sprayer in 2001 to 15 to 20 bushes per farm. In 2002, treatment areas were 0.5 ha or greater, and CPPU was applied using commercial airblast sprayers. Spray volumes ranged from 280 to 560 L/ha. Target application dates were 12 to 15 days after bloom. A total of five commercial farms were involved in this test. Table 1 summarizes dates of bloom and CPPU application for each farm. A brief characterization of each farm site follows:

*Farm 1:* This farm, located in southwestern Appling Co., consists of more than 110 ha of blueberries. Bushes used in this experiment were from 5 to 8 years old, and were considered very vigorous. Generally, row width at the farm was 3 to 3.5 m, and plant spacing was 1.2 m. The farm had raised beds, used a herbicide strip, had cultivated middles, and had overhead irrigation. The 'Climax' and 'Tifblue' blueberries utilized

TABLE 1. Dates of 50% bloom and dates of CPPU application for the rabbiteye blueberry cultivars 'Climax' and 'Tifblue' on five commercial farms in south Georgia during 2001 and 2002.

| Location | 2001 | | 2002 | |
|---|---|---|---|---|
| | Date of 50% bloom | Date of CPPU application | Date of 50% bloom | Date of CPPU application |
| Climax | | | | |
| Farm 1 | March 2 | March 14 | March 21 | April 5 |
| Farm 2 | March 3 | March 13 | March 22 | April 6 |
| Farm 3 | February 26 | March 13 | --- | --- |
| Farm 4a | March 1 | March 14 | March 22 | April 4 |
| Farm 4b | --- | --- | March 24 | April 8 |
| Farm 5 | --- | --- | March 23 | April 5 |
| Tifblue | | | | |
| Farm 1 | March 12 | March 27 | March 27 | April 6 |
| Farm 2 | March 10 | March 26 | March 26 | April 6 |
| Farm 3 | March 11 | March 27 | --- | --- |
| Farm 4a | March 12 | March 27 | March 28 | April 9 |
| Farm 4b | --- | --- | March 26 | April 8 |
| Farm 5 | --- | --- | March 28 | April 8 |

at this farm were in separate plantings, and each was grown with the variety Premier. Control and CPPU treatments were applied during both years at this farm. Freezes on March 8, 2001 and February 28, 2002 caused little or no damage to flowers at this farm.

*Farm 2:* This farm, located in eastern Bacon Co., consists of more than 10 ha of blueberries. Bushes used in this experiment were more than 10 years old, and were considered moderately vigorous. Generally, row width at the farm was 3.6 m, and plant spacing was 1.8 m. The farm had raised beds, used a herbicide strip, had grass middles, and had no irrigation. The 'Climax' and 'Tifblue' blueberries utilized at this farm were in the same planting. Control and CPPU treatments were applied during both years at this farm. Freezes on March 8, 2001 and February 28, 2002 caused slight to moderate flower damage at this farm.

*Farm 3:* This farm, located in central Appling Co., consists of more than 24 ha of blueberries. Bushes used in this experiment were more than 12 years old, and had a low degree of vigor. Generally, row width at the farm was 3.6 m, and plant spacing was 1.8 m. The farm had raised beds, used a herbicide strip, had cultivated middles, and had subsurface irrigation. The 'Climax' and 'Tifblue' blueberries utilized at this farm were in the same planting. Control and CPPU treatments were applied only in 2001 at this farm. A freeze on March 8, 2002 caused only slight flower damage at this farm.

*Farm 4:* This farm, located in northern Pierce Co., consists of more than 40 ha of blueberries. Bushes used in this experiment were more than 14 years old, and were considered moderate to highly vigorous. Generally, row width at the farm was 3.6 m, and plant spacing was 1.5 m. The farm had flat beds, used a herbicide strip, had cultivated middles, and had subsurface irrigation. The 'Climax' and 'Tifblue' blueberries utilized at this farm were in the same planting. Control and CPPU treatments were applied to one site during 2001 and two sites during 2002 at this farm. Freezes on March 8, 2001 and February 28, 2002 caused moderate to severe flower damage at this farm.

*Farm 5:* This farm, located in Ware Co., consists of more than 40 ha of blueberries. Bushes used in this experiment were more than 12 years old, and had a moderate to high degree of vigor. Generally, row width at the farm was 3.6 m, and plant spacing was 1.8 m. The farm had raised beds, had cultivated middles, and overhead irrigation. The 'Climax' and 'Tifblue' at this farm were in the same planting, along with the cultivars Brightwell and Premier. Control and CPPU treatments were applied both years at this farm; however, only data from 2002 were available. A freeze on February 28, 2002 caused slight to moderate flower damage at this farm.

Data taken from treatments at all farms consisted of fruit set and berry size. Prior to treatments, branches were tagged and flower bud numbers were determined. The average number of flowers per bud was also determined for 100 buds for each treatment at each farm. A total of 12 to 24 plants was tagged for each treatment at each site. Fruit set was calculated from the flower bud counts and subsequent berry counts. Berry size was determined at the beginning of commercial harvest for each cultivar at each site. Six to twelve samples of 50 ripe berries were randomly taken from tagged bushes and were weighed.

## RESULTS AND DISCUSSION

CPPU increased fruit set of 'Climax' at all commercial farms as compared to the control treatment during 2001 (Table 2). In fact, across all farms, CPPU more than doubled fruit set during the first year (27.7% for control versus 60.0% for CPPU alone). Fruit set of control plots over all were better for 'Climax' in 2002 than in 2001. However, there were still positive increases in fruit set due to CPPU in 2002. Fruit set of 'Climax' control plants varied across the 5 farms each year. Some of the control variability could be attributed to the degree of freeze damage flowers at each farm experienced. The greatest response to CPPU during both years for 'Climax' was on Farm 2, where the resulting fruit set was 50 to 82%, compared to only 21 to 27% for the control.

In general, CPPU applications increased berry size of 'Climax' as compared to control plants, with the greater increases in size observed during 2002 (Table 2). In fact, in 2002, 'Climax' fruit size across all farms was increased by nearly 17% when treated with CPPU. The greatest increase in berry size from CPPU was around 22%. Increases in size due to CPPU were observed for both early and later harvest berries (data not shown). CPPU sometimes delayed maturity by a few days (data not shown), but the degree of delay did not appear to be of much consequence in most instances.

For 'Tifblue', CPPU increased fruit set to a degree on three farms during 2001 and 2002 (Table 3). The increases in fruit set of the CPPU treatment over the control were less than that observed for 'Climax'. There was a slight decrease in fruit set at Farm 1 caused by CPPU during 2001, but the opposite response occurred during 2002. One of the most pronounced increases of 'Tifblue' fruit set due to CPPU was on Farm 3 in 2001, where the bloom time separation between the cultivars

TABLE 2. Fruit set and berry size of 'Climax' rabbiteye blueberry in response to the growth regulator CPPU on five commercial farms in Georgia during 2001 and 2002.

| Location | 2001 | | 2002 | |
|---|---|---|---|---|
| | Control | CPPU | Control | CPPU |
| | Fruit set (%) [z] | | | |
| Farm 1 | 40.1 ± 3.9 | 60.2 ± 5.9 | 60.8 ± 3.1 | 62.7 ± 3.7 |
| Farm 2 | 20.9 ± 3.6 | 81.7 ± 2.2 | 27.3 ± 3.7 | 52.3 ± 6.8 |
| Farm 3 | 34.1 ± 6.4 | 66.5 ± 6.6 | --- | --- |
| Farm 4a | 15.5 ± 2.3 | 31.6 ± 5.6 | 76.6 ± 4.4 | 81.1 ± 4.0 |
| Farm 4b | --- | --- | 43.6 ± 2.0 | 59.1 ± 2.6 |
| Farm 5 | --- | --- | 64.5 ± 5.0 | 74.9 ± 4.3 |
| Average all farms | 27.7 | 60.0 | 54.6 | 66.0 |
| | Berry size (g/50 berries) [y] | | | |
| Farm 1 | 53.2 ± 1.8 | 53.2 ± 1.9 | 48.2 ± 1.9 | 59.3 ± 1.0 |
| Farm 2 | 65.6 ± 1.2 | 76.8 ± 1.9 | 65.0 ± 1.7 | 74.0 ± 2.6 |
| Farm 3 | 49.8 ± 0.8 | 60.6 ± 0.8 | --- | --- |
| Farm 4a | 69.4 ± 2.3 | 71.2 ± 1.4 | 61.0 ± 2.1 | 72.0 ± 1.6 |
| Farm 4b | --- | --- | 59.2 ± 1.5 | 72.0 ± 0.7 |
| Farm 5 | --- | --- | 58.0 ± 0.8 | 63.5 ± 1.7 |
| Average all farms | 59.5 | 65.5 | 58.3 | 68.2 |

[z] Values are means ± standard error with n = 12 or n = 24.
[y] Values are means ± standard error with n = 6 or n = 12.

was considerable. 'Tifblue' at Farm 4a had extremely poor fruit set due to severe freeze damage during 2001. The poor fruit set observed for 'Tifblue' on Farm 4b during 2002 is not readily explained. This field had been heavily pruned (hedged) in the past two years, and perhaps this had some influence. The degree of fruit set for the 'Tifblue' control treatment overall (33 to 39%), is greater than growers experience in many years. 'Tifblue' fruit set can be as low as 10% when relying only on pollination, especially in a 'Tifblue/Climax' mix (Lyrene and Crocker, 1983; Lyrene and Goldy, 1983; NeSmith and Krewer, 1997a; NeSmith et al., 1999).

TABLE 3. Fruit set and berry size of 'Tifblue' rabbiteye blueberry in response to the growth regulator CPPU on five commercial farms in Georgia during 2001 and 2002.

| Location | 2001 | | 2002 | |
|---|---|---|---|---|
| | Control | CPPU | Control | CPPU |
| Fruit set (%) [z] | | | | |
| Farm 1 | 58.0 ± 6.1 | 41.1 ± 2.3 | 66.2 ± 8.9 | 76.9 ± 5.1 |
| Farm 2 | 32.4 ± 2.9 | 46.8 ± 5.7 | 24.4 ± 2.4 | 39.6 ± 4.7 |
| Farm 3 | 33.5 ± 8.1 | 51.6 ± 9.2 | --- | --- |
| Farm 4a | 7.8 ± 1.3 | 16.9 ± 4.2 | 21.5 ± 2.9 | 36.5 ± 4.6 |
| Farm 4b | --- | --- | 11.8 ± 2.0 | 10.1 ± 1.5 |
| Farm 5 | --- | --- | 70.0 ± 4.8 | 74.4 ± 3.6 |
| Average all farms | 32.9 | 39.1 | 38.8 | 47.5 |
| Berry size (g/50 berries) [y] | | | | |
| Farm 1 | 77.2 ± 1.9 | 75.8 ± 1.3 | 58.8 ± 1.2 | 63.0 ± 1.1 |
| Farm 2 | 69.6 ± 1.4 | 66.2 ± 1.4 | 61.3 ± 1.1 | 74.0 ± 1.7 |
| Farm 3 | 66.8 ± 2.2 | 62.8 ± 1.3 | --- | --- |
| Farm 4a | 67.8 ± 2.7 | 64.4 ± 2.1 | 64.0 ± 1.5 | 76.0 ± 1.4 |
| Farm 4b | --- | --- | 56.7 ± 1.6 | 73.3 ± 0.9 |
| Farm 5 | --- | --- | 73.2 ± 1.6 | 79.2 ± 1.9 |
| Average all farms | 70.3 | 67.3 | 62.8 | 73.1 |

[z] Values are means ± standard error with n = 12 or n = 24.
[y] Values are means ± standard error with n = 6 or n = 12.

There was essentially no difference in berry size among treatments for 'Tifblue' during 2001; however, during 2002 berry size was increased by CPPU application for this cultivar (Table 3). In fact, during 2002 berry size increases for 'Tifblue' resulting from CPPU were similar to that for 'Climax'. The largest increase in berry size (nearly 30%) for all locations occurred at Farm 4b during 2002. The degree of berry size increase during 2002 varied among Farms. The variation in berry size response to CPPU from farm-to-farm and from year-to-year could be related to application timings. Additional research is needed to optimize CPPU applications in relation to bloom time. As with 'Climax',

CPPU tended to cause a slight delay in maturity of 'Tifblue' in some instances (data not shown).

## CONCLUSIONS AND GROWER BENEFITS

In general, the data from these field trials indicate that CPPU can enhance fruit set and increase berry size of rabbiteye blueberries. The effects were more dramatic with 'Climax' than with 'Tifblue', although, both cultivars did display positive responses to the growth regulator. These large-scale field trials suggest CPPU will be beneficial in rabbiteye blueberry production, especially under conditions of poor pollination. Additional research with other rabbiteye and highbush blueberry cultivars, as well as experiments concerning CPPU rates and application timings are needed.

## LITERATURE CITED

Antognozzi, E., F. Famiani, A. Palliotti, and A. Tombesi. 1993a. Effects of CPPU (cytokinin) on kiwifruit productivity. Acta Hort. 329: 150-152.

Antognozzi, E., P. Proietti, and M. Boco. 1993b. Effect of CPPU (cytokinin) on table olive cultivars. Acta Hort. 329: 153-155.

Greene, D.W. 1989. CPPU influences 'McIntosh' apple crop load and fruit characteristics. HortScience 24: 94-96.

Greene, D.W. 1993. A comparison of the effects of several cytokinins on apple fruit set and fruit quality. Acta Hort. 329: 144-146.

Looney, N.E. 1993. Improving fruit size, appearance, and other aspects of fruit crop "quality" with plant bioregulating chemicals. Acta Hort. 329: 120-127.

Lyrene, P.M. and T.E. Crocker. 1983. Poor fruit set on rabbiteye blueberries after mild winters: Possible causes and remedies. Proc. Fla. State Hort. Soc. 96: 195-197.

Lyrene, P.M. and R.G. Goldy. 1983. Cultivar variation in fruit set and number of flowers per cluster for rabbiteye blueberry. HortScience 18: 228-229.

NeSmith. D.S. 2002. Response of rabbiteye blueberry (*Vaccinium ashei* Reade) to the growth regulators CPPU and gibberellic acid. HortScience 37:666-668.

NeSmith, D.S. and G. Krewer. 1992. Flower bud stage and chill hours influence the activity of $GA_3$ applied to rabbiteye blueberry. HortScience 27: 316-318.

NeSmith, D.S. and G. Krewer. 1997a. Response of rabbiteye blueberry (*Vaccinium ashei*) to gibberellic acid rate. Acta Hort. 446: 337-342.

NeSmith, D.S. and G. Krewer. 1997b. Fruit set of eight rabbiteye blueberry (*Vaccinium ashei* Reade) cultivars in response to gibberellic acid application. Fruit Var. J. 51: 124-128.

NeSmith, D.S. and G. Krewer. 1999. Effect of bee pollination and $GA_3$ on fruit size and maturity of three rabbiteye blueberry cultivars with similar fruit densities. HortScience. 34: 1106-1107.

NeSmith, D.S., G. Krewer, and O.M. Lindstrom. 1999. Fruit set of rabbiteye blueberry (*Vaccinium ashei*) after subfreezing temperatures. J. Amer. Soc. Hort. Sci. 124: 337-340.

NeSmith, D.S., G. Krewer, M. Rieger, and B. Mullinix. 1995. Gibberellic acid-induced fruit set of rabbiteye blueberry following freeze and physical injury. HortScience 30: 1241-1243.

Reynolds, A.G., D.A. Wardle, C. Zurowski, and N.E. Looney. 1992. Phenylureas CPPU and thidiazuron affect yield components, fruit composition, and storage potential of four seedless grape selections. J. Amer. Soc. Hort. Sci. 117: 85-89.

Sugiyama, N. and Y.T. Yamaki. 1995. Effects of CPPU on fruit set and fruit growth in Japanese persimmon. Scientia Hort. 60: 337-343.

# Susceptibility
# of Southern Highbush Blueberry Cultivars
# to Botryosphaeria Stem Blight

Barbara J. Smith

**SUMMARY.** Stem blight, caused by the fungus *Botryosphaeria dothidea*, is a destructive disease of rabbiteye (*Vaccinium ashei* Reade) and highbush (*V. corymbosum* L.) blueberries in the southeastern United States. The susceptibility of 20 southern highbush, two rabbiteye, and two highbush cultivars were compared using a detached stem assay. Fresh isolates of *B. dothidea* obtained from infected southern highbush blueberry plants were used as inoculum in this study. Succulent, partially-hardened stems were surface disinfected, rinsed in sterile distilled water, wounded by scraping away a section of bark, and inoculated by covering the wound with a mycelial agar block of *B. dothedia* and securing with parafilm wrap. The base of each stem was inserted into moistened, sterilized sand in a 150 mm × 25 mm tissue culture tube and incubated at 25°C, 100% RH for 30 days. Lesion length was measured after 15 days incubation. Cultivars with the shortest mean lesion length were classified as relatively resistant and included 'Pearl River', 'Emerald', 'Star', 'Sharpblue', 'Elliott', 'Misty', 'Bluecrisp', 'Darrow', 'Southmoon', 'Ozarkblue', 'Sapphire', and 'Brightwell'. Cultivars with the longest le-

Barbara J. Smith is Research Plant Pathologist, USDA-ARS Small Fruit Research Station, P.O. Box 287, Poplarville, MS 39740 (E-mail: BarbaraSmith@ars.usda.gov). The author thanks Wanda S. Elliott for technical assistance with this project.

[Haworth co-indexing entry note]: "Susceptibility of Southern Highbush Blueberry Cultivars to Botryosphaeria Stem Blight." Smith, Barbara J. Co-published simultaneously in *Small Fruits Review* (Food Products Press, an imprint of The Haworth Press, Inc.) Vol. 3, No. 1/2, 2004, pp. 193-201; and: *Proceedings of the Ninth North American Blueberry Research and Extension Workers Conference* (ed: Charles F. Forney, and Leonard J. Eaton) Food Products Press, an imprint of The Haworth Press, Inc., 2004, pp. 193-201. Single or multiple copies of this article are available for a fee from The Haworth Document Delivery Service [1-800-HAWORTH, 9:00 a.m. - 5:00 p.m. (EST). E-mail address: docdelivery@haworthpress.com].

sions at 15 days were classified as relatively susceptible and included 'Legacy', 'Gulf Coast', 'Cooper', 'Georgiagem', 'O'Neal', 'Reveille', 'Jubilee', and 'Magnolia'. *[Article copies available for a fee from The Haworth Document Delivery Service: 1-800-HAWORTH. E-mail address: <docdelivery@haworthpress.com> Website: <http://www.HaworthPress.com>]*

**KEYWORDS.** *Vaccinium ashei* Reade, *Vaccinium corymbosum* L.

## *INTRODUCTION*

Stem blight, caused by the fungus *Botryosphaeria dothidea* (Moug. LFr.) Ces. and De Not (anamorph *Fusicoccum aesuli* Corda.) is a widespread and destructive disease of highbush (*Vaccinium corymbosum* L.) and rabbiteye blueberries (*V. ashei* Reade) grown in the southeastern United States (Milholland, 1995). Losses are most severe in young fields where plants often become infected and die in the first two years (Milholland, 1972). In older fields stem blight is most commonly seen as a dieback of one or more canes in a bush; however, mature bushes of susceptible cultivars, such as 'Tifblue', sometimes die from stem blight infection (Smith, 1997). Creswell and Milholland (1987) reported a 23% incidence of stem blight in North Carolina in 1985.

Southern highbush blueberry cultivars are hybrids between *V. corymbosum* and various blueberry species native to the southeastern US. For example the cultivar, 'Gulf Coast', has a genetic composition of 72% *V. corymbosum*, 25% *V. darrowi*, and 3% *V. angustifolium* (Lang, 1993). These cultivars are being released for production in the southeastern US because of their low winter chill requirement, late spring bloom, and early fruit production. The area planted with these cultivars is steadily increasing; however, little is known about their susceptibility to diseases common in the southeastern US.

*Botryosphaeria dothidea* infects blueberry stems through wounds such as those caused by herbicide injury, winter injury, pruning, and mechanical harvesters (Milholland, 1972; 1995). Drought stress has been shown to predispose plants to stem blight (Cline, 2002; Milholland, 1972). Fields that are irrigated regularly during dry periods have a much lower incidence of bush death and stem blight than fields that are not. Fungicides are not effective for the control of this disease in the field even though fungicide dip treatments are useful for propagation (Cline and Milholland, 1992). The recommended control for stem blight is to prune out diseased canes during the winter and avoid wounding the plants.

Stem blight resistance is a primary objective of most southern blueberry breeding programs (Ballington et al., 1993; Cline et al., 1993; Gupton and Smith, 1989), and selection of resistant genotypes generally relies on natural infection of seedlings in field trials (Buckley,1990; Creswell and Milholland, 1987). A detached stem assay showed good correlation with natural field infections and with artificially inoculated plants in the field, greenhouse, and laboratory of rabbiteye blueberry cultivars (Smith, 1997). The objective of this study was to determine the relative susceptibility of southern highbush cultivars to stem blight using the detached stem assay.

## MATERIALS AND METHODS

Since *B. dothidea* isolates rapidly lose virulence in culture (Cline et al., 1993), fresh isolates were obtained from symptomatic 2- to 4-year-old southern highbush blueberry cultivars growing in research plots at Poplarville, MS about one month before the initiation of the study. Five mm internal stem pieces were cut from the edge of lesions, surface sterilized, rinsed in sterile distilled water, and placed on acidified potato dextrose agar (A-PDA). Fungal cultures were maintained on A-PDA and transferred every 14 to 21 d. Inoculum consisted of 2 mm square mycelial blocks cut from 10- to 15-d-old A-PDA cultures. Two virulent isolates, PR3 and CP5, were selected from six isolates tested for pathogenicity on detached stems of the susceptible rabbiteye cultivar, 'Tifblue' (data not shown).

Assay stems were collected from 2- to 4-year-old southern highbush, rabbiteye, and low-chill highbush cultivars (Table 1) established at Poplarville on Ruston fine sandy loam on raised beds mulched with pine bark. Fifteen stems each of six or seven cultivars were collected on four dates within the same week for a total of 24 cultivars. For comparison of collection date, stems of 'Biloxi' and 'Magnolia' were collected on the first and last collection date. Succulent, partially hardened-off stems, 150 mm or more in length, were cut from the bushes, immersed in water, and held at 4°C overnight (Creswell and Milholland, 1987). All leaves except the terminal three were removed from each stem, and the stem was disinfected by immersing for 15 min in a 10% bleach solution followed by three rinses in sterile distilled water. Six stems of each

TABLE 1. Blueberry cultivars assayed for susceptibility to stem blight.

| Cultivar | Type[z] | Date and location of release |
|---|---|---|
| Biloxi | SHB | 1998 USDA Poplarville, Mississippi |
| Bluecrisp | SHB | 1997 Florida |
| Cooper | SHB | 1987 USDA Poplarville, Mississippi |
| Emerald | SHB | 1999 Florida |
| Georgiagem | SHB | 1987 Georgia & USDA |
| Gulf Coast | SHB | 1987 USDA Poplarville, Mssissippi |
| Jubliee | SHB | 1995 USDA Poplarville, Mississippi |
| Legacy | SHB | 1993 USDA & New Jersey |
| Magnolia | SHB | 1995 USDA Poplarville, Mississippi |
| Misty | SHB | 1989 Florida |
| O'Neal | SHB | 1987 North Carolina |
| Ozarkblue | SHB | 1996 Arkansas |
| Pearl River | SHB | 1995 USDA Poplarville, Mississippi |
| Reveille | SHB | 1990 North Carolina |
| Santa Fe | SHB | 1999 Florida |
| Sapphire | SHB | 1998 Florida |
| Sharpblue | SHB | 1976 Florida |
| Southmoon | SHB | 1996 Florida |
| Star | SHB | 1996 Florida |
| Windsor | SHB | 1999 Florida |
| Brightwell | RE | 1983 Georgia & USDA |
| Tifblue | RE | 1955 Georgia & USDA |
| Darrow | HB | 1965 USDA & New Jersey |
| Elliott | HB | 1973 USDA |

[z] SHB = Southern Highbush (hybrid of *Vaccinium corymbosum* L. and various *Vaccinium* species native to the southeastern US, RE = Rabbiteye (*V. ashei* Reade), HB = Highbush (*Vaccinium corymbosum* L.)

cultivar were inoculated with each of the two isolates. Three stems of each cultivar inoculated with A-PDA served as the non-inoculated control treatment. Within each cultivar, stems were randomly assigned to each inoculum treatment. Each stem was wounded ~75 mm from the terminal by using a scalpel to remove a 2 mm × 4 mm section of bark. The inoculum block was secured to the wounded site with a strip of laboratory film (Parafilm "M", American National Can, Menasha, WI). The base of each inoculated stem was then inserted about 25 mm deep into moistened, sterilized sand in 150 mm × 25 mm tissue culture tubes. The stems were incubated at 25°C in a moist chamber at 100 % RH for 30 d. Sterile distilled water was added to each tube as needed to keep the sand moist. The parafilm and inoculum block were carefully removed after 15 d. Disease development was rated at 15 and 30 d by measuring lesion length.

Data were analyzed with an analysis of variance using the "proc glm" command of the statistical analysis software SAS 8.2 (SAS Institute Inc., 1999). Least Significant Difference (LSD) values were calculated to compare reactions of individual genotypes to pathogen isolates.

## *RESULTS*

The mean lesion lengths on wounded blueberry stems 15 and 30 d after inoculation with *B. dothedia* are listed in Tables 2 and 3. There was a significant difference between the two fungal isolates in mean lesion length with CP5 causing longer lesions (41.7 and 91.8 mm at 15 and 30 d) than PR3 (30.6 and 79.5 mm). However, there was not a significant interaction between cultivars and isolates. Cultivars with the shortest mean lesion length for both isolates at 15 d were classified as relatively resistant and included 'Pearl River', 'Emerald', 'Star', 'Sharpblue', 'Elliott', 'Misty', 'Bluecrisp', 'Darrow', 'Southmoon', 'Ozarkblue', 'Sapphire', and 'Brightwell'. Those cultivars with the longest lesions after 15 d incubation were considered relatively susceptible and included 'Legacy', 'Gulf Coast', 'Cooper', Georgiagem, 'O'Neal', 'Reveille', 'Jubilee', and 'Magnolia'.

There were no differences in lesion length of 'Biloxi' and 'Magnolia' stems collected on the first and last collection dates at either the 15 or 30 d rating or with either isolate (Tables 2 and 3). This indicates that the four stem collection dates in this study did not affect lesion length. All the collection dates for this study were within a week of each other and care was taken to collect stems at a similar growth stage.

The mean lesion length of all non-inoculated stems (those inoculated with the agar block) was 5.0 mm after 15 d indicating that the wounding did not cause the lesions. By 30 d the mean lesion length of the non-inoculated stems (37.6 mm) indicates that secondary infection had occurred during incubation in the warm humid environment. The lesions on the non-inoculated stems usually did not occur at the wound site, but rather began at buds suggesting that fungal spores present in the bud scale crevices may not have been killed when the stem was disinfested. Due to the presence of lesions on the non-inoculated stems after 30 d incubation, the lesion development on some of the inoculated stems may be due to secondary infections rather than the inoculation at the wound

TABLE 2. Mean lesion length (mm) on detached blueberry stems 15 d after wound inoculation with two isolates, PR3 and CP5, of the stem blight pathogen, *Botryosphaeria dothidea.*

| Cultivar | PR3[z] | CP5[z] | Mean[y] | Control[x] |
|---|---|---|---|---|
| Pearl River | 5.7 ef[w] | 5.2 i | 5.5 j | 6.3 |
| Emerald | 5.2 f | 6.2 i | 5.7 j | 3.7 |
| Star | 9.3 def | 5.7 i | 7.5 j | 4.7 |
| Sharpblue | 10.0 ef | 8.0 i | 9.0 ij | nt [v] |
| Elliott | 7.5 ef | 12.8 hi | 10.2 hij | 3.3 |
| Misty | 12.3 def | 9.0 i | 10.7 hij | 5.3 |
| Bluecrisp | 20.3 cdef | 5.7 i | 13.0 ghij | 3.0 |
| Darrow | 26.3 cdef | 8.3 i | 17.3 fghij | 4.0 |
| Southmoon | 17.7 cdef | 19.3 ghi | 18.5 fghij | 3.3 |
| Ozarkblue | 6.5 ef | 38.7 defghi | 22.6 efghij | 3.3 |
| Sapphire | 5.5 f | 39.8 defghi | 22.7 efghij | nt |
| Brightwell | 16.0 cdef | 36.8 efghi | 26.4 efghij | nt |
| Tifblue | 18.8 cdef | 48.0 defgh | 33.4 efgh | 4.7 |
| Windsor | 37.0 bcde | 30.3 fghi | 33.7 defgh | 3.3 |
| Biloxi (1) | 20.5 cdef | 55.0 cdefg | 37.8 cdefg | 4.3 |
| Biloxi (2) | 28.0 cdef | 47.6 defgh | 37.8 cdef | 3.0 |
| Santa Fe | 39.5 bcd | 39.8 defghi | 39.7 cdefg | 3.3 |
| Magnolia (1) | 33.3 bcdef | 60.5 cdef | 46.9 bcde | 25.3 |
| Magnolia (2) | 23.8 cdef | 86.5 abc | 55.2 bcd | 3.3 |
| Jubilee | 36.5 bcdef | 76.0 bcd | 56.3 bcd | 4.0 |
| Reveille | 45.7 bc | 68.7 bcde | 57.2 bcd | 6.7 |
| O'Neal | 61.2 b | 55.2 cdefg | 58.2 bc | 4.0 |
| Georgiagem | 61.5 b | 60.0 cdef | 60.8 bc | 3.3 |
| Cooper | 61.2 b | 74.0 bcde | 67.6 b | 4.0 |
| Gulf Coast | 98.3 a | 99.2 ab | 98.8 a | 4.0 |
| Legacy | 96.3 a | 119.3 a | 107.8 a | 5.0 |
| Column mean | 31.5 | 43.2 | 36.3 | 5.0 |
| LSD (0.05) | 31.0 | 38.1 | 24.5 | ns[u] |

[z] Mean lesion length of six stems.
[y] Mean lesion length of two isolates (12 stems).
[x] Mean lesion length of three stems inoculated with agar block.
[w] Means followed by the same letter within columns are not significantly different (P ≤ 0.05).
[v] Not tested.
[u] Not significant.

TABLE 3. Mean lesion length (mm) on detached blueberry stems 30 d after wound inoculation with two isolates, PR3 and CP5, of the stem blight pathogen, *Botryosphaeria dothidea*.

| Cultivar | PR3[z] | CP5[z] | Mean[y] | Control[x] |
|---|---|---|---|---|
| Sharpblue | 29.6 hi[w] | 8.8 h | 19.2 i | nt[v] |
| Emerald | 7.2 i | 45.5 fgh | 26.3 hi | 3.7 e |
| Pearl River | 19.3 i | 39.2 gh | 29.2 hi | 6.3 e |
| Star | 28.8 hi | 33.2 gh | 31.0 hi | 8.3 de |
| Southmoon | 45.5 ghi | 58.0 efg | 51.8 gh | 3.3 e |
| Misty | 30.3 hi | 76.2 cdefg | 53.2 gh | 5.7 e |
| Sapphire | 67.5 fgh | 68.0 defg | 67.8 fg | nt |
| Ozarkblue | 40.5 ghi | 95.7 abcde | 68.1 fg | 3.3 e |
| Tifblue | 35.2 ghi | 108.0 abcd | 71.6 fg | 62.7 abcde |
| Jubilee | 90.3 cdef | 66.7 defg | 78.5 efg | 56.3 bcde |
| Elliott | 67.8 fgh | 95.2 abcde | 81.5 efg | 84.3 abc |
| Magnolia (1) | 69.2 efgh | 95.7 abcde | 82.4 efg | 46.7 cde |
| Darrow | 73.0 defg | 97.7 abcde | 85.3 ef | 51.0 cde |
| Windsor | 90.2 cdef | 93.3 abcde | 91.8 ef | 3.3 e |
| Santa Fe | 93.5 bcdef | 90.7 bcdef | 92.1 def | 3.3 e |
| Magnolia (2) | 65.0 fgh | 129.7 ab | 97.3 cdef | 3.3 e |
| Bluecrisp | 101.3 abcdef | 107.2 abcd | 104.2 bcde | 4.0 e |
| Biloxi (2) | 118.8 abc | 87.0 bcdef | 104.4 bcde | 53.0 cde |
| Biloxi (1) | 110.2 abcde | 109.0 abcd | 109.6 abcde | 54.3 bcde |
| Gulf Coast | 113.7 abcd | 126.0 ab | 119.8 abcde | 4.0 e |
| O'Neal | 113.5 abcd | 126.5 ab | 120.0 abcde | 9.3 de |
| Cooper | 123.2 abc | 120.7 abc | 121.9 abcde | 34.0 cde |
| Reveille | 119.3 abc | 124.5 ab | 121.9 abcde | 135.0 a |
| Georgiagem | 125.3 abc | 124.7 ab | 125.0 abc | 82.0 abcd |
| Legacy | 134.0 ab | 132.0 ab | 133.0 ab | 127.7 ab |
| Brightwell | 140.8 a | 138.6 a | 139.7 a | nt |
| Column mean | 91.8 | 79.5 | 85.6 | 37.6 |
| LSD (0.05) | 41.9 | 47.3 | 31.4 | 74.1 |

[z] Mean lesion length of six stems.
[y] Mean lesion length of two isolates (12 stems).
[x] Mean lesion length of three stems inoculated with agar block.
[w] Means followed by the same letter within columns are not significantly different ($P \leq 0.05$).
[v] Not tested.

site and therefore, only the lesion length after 15 d incubation was used to rate cultivar susceptibility.

## DISCUSSION AND GROWER BENEFITS

Ten of the 20 southern highbush cultivars, both highbush cultivars, and one of the rabbiteye cultivars were classified as relatively resistant to stem blight based on the 15-d measurement of lesion length on detached stems. This confirms field observations and previous reports that stem blight resistance is present among commercial blueberry cultivars (Ballington et al., 1993; Buckley, 1990; Gupton and Smith, 1989). Buckley (1990) indicated that stem blight resistance in highbush cultivated blueberries comes from lowbush blueberry (*V. angustifolium*). Stem blight resistance has also been identified in rabbiteye cultivars (Gupton and Smith, 1989; Smith, 1997) and in *V. elliotti* (Ballington et al., 1993).

'Tifblue' was the most susceptible rabbiteye cultivar in an earlier trial (Smith, 1997). Since its lesion length in this trial was near the average for all cultivars, it and the southern highbush cultivars, 'Windsor', 'Biloxi', and 'Santa Fe', were classified as tolerant. The inclusion of 'Tifblue' in the tolerant class in this trial of mostly southern highbush cultivars supports previous reports that rabbiteye cultivars in general are more resistant to stem blight than highbush cultivars (Ballington et al., 1993; Milholland, 1995). 'O'Neal' was classified as relatively susceptible to stem blight in this trial; however, it is considered field resistant (Milholland, 1995). It is not surprising that some discrepancies may occur between the detached stem assay and field observations since wound inoculation with fresh, virulent isolates is a harsh test for stem blight resistance (Cline et al., 1993). However, screening procedures using this type of technique facilitates mass selection of resistant genotypes. Breeders and growers can use the information generated from this and similar screening procedures to choose appropriate cultivars for parent lines and for planting in areas where stem blight is known to be a problem.

## LITERATURE CITED

Ballington, J.R., S.D. Rooks, R.D. Milholland, W.O. Cline, and J.R. Meyers. 1993. Breeding blueberries for pest resistance in North Carolina. Acta Hort. 346:87-94.

Buckley, B. 1990. Occurrence of stem blight resistance in blueberry. PhD thesis. North Carolina State Univ., Raleigh, NC.

Cline, W.O. 2002. Stem blight of blueberry. Fruit disease information note 9. North Carolina State Univ. http://www.ces.ncsu.edu/depts/pp/notes/Fruit/fdin009/fdin009. htm

Cline, W.O. and R.D. Milholland. 1992. Root dip treatments for controlling blueberry stem blight caused by *Botryosphaeria dothidea* in container-grown nursery plants. Plant Dis. 76:136-138.

Cline, W.O., R.D. Milholland, S.D. Rooks, and J.R. Ballington. 1993. Techniques for breeding for resistance to blueberry stem blight caused by *Botryosphaeria dothidea*. Acta Hort. 346:107-109.

Creswell, T.C. and R.D. Milholland. 1987. Responses of blueberry genotypes to infection by *Botryosphaeria dothidea*. Plant Dis. 71:710-713.

Gupton, C.L. and B.J. Smith. 1989. Inheritance of tolerance to stem blight in *Vaccinium* species. HortScience 24:748.

Lang, G.A. 1993. Southern highbush blueberries: Physiological and cultural factors important for optimal cropping of these complex hybrids. Acta Hort. 346:72-80.

Milholland, R.D. 1972. Histopathology and pathogenicity of *Botryosphaeria dothidea* on blueberry stems. Phytopathology 62:654-660.

Milholland, R.D. 1995. *Botryosphaeria* stem blight. Pp. 10-11, In: F.L. Caruso and D.C. Ramsdell (eds.). Compendium of blueberry and cranberry diseases, APS Press. St. Paul, MN.

SAS Institute Inc. 1999. SAS/STAT user's guide. Release 8.02. SAS Institute Inc., Cary, NC.

Smith, B.J. 1997. Detached stem assay to evaluate the severity of stem blight of rabbiteye blueberry (*Vaccinium ashei*). Acta Hort. 446:457-464.

# Post-Harvest Hedging and Pruning of Three Year Pruning Trial on 'Climax' and 'Tifblue' Rabbiteye Blueberry

Gerard Krewer
Danny Stanaland
Scott NeSmith
Ben Mullinix

**SUMMARY.** Control of bush height is a major problem in rabbiteye blueberries. Eight-year-old, moderately vigorous *Vaccinium ashei* cv. Climax, were roof-top hedged in early August to a height of about 2.2 m (7 ft). Fourteen-year-old, low vigor 'Tifblue' were roof-top hedged to a height of about 2.8 m (9 ft) 0, 2, 4, and 6 weeks after harvest. Average

Gerard Krewer is Professor, Horticulture Department, University of Georgia, Tifton Campus, Tifton, GA 31793.

Danny Stanaland is County Agent, Bacon County Extension Service, Alma, GA 31510.

Scott NeSmith is Professor, Horticulture Department, University of Georgia, Griffin Campus, Griffin, GA 30223.

Ben Mullinix is Statistician, Statistical Services, University of Georgia, Tifton Campus, Tifton, GA 31793.

The authors would like to thank MBG Marketing, Inc. and Wade Farms for support of this research project.

[Haworth co-indexing entry note]: "Post-Harvest Hedging and Pruning of Three Year Pruning Trial on 'Climax' and 'Tifblue' Rabbiteye Blueberry." Krewer, Gerard et al. Co-published simultaneously in *Small Fruits Review* (Food Products Press, an imprint of The Haworth Press, Inc.) Vol. 3, No. 1/2, 2004, pp. 203-212; and: *Proceedings of the Ninth North American Blueberry Research and Extension Workers Conference* (ed: Charles F. Forney, and Leonard J. Eaton) Food Products Press, an imprint of The Haworth Press, Inc., 2004, pp. 203-212. Single or multiple copies of this article are available for a fee from The Haworth Document Delivery Service [1-800-HAWORTH, 9:00 a.m. - 5:00 p.m. (EST). E-mail address: docdelivery@haworthpress.com].

Digital Object Identifer: 10.1300/J301v03n01_20

bush yield on 'Climax' was 5.3 kg (11.6 lbs) for the unpruned control, 4.6 kg (10.1 lbs) for the moderate roof-top hedged, and 4.8 kg (10.5 lbs) for the winter cane renewal pruning treatment. There were no significant differences among yields but there was a trend toward lower yields on the first harvest of 'Climax' with roof-top hedging compared to unpruned or cane renewal pruned. Average bush yield on 'Tifblue' was 2.9 kg (6.4 lbs) for the unpruned control, 2.5 kg (5.5 lbs) for pruning 0 weeks after harvest, 3.3 kg (7.3 lbs) for pruning two weeks after harvest, 2.8 kg (6.1 lbs) for pruning four weeks after harvest, and 2.8 kg (6.3 lbs) for pruning six weeks after harvest. 'Tifblue' that were roof-top hedged two weeks after harvest tended to produce the highest yields. Bushes that were roof-top hedged were much easier to hand harvest. *[Article copies available for a fee from The Haworth Document Delivery Service: 1-800-HAWORTH. E-mail address: <docdelivery@haworthpress.com> Website: <http://www.HaworthPress.com>* © *2004 by The Haworth Press, Inc. All rights reserved.]*

**KEYWORDS.** *Vaccinium ashei*, blueberry pruning, roof-top hedging

## *INTRODUCTION*

Pruning of large rabbiteye bushes has become a major concern to growers in the Southeastern United States. Without pruning, rabbiteye bushes often grow to heights of 4 m or more (13 ft) in 10 years making hand harvest very difficult. Moderate post-harvest roof-top hedging of blueberries has become a common practice in the production of northern and southern highbush blueberries in North Carolina since they ripen early and pruning can be conducted in June. However, even June pruning causes a slight depression of yields on most cultivars the next year, so this practice is normally done biennially (Mainland, 1989). With rabbiteye blueberries, research from North Carolina indicates that 'Tifblue' blueberry must be hedged by 15 July to maintain yields (Mainland, 1989). This usually means sacrificing part of the current season's crop. Because South Georgia has about one month longer growing season than Eastern North Carolina, research is needed to determine if post-harvest rooftop hedging can be used with out yield loss. Early August is a practical time for growers to perform this practice, since all cultivars have finished bearing by this time.

Cane renewal pruning has been used successfully by a number of growers in Georgia. However, little experimental data is available on the effect of this pruning system on the yields of rabbiteyes. In this study, we collected information on both pruning systems and on a combination of rooftop hedging and cane renewal in one year.

## MATERIALS AND METHODS

*Hedging and cane renewal with 'Climax'.* Eight-year-old drip irrigated 'Climax' rabbiteye blueberries bushes growing on a commercial farm near Alma, GA were used in the experiment. All plants received pre- and post-harvest fertilization totaling 32.9 kg (72.5 lbs) of N, 48.8 kg (107.5 lbs) of phosphate and 22.7 kg (50 lbs) of potash per acre. Pre-harvest fertilization was 226 kg (500 lbs) per acre of 10-10-10 and post-harvest fertilization was 56.7 kg (125 lbs) of 18-46-0. Rows were selected and a pruning treatment applied to each row, since initially a mechanical peach hedger was used on the entire row. Within each row, four replications of 10 bushes each were selected at random. Analysis was completely randomized design. Treatments were (1) Unpruned control, (2) August moderate roof-top hedge to about 2.1 m (7 ft) at the apex each year removing about 15 to 45 cm (6 to 18 in) of canopy, (3) Winter cane renewal removing about 20% of the canopy in 1999 and 2001, and (4) August moderate roof-top hedge each year plus cane renewal in 2001. Since treatment four was only conducted the last year, rooftop hedge data for 1999 and 2000 was a combination of treatments two and four representing 80 bushes.

At the time of first hedging on 1 August 1998, the tallest bushes were about 2.6 m (8.5 ft) in height and the shortest bushes about 2.2 m (7 ft) in height. Bushes were hedged to 2 to 2.3 m (6.5 to 7.5 ft) in height at the apex with a mechanical peach hedger set at an angle of 55 degrees from the horizontal. The year of first hedging (1998), shoots of varied lengths up to 61 cm (24 inches) were removed by the hedger from the sloping 'roof' portion of the hedge cut. However, the length of most of the shoots cut were much shorter. The cane renewal treatment was applied about 1 Mar. 1999 by removing one or two canes with lopping shears within 46 cm (18 inches) of ground level. Since the 'Climax' cultivar has few canes, this was about 20% of the canopy.

Flower bud development stage on the 1998 spring growth and 1998 fall regrowth after hedging was rated using the Spiers scale by examining 30 shoots of each type on 3 March 1999 (Spiers, 1978).

Fruit were harvest by hand on 9 June 1999 and by machine on 22 June 1999. It is estimated that about 80% of the fruit were harvested in these two pickings. On the 9 June harvest, 100 berries per sample (replication) were selected at random and weighted on a gram scale. All fruit were taken to the packing house and weighed.

On 17 August 1999, the moderate roof-top hedging treatment was rehedged with a hand held sickle bar hedger to a uniform height of 2.2 m (7 ft) at the apex. It is estimated from one to three percent of the canopy was removed with some shoots removed being up to 17.8 cm (7 in) in length, but many only 5.1 to 7.6 cm (2 to 3 in) in length. This is due in part to the variation in pruning height (about 30 cm (1 ft) variation) produced by the mechanical hedger in year one.

In 2000 and 2001, bushes were fertilized as in 1999. Fruit were hand harvested on 9 June 2000 and machine harvested on 23 June 2000. One hundred berries per replication were selected at random and weighted on a gram scale to obtain berry weight.

On 3 August 2000, the moderate rooftop hedging treatment was rehedged with a hand held sickle bar hedger to a uniform height of 2.2 m (7 ft) at the apex. Hedging angle was about 55 degrees from the horizontal. It is estimated that about 10% of the canopy was removed by roof-top hedging. Cane renewal pruning was conducted on 22 February 2001 by cutting one or two canes with a small chainsaw near ground level. Most of these 'Climax' bushes had about five major canes and several smaller canes from the 1999 pruning. It is estimated that 20% of the canopy was removed in this process. For the first time in the study, cane renewal pruning plus rooftop hedging was implemented (Treatment 4). Since these bushes had not been previously pruned, cutting one or two massive canes removed an estimated 25%-30% of the canopy. Yield and fruit size data were analyzed using Proc MIXED (Littell et al., 1996).

*Timing of post-harvest hedging 'Tifblue'.* In 1997, a post-harvest hedging pruning experiment was initiated at the University of Georgia's Blueberry Research Farm near Alapaha, GA. Plants used were 14 year-old, non-irrigated 'Tifblue' rabbiteye blueberries that had not been previously pruned. The treatments applied were roof-top hedging of the bushes at different intervals after harvest. Roof-top hedging angled from 1.5 m (5 ft) (lower canopy) to 2.7 m (9 ft) (canopy apex) was performed on bushes. Approximately 46 to 61 cm (1.5 to 2 ft) of canopy was removed from the bush the first year. The treatments were no hedging and hedging applied 0, 2, 4, and 6 weeks post-harvest. Light hedging 5 to 30 cm (2 to 12 in removed) to maintain height was again conducted in 1998 and 1999 at the same time intervals. There were four bushes per plot, with five replications of each treatment. Yield and fruit size were determined each year. In 1998, yield was determined by harvesting the two center bushes of each plot two times by hand. In 1999 and 2000, yield was obtained by mechanically harvesting fruit twice from all four bushes in each plot. Yield and fruit size data were subjected to analysis of variance procedures. All plots received standard fertility and fungicide applications each year.

## RESULTS

*Hedging and cane renewal with 'Climax'.* Regrowth after moderate roof-top hedging in 1998 averaged 12.7 to 17.8 cm (5 to 7 in) in length. This fall regrowth produced flower buds which were delayed in development in the spring of 1999 compared to flower buds produced on the spring 1998 wood. Mean flower bud development on 3 March 1999 was 4.41 (Spiers, 1978) on the Spiers scale for the 1998 spring wood and 3.21 for the 1998 fall regrowth after pruning (data not shown). This 1.2 difference in the flower bud development scale represents over a one week delay in bloom. However, most of blooms on the fall regrowth after pruning failed to set fruit, probably due to an infestation of thrips, which attacked late-season flowers in 1999.

Fruit on hedged bushes was at a maximum height of about 2 m (6.5 ft) at harvest, while control fruit was up to 2.4 m (8 ft) high at harvest. Hand pickers preferred the bushes that were moderately roof-top hedged, since they could reach nearly all the fruit with both hands. On the taller bushes of the control and cane renewal treatment, the tops of the bushes had to be picked by holding a cane down with one hand and picking with the other.

Berry weight was slightly smaller on the roof-top hedge treatment than the control on the first harvest (Table 1). There was a trend for the cane renewal pruned treatment to produce higher yields on the first harvest date (Table 1). On second harvest date, cane renewal treatment had the lowest yields. Total yield was not significantly different between treatments. Roof-top hedging in early August, appeared to improve light penetration into the lower canopy and maintained yields, even though bush size was reduced.

Bush growth in the spring of 1999 was limited, so the total amount of wood removed by rooftop pruning was estimated to be about 3% of the canopy. Post-harvest hedging brought the fruit bearing apex of the canopy down to 2.2 m (7 ft) in height after hedging. A regrowth from the hedging cuts of about 15 cm (6 in) occurred after hedging in 1999.

**TABLE 1.** Effect of pruning treatments on 'Climax' yield and fruit weight in 1999

| Treatment (year) | Yield per bush (kg) | | | Berry weight (g) |
|---|---|---|---|---|
| | 1st harvest | 2nd harvest | Total | |
| Control | 1.9a[z] | 2.0a | 3.9a | 1.40a |
| Rooftop hedge (98) | 1.8a | 1.9a | 3.7a | 1.27b |
| Cane renewal (99) | 2.7a | 1.4b | 4.1a | 1.35ab |

[z]Means in a column with the same letter are not significantly different at the *P* < 0.05 level according to Fisher's LSD mean separation test.

Fruit set in the spring of 2000 was good on both the spring wood and the later blooming fall wood since thrips were controlled in 2000. Height to the top of the bearing canopy on the control at harvest was up to 2.7 m (9 ft), while the top of the bearing canopy on the roof-top hedged was just over 2.2 m (7 ft).

In 2000, there was no significant difference in the yield between treatments on the first harvest date. However, there was a trend toward the control and cane renewal having the highest yield (Table 2). On the second harvest date the control had a significantly higher yield than the cane renewal treatment, but was not significantly different than the rooftop hedge. Total yield was not significantly different between treatments. However, there was a trend for slightly higher yields on the control than the cane renewal or roof-top hedge.

On the first harvest date berry weight was significantly greater on the cane renewal than the roof-top hedge treatment. On the second harvest date there was no significant difference in berry weights between treatments.

Roof-top hedged bushes grew about 15 cm (6 in) above the hedging cuts in the fall of 1999 and about 25 cm (10 in) in the spring and summer prior to 3 August 2000 roof-top hedging. An average of about 41 cm (16 in) of canopy was removed. It is estimated this was about 10% of the canopy. The tops of control and cane renewal pruned bushes grew about 15 to 20 cm (6 to 8 in) between bud break and 3 August 2000. Regrowth after roof-top hedging in the fall of 2000 averaged about 13 cm (5 in) followed by vigorous spring growth. The net result was for all treatments to produce bushes of similar height at harvest 2001, but with a lower fruiting canopy on the roof-top hedged treatment. At harvest, the

TABLE 2. Effect of pruning treatments on 'Climax' yield and fruit weight in 2000

| Treatment (year) | Yield per bush (kg) | | | Berry weight (g) | |
|---|---|---|---|---|---|
| | 1st harvest | 2nd harvest | total | 1st harvest | 2nd harvest |
| Control | 3.3a[z] | 3.4a | 6.7a | 1.18ab | 0.88a |
| Rooftop hedge (98,99) | 2.2a | 2.9ab | 5.1a | 1.13b | 0.95a |
| Cane renewal (99) | 3.2a | 2.5b | 5.7a | 1.22a | 0.86a |
| LSD | 1.6 | 0.76 | 1.67 | 0.08 | 0.11 |

[z]Mean separation within columns using LSD option ($P < 0.05$) in Proc MIXED (Littell, 1996).
Means in a column with the same letter are not significantly different at the $P < 0.05$ level according to Fisher's LSD mean separation test.

control and cane renewal treatments had fruit up to 2.7 to 3 m (9 to 10 ft) in height, while the roof-top hedged treatment averaged about 2.3 m (7.5 ft). This allowed efficient, two-handed harvest of the roof-top pruned bushes, while the tops of the control and cane renewal pruned bushes were harvested only by pickers using one hand to hold down the canes and picking with their other second hand.

There was no significant difference in yield between treatments on the first, second, and fourth harvest. On the third harvest, yields were lower on the cane renewal and roof-top plus cane renewal than the control and the roof-top hedge. Total yield was not significantly different between treatments. There was also no significant difference in berry weight on the second harvest (Table 3).

Average bush yield over three years was 5.3 kg (11.6 lbs) for the control, 4.6 kg (10.1 lbs) for the moderate roof-top hedging, 4.8 kg (10.5 lbs) for the cane renewal, and 3.4 kg (7.71 lbs) for the rooftop plus cane renewal (one year's data only).

*Timing of post-harvest hedging 'Tifblue'.* Total yields from 1998, 1999, and 2000 are presented in Table 4. There were no significant yield differences between treatments in any of the years at either the first harvest (data not shown) or for the season total. There were also no significant differences in fruit size among treatments in any year (data not shown). In 1998, there was a spring freeze which damaged some of the flowers and reduced fruit set for all treatments. In 1999, fruit set was greatly reduced for all treatments following a severe thrips infestation. In 2000, there were no unusual freeze or insect problems. In 1999 and

TABLE 3. Effect of pruning treatments on 'Climax' yield and fruit weight in 2001

| Treatment (year) | Yield per bush (kg) | | | | | Berry weight (g) |
| | 1st harvest | 2nd harvest | 3rd harvest | 4th harvest | total | 2nd harvest |
| --- | --- | --- | --- | --- | --- | --- |
| Control | 1.9a[z] | 1.1a | 1.3a[y] | 0.8a | 5.1a | 1.37a |
| Rooftop hedge (98,99,00) | 1.8a | 0.8a | 1.5a | 0.8a | 4.9a | 1.39a |
| Cane renewal (99,01) | 2.2a | 1.0a | 0.8b | 0.7a | 4.7a | 1.37a |
| Rooftop (00) + cane renewal (01) | 1.3a | 0.9 | 0.7 | 0.5b | 3.4a | 1.40a |
| LSD | 1.06 | 0.47 | 0.37 | 0.25 | 1.8 | 0.11 |

[z]Mean separation within column using LSD option ($P < 0.05$) in Proc MIXED (Littell, 1996).
[y]Assignment of letters for comparing treatments and between line 1 (control) and line 3 (cane renewal), and between line 2 (rooftop hedge) and line 4 (combination).

TABLE 4. Yield from 1998, 1999, and 2000 for mature 'Tifblue' rabbiteye blueberry bushes in response to post-harvest hedging (PHH). Treatments were (1) no hedging; (2) hedging immediately after harvest; (3) hedging 2 weeks after harvest; (4) hedging 4 weeks after harvest; and (5) hedging 6 weeks after harvest.

| Hedging treatment | Total yield | | |
| | 1998 | 1999 | 2000 |
| --- | --- | --- | --- |
| | -------------------kg/bush------------------- | | |
| No hedging | 4.5 | 1.1 | 3.1 |
| 0 weeks PHH | 3.6 | 0.9 | 3.0 |
| 2 weeks PHH | 4.3 | 1.6 | 4.0 |
| 4 weeks PHH | 4.2 | 0.9 | 3.2 |
| 6 weeks PHH | 4.4 | 1.0 | 3.1 |

[z]There were no significant differences ($P < 0.05$) in yields in a given year.

2000, the plots were machine harvested, but there were no apparent differences in fruit quality among treatments.

## DISCUSSION AND GROWER BENEFITS

During the first or second harvests of 'Climax' in 2000 and 2001, which was conducted with the help of commercial field Hispanic workers, it was observed that fruit above 2.7 m (7 ft) in height was picked only with great difficulty, it was necessary for the worker to pull the tall canes down with one hand and pick with the other hand. In commercial areas of the field, this fruit was not harvested by the hand pickers, resulting in a significant amount of the fruit left in the top of the bush until machine harvest at the end of the hand picking season. This probably lowered the quality of the machine picked fruit. During machine harvest, the control and cane renewal pruned bushes protruded above the top of the harvester, causing a compression of the canopy going into the harvester. This type of situation is associated with a loss of berry quality and harvester efficiency.

Although roof-top hedging produced a desirable hand picking height, there was a trend toward slightly decreased first harvest compared to the control and cane renewal with 'Climax'. Control bushes produced an average of 2.3 kg (5.2 lbs) of fruit on the first harvest, while roof-top hedged produced 2 kg (4.32 lbs) and cane renewal pruned produced 2.7 kg (5.99 lbs). Growers harvesting fresh market blueberries in a typical declining price situation in early June need to consider yields at first harvest in their economic analysis, especially in a year with low frozen berry prices. However, rooftop hedging appears to increase hand harvest and machine harvest efficiency and can be performed at relatively low cost compared to hand pruning, so this may off-set the slight yield loss. With 'Climax', a combination of post-harvest, rooftop hedging (which removed about 10% of the canopy in August 2000) and cane renewal (which removed 25%-30% of the canopy in February 2001) showed a trend toward depressed yields on both first and total harvest. This treatment obviously removed too many flower buds. Although some cane renewal or shoot thinning will probably be necessary to prevent the formation of 'crow's feet'(multiple branches from the same area of cane) following moderate rooftop hedging, care should be taken not to remove an excessive number of flower buds in the process.

'Tifblue' did not have a trend toward reduced yield on the first harvest with roof-top hedging. The data suggest that moderate post-harvest rooftop hedging can be accomplished without loss of yield, at least up through the fourth year after initial hedging. The plants that were hedged were much more manageable in terms of routine cultural practices such as mowing, herbicide applications, and fertilization. The timing of post-harvest hedging did not significantly influence response; however, there was a trend in 1999 and 2000 toward slightly higher yields when hedging was performed 2 weeks post-harvest.

It appears from this study that in South Georgia (latitude 31°), August, moderate, roof-top, post-harvest hedging of rabbiteye blueberries is a viable method of controlling plant height and allowing more sunlight to reach the interior of the canopy. Undoubtably, over time, some additional cane renewal pruning or detailed pruning will be necessary to keep the bushes open enough for good internal bearing. However, the cost of mechanical post-harvest hedging is very low compared to using cane renewal pruning as the primary pruning method. Moderate, post-harvest roof-top hedging should be a significant cost reduction management tool for the South Georgia blueberry grower.

## LITERATURE CITED

Littell, R.C., G.A. Milliken, W.W. Stroup, and R.D. Wolfinger. 1996. SAS System for mixed models. SAS Institute, Inc. Cary, NC.

Mainland, C.M. 1989. Pruning blueberries. Proc. 23rd Annu. Open House Southeast Blueberry Council. pp. 10-15.

Spiers, J.M. 1978. Effect of stage of bud development on cold injury in rabbiteye blueberry. J. Amer. Soc. Hort. Sci. 103: 452-455.

# Fungal Pathogens Associated with Blueberry Propagation Beds in North Carolina

## Bill Cline

**SUMMARY.** Death of blueberry cuttings in commercial rooting beds was observed due to abiotic and biotic causes. Abiotic causes included poor watering practices, water quality, rooting medium, and inadequate drainage due to poor rooting bed design. Biotic causes were attributable to fungi and included (1) non-pathogenic Basidiomycetes colonizing unsterilized rooting media, (2) airborne or rain-splashed pathogens infecting individual cuttings (*Botryosphaeria, Pestalotia* and other sp.), and (3) *Cylindrocladium* sp. that spread radially from the initial infection, producing circular dead spots in rooting beds. Re-use of *Cylindrocladium*-infested media resulted in complete loss of cuttings. Methyl bromide fumigation was successfully used to sanitize infested media. *[Article copies available for a fee from The Haworth Document Delivery Service: 1-800-HAWORTH. E-mail address: <docdelivery@haworthpress.com> Website: <http://www.HaworthPress.com> © 2004 by The Haworth Press, Inc. All rights reserved.]*

Bill Cline is Researcher/Extension Specialist, Plant Pathology Department, North Carolina State University, NCSU Horticultural Crops Research Station, 3800 Castle Hayne Road, Castle Hayne, NC 28429 (E-mail: bill_cline@ncsu.edu).

This research was supported in part by funding from the North Carolina Blueberry Council, Inc.

The author would like to thank Benny Bloodworth and Terry Bland for technical assistance.

[Haworth co-indexing entry note]: "Fungal Pathogens Associated with Blueberry Propagation Beds in North Carolina." Cline, Bill. Co-published simultaneously in *Small Fruits Review* (Food Products Press, an imprint of The Haworth Press, Inc.) Vol. 3, No. 1/2, 2004, pp. 213-219; and: *Proceedings of the Ninth North American Blueberry Research and Extension Workers Conference* (ed: Charles F. Forney, and Leonard J. Eaton) Food Products Press, an imprint of The Haworth Press, Inc., 2004, pp. 213-219. Single or multiple copies of this article are available for a fee from The Haworth Document Delivery Service [1-800-HAWORTH, 9:00 a.m. - 5:00 p.m. (EST). E-mail address: docdelivery@haworthpress.com].

Digital Object Identifer: 10.1300/J301v03n01_21

**KEYWORDS.** Disease, rooting, *Cylindrocladium*, hardwood, softwood, fumigation, sanitation

## INTRODUCTION

Blueberry plants (*Vaccinium corymbosum* L., *Vaccinium ashei* Reade) used to establish commercial fields in North Carolina are most often propagated by the grower or by local nurseries, using hardwood or softwood cuttings stuck in outdoor ground beds with or without shade. Beds are filled 15-18 cm in depth with pine bark, sawdust, peat: sand, or combinations of these well-drained media, then watered as needed with sprinklers or mist nozzles. Hardwood 'whips' have traditionally been taken in early spring, cut into 10-13 cm lengths, and refrigerated until stuck in April (Mainland, 1966). Softwood propagation consists of collecting semi-hardened leafy shoots in August for propagation under intermittent mist (Mainland and Bland, 1993; Spangler and Sneed, 1973). Once rooted, both types of cuttings are usually left in the rooting bed until February or March, then transplanted directly to the field.

Observation of failed propagation attempts in North Carolina suggests a broad range of abiotic causes: (1) Poor maintenance of beds, poor water management, or improper handing of cuttings during collection and storage all may damage cuttings beyond recovery. (2) Badly designed beds with inadequate drainage are doomed to failure. In coastal NC, bed drainage must allow for sudden showers of 3-5 inches, and ideally will accommodate even the heaviest rainfalls (10+ inches from tropical storms and hurricanes) without drowning the cuttings. For this reason, most beds are built atop an 18-24 inch layer of sand. (3) Even with a well-designed bed, the rooting medium must drain quickly yet maintain consistent moisture and aeration throughout the season. Aged pine sawdust or milled pine bark is the standard, but hardwood sawdust or bark cannot be used because it decays too readily and forms impenetrable crusts or layers that shed water. Most peat-based rooting media also retain too much water during heavy rains. Even the preferred media of pine sawdust can be too old or decayed to be useable. (4) High levels of bicarbonates, sodium, chlorine, and iron in well water have been observed to kill entire propagation beds through mineral accumulation in the rooting medium and on the cuttings.

Biotic causes of rooting failures are mainly fungal. Non-pathogenic saprophyte fungi may colonize rooting media, producing a mycelial mat that interferes with bed drainage. Saprobes sometimes produce mushrooms, which are fruiting structures (basidiocarps), in propagation beds. Pathogenic fungi may infect and kill cuttings in the rooting bed, or may colonize plants that survive and carry the disease to the field (Cline, 1998). Of particular interest is the plant pathogenic genus *Cylindrocladium*, which contains species known to be pests in plant propagation beds. *Cylindrocladium (Calonectria) crotalarie* (Loos) Bell & Sob. has been reported to cause stem, leaf, and root rot of blueberry seedlings, and the histopathology of this disease has been investigated (Milholland, 1974a; 1974b). The fungus may spread by conidia, chlamydospores, or microsclerotia and is known to infest soil and potting/rooting media. The taxonomy of *Cylindrocladium* has undergone revision since Dr. Milholland's work was conducted, and the specific epithet is now putatively *Calonectria ilicicola* Boedijin & Reitsma (imperfect stage: *Cylindrocladium parasiticum* Crous, Wingfield & Alfenas) (Crous, 2002).

The following studies were undertaken to document and evaluate reasons for propagation failures, to investigate the epidemiology of *Cylindrocladium* sp. and other pathogens in blueberry propagation beds, and to determine whether methyl bromide fumigation commonly used by growers was capable of sanitizing infested rooting media for reuse as rooting media, potting mix or in high-density field plantings.

## MATERIALS AND METHODS

*Field investigations.* Propagation beds with reported problems were visited over a 4 yr period. Dead areas were mapped or photographed as needed to document patterns of plant death within each bed. Causes were determined by visual inspection, and observations were recorded. Where needed, water, media, or plant samples were collected for analysis. Disease isolations were made on acidified potato-dextrose agar (aPDA, Difco Laboratories, Detroit, MI) from individual symptomatic plants or from declining and/or symptomless plants growing at the margin of dead areas.

*Greenhouse studies.* Rooting media known to be infested with *Cylindrocladium* sp. was collected from two grower sites, and samples were retained for later quantification. Half of each sample was fumigated with methyl bromide. Infested vs. fumigated media from each site was used to grow rooted 'O'Neal' blueberry cuttings in 4-inch pots the following spring. Bushes were grown for 6 months in the greenhouse. Symptoms were evaluated 6 months after planting and all stems were cultured on a PDA. The experiment was conducted once in 2001 and again in 2002. The 2002 repeat of the study used the same rooting media samples as in 2001, after samples had weathered outdoors for one year. Greenhouse data was analyzed using PROC ANOVA (SAS version 6.12, SAS Institute, Cary, NC) and results are presented in Tables 1 and 2.

## RESULTS AND DISCUSSION

*Field observations.* During the survey period (1998-2002), problems were reported and investigated in twenty-three propagation beds where some or all cuttings failed to root. A wide variety of causes was documented: Water quality (6); overwatering (4); disease (4); faulty bed design (2); poor rooting media (3); mishandling of cuttings (2), and spray injury (1).

Abiotic problems were complicated by a shortage of the preferred rooting medium. Most growers were using a coarse pine sawdust recovered from huge, old, abandoned piles left behind by portable sawmills common in the 1950s and 1960s. Digging a short way into these piles reveals a treasure of essentially sterile, reddish-brown sawdust with coarse (2-4 mm) particles, few fine particles and a consistency similar to medium-grade vermiculite or perlite. As these piles have become scarcer and smaller, growers have resorted to re-using the 'dust,' or trying to use less-than-perfect remnants of piles they previously passed over. Also, as these sawdust piles age or are disturbed, they become de-

TABLE 1. Effect of fumigation and media source on incidence of *Cylindrocladium* stem and root rot of blueberry, 2001 (N = 12).

| Source and Treatment | Survival (%) | Foliage[x] (1-4) | Stem color[y] (1-3) | # Shoots | Max. shoot length (cm) | a PDA recovery (%) |
|---|---|---|---|---|---|---|
| Site A Fumigated | 92 ab[z] | 1.2 a | 1.2 a | 3.1 a | 6.5 b | 8 a |
| Site B Fumigated | 100 a | 1.3 a | 1.1 a | 4.4 b | 6.5 b | 0 a |
| Site B Non-fumigated | 58 c | 3.2 c | 2.5 b | 2.2 a | 2.6 c | 67 b |
| Site C Fumigated | 100 a | 1.0 a | 1.0 a | 4.5 b | 10.3 a | 0 a |
| Site C Non-fumigated | 75 bc | 2.3 b | 2.5 b | 2.5 a | 5.3 b | 92 b |

[z] Means in columns followed by the same letter are not significantly different ( $P \le 0.05$ ).
[xy] Foliage and stem color rated as: 1 = dark green, 2 = light green, 3 = chlorotic, 4 = severely chlorotic or necrotic.

TABLE 2. Effect of fumigation and media source on incidence of *Cylindrocladium* stem and root rot of blueberry, 2002 (N = 12).

| Source and Treatment | Survival (%) | Foliage[x] (1-3) | Stem color[y] (1-3) | # Shoots | Max. shoot length (cm) | a PDA recovery (%) |
|---|---|---|---|---|---|---|
| Site A Fumigated | 100 a | 1.0 a | 1.0 a | 4.3 ab | 15.7 a | 0 a |
| Site B Fumigated | 100 a | 1.2 a | 1.1 a | 3.4 bc | 8.3 cd | 0 a |
| Site B Non-fumigated | 100 a | 2.0 b | 1.8 b | 3.1 c | 5.6 d | 83 b |
| Site C Fumigated | 100 a | 1.2 a | 1.2 a | 4.4 a | 11.9 b | 0 a |
| Site C Non-fumigated | 92 a | 1.7 b | 1.3 ab | 3.9 abc | 9.3 bc | 75 b |

[z] Means in columns followed by the same letter are not significantly different ($P \leq 0.05$).
[xy] Foliage and stem color rated as: 1 = dark green, 2 = light green, 3 = chlorotic, 4 = severely chlorotic or necrotic.

cayed or contaminated with surrounding soil. Re-used sawdust (from the previous year's rooting bed) may contain not only pathogens but also accumulated minerals from the irrigation water (bicarbonates, sodium, iron) and has a less desirable consistency, with more silty fine particles and a tendency to become waterlogged.

Fungal genera isolated from cuttings included *Cylindrocladium, Botryosphaeria, Pestalotia, Alternaria, Trichoderma,* and others that were unidentified. Only *Cylindrocladium* was consistently associated with plant-to-plant spread of disease, as evidenced by patterns observed. In beds with fresh (not previously used) rooting media, symptoms associated with *Cylindrocladium* were most often encountered as a circular area of dead plants 0.3-1.2 m in diameter, with infections spreading radially from a point source. In older rooting beds where media had been stirred up and reused, disease occurred throughout the bed. When re-used media from infested beds was not fumigated, it was not unusual for all cuttings in the rooting bed to die.

*Greenhouse studies.* Fumigation of media significantly increased survival of plants potted in infested media in 2001 (Table 1). Foliage color, stem color, number of shoots, and shoot length were also increased. *Cylindrocladium* was recovered from stems of plants grown in non-fumigated media from sites B and C at levels of 67% and 92%, respectively, and from only one plant in fumigated media. In 2002, plant survival was not significantly affected as all plants save one survived (Table 2). Foliage and stem color were improved by fumigation, but there was no response in number of shoots or shoot length. However, *Cylindrocladium* was still present in stems of plants grown in non-fumigated media, and was recovered from most. None of the plants grown in fumigated media in 2002 was found to be infected.

## CONCLUSIONS

Most blueberry rooting bed failures in NC are due to preventable abiotic causes that can be addressed by educating growers on the importance of water testing, bed design, choice of rooting medium, and proper handing of hardwood and softwood cuttings. Re-use of rooting media is not advised. *Cylindrocladium* is the primary fungal cause of death in both hardwood and softwood blueberry rooting beds in NC. The source of initial inoculum is not known, but once infested, rooting beds serve as disease reservoirs, and catastrophic losses can occur under warm, humid conditions maintained in rooting beds. Inoculum levels in used, infested media may decrease over time if the material is allowed to dry out and weather, and this will allow survival of potted plants even if the material is not fumigated prior to use as a potting mix. However, plants will continued to become infected, with possible later consequences. Where methyl bromide was used to sanitize media, plants were shown to be free of the fungus, except for a single plant which may have been cross-contaminated by adjacent infected plants during routine greenhouse watering of randomized pots.

## GROWER BENEFITS

Use of new, clean, uninfested rooting media is essential for propagation. Media should not be re-used unless thoroughly fumigated to destroy fungi. Even with the use of disease-free rooting media, growers must attend to all aspects of propagation (water quality, media quality, bed drainage, collection and handling of cuttings) in order to produce plants successfully.

## LITERATURE CITED

Cline, W. O. 1998. Fungal pathogens colonizing first-year rooted blueberry cuttings. P. 122, In: Proc. 8th North Amer. Blueberry Res. Ext. Workers Conf., May 27-29, Wilmington, NC. North Carolina State Univ., Raleigh, NC.

Crous, P. W. 2002. Taxonomy and pathology of *Cylindrocladium (Calonectria)* and allied genera. APS Press, St. Paul, MN.

Mainland, C. M. 1966. Propagation and planting. Pp. 111-131, In: P. Eck and N. F. Childers, (eds.). Blueberry culture. Rutgers University Press, New Brunswick, NJ.

Mainland, C. M. and W. T. Bland. 1993. Propagation techniques for highbush, southern highbush and rabbiteye softwood or hardwood cuttings. Pp. 1-8, In: Proc. 27th Ann. Open House, Southeastern Blueberry Council, Inc. N. C. Coop. Ext. Ser., January 31, Elizabethtown, NC.

Milholland, R. D. 1974a. Stem and root rot of blueberry caused by *Calonectria crotolariae*. Phytopathology 64:831-834.

Milholland, R. D. 1974b. Histopathology of *Calonectria crotolariae* on highbush blueberry. Phytopathology 64:1228-1231.

Spangler, R. L. and R. E. Sneed. 1973. Intermittent mist propagation. Circ. 506 (revised), N. C. Agr. Ext. Ser., Raleigh, NC.

# *Vaccinia gloriosa*

## S. P. Vander Kloet

**SUMMARY.** The genus *Vaccinium* contains not only such well known sections as blueberries, cranberries, lingon berries, and bilberries but also a vast array of tropical epiphytes. Indeed at the equator, one can find terrestrial shrubs on coastal sand dunes; lianas, vines and epiphytes in primary and secondary forests, and epipetric shrubberies at 3200 msm and above. Unfortunately, DNA sequences derived from two chloroplast genes and one nuclear ribosomal gene strongly favour a polyphyletic interpretation for the genus *Vaccinium* viz.: (1) the *Bracteata* Clade; (2) the *Myrtillus* Clade; and (3) the *Vaccinium* Clade. Prior to dismembering *Vaccinium sensu lato*, the users, especially the horticultural community, ought to be consulted. Given that *V. uliginosum* is the type for the genus, what elements ought to be conserved. Conversely, perhaps the boundaries of *Vaccinium* ought to be expanded to include all the currently recognized segregate genera such as *Agapetes* and *Cavendishia*. *[Article copies available for a fee from The Haworth Document Delivery Service: 1-800-HAWORTH. E-mail address: <docdelivery@haworthpress.com> Website: <http://www.HaworthPress.com> © 2004 by The Haworth Press, Inc. All rights reserved.]*

**KEYWORDS.** *Vaccinium*, blueberry, taxonomy

The genus *Vaccinium* was erected by Linnaeus in 1737 to accommodate those monogynous plants with 8 stamens and an inferior 4-loculed

S. P. Vander Kloet is University Botantist, K.C. Irving Research Centre, Acadia University, Wolfville, NS B4P 2R6.

[Haworth co-indexing entry note]: "*Vaccinia gloriosa*." Vander Kloet, S. P. Co-published simultaneously in *Small Fruits Review* (Food Products Press, an imprint of The Haworth Press, Inc.) Vol. 3, No. 3/4, 2004, pp. 221-227; and: *Proceedings of the Ninth North American Blueberry Research and Extension Workers Conference* (ed: Charles F. Forney, and Leonard J. Eaton) Food Products Press, an imprint of The Haworth Press, Inc., 2004, pp. 221-227. Single or multiple copies of this article are available for a fee from The Haworth Document Delivery Service [1-800-HAWORTH, 9:00 a.m. - 5:00 p.m. (EST). E-mail address: docdelivery@haworthpress.com].

ovary that ripened into a berry containing a few small seeds. Initially the genus had four species; at last count about 500 species had been assigned to *Vaccinium* (Sleumer, 1967; Vander Kloet, 1988), most of whom are found in the old world tropics not because of habitat abundance and rapid speciation but because of differing views on the stability of androecial features. Seminal here were the opinions of Wright (1850) and Hooker (1854) both of whom dismissed staminal features such as the absence of tubules, spurs, or awns as minor variations as opposed to Klotzsch (1851) who regarded androecial features as pivotal in erecting segregate genera in the Vaccinieae.

Currently the characters that define *Vaccinium* L. are rather vague and indistinct, viz., small 4- or 5-merous flowers and a 4- or 5-loculed inferior ovary (Stevens, 1972). Such a definition fits taxa not now included in *Vaccinium* such as *Diogenesia* Sleumer, *Disterigma* (Klotzsch) Niedenzu, and *Sphyrospermum* Poepp and Endlich. and should exclude others such as *V. barandanum* Vidal, *V. chapaense* Merrill, and *V. cylindraceum* Smith.

Nevertheless some effort has been made to limit the morphological diversity of *Vaccinium* by transferring *V. poasanum* Donnell Smith to *Symphysia* C.B. Presl (Vander Kloet et al., 2003). Conversely transferring *Symphysia racemosa* (Vahl) Stearn to *Vaccinium* as proposed by Wilbur and Luteyn (1978) significantly increases the character suite that defines the genus since *S. racemosa* is pleiomerous and develops the calyx tube prior to anthesis (Vander Kloet, 1985). Adding characters such as these to *Vaccinium* is tantamount to making the genus a dumping ground for any Vaccinieae of uncertain affinity. Not that generic boundaries should be inelastic, but candidate taxa ought to at least have the character unique to the genus in which they are about to be submerged. At the moment new taxa are assigned to *Vaccinium* not because they possess the characters referred to in the protolog but rather because they bear some resemblance to at least one or more species already placed therein. Such a pragmatic approach is an abomination in the eyes of cladists who argue that candidate taxa ought at least to have the character unique to the genus in which they are about to be submerged.

Phylogenetic analyses of Vaccinieae by Kron and her colleagues (see Kron et al. 2002 and references therein) using molecular data from ribosomal and chloroplast DNA strongly favour a polypheletic interpretation for *Vaccinium* (Figure 1). Most of these analyses suggest three weakly supported clades although each clade has branches with strong support, viz., (1) the *"myrtillus–uliginosum"* clade; this clade not only contains the four species that Linnaeus used to formulate the genus but

FIGURE 1. Strict concensus tree obtained from the parsimony analysis of *mat*
K data of 93 species of *Vaccinieae* ex. (Kron et al. 2002. Amer. J. Bot. 89: 331,
reprinted with permission.)

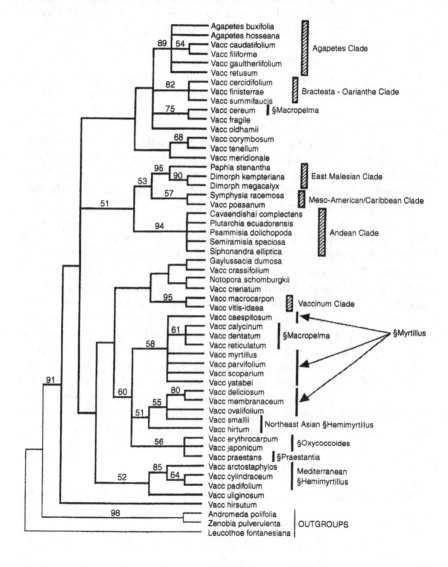

also the type of the genus *V. uliginosum* L. (VanderKloet, 1981); (2) The "Gonwanaland" clade; this clade comprises a group of segregate genera from the Caribbean, Central America, South America and the Pacific Rim; (3) The "Laurasian" clade; this clade contains *Agapetes* D. Don as well as related *Vaccinia* from SE Asia, North America and the Caribbean.

In terms of gross morphology, these three clades can be characterized as follows: the *uliginsoum-myrtillus* group has its inflorescence often reduced to a single flower in the axils of the lowermost leaves of the leafy shoot; the pedicel is often continuous with the calyx tube, and the berry is usually 4- or 5-locular.

The second clade comprises segregate genera that exhibit some of the showiest flowers of the tribe either borne singly in leaf axils or in elaborate inflorescences. The stamens are frequently dimorphic and show some cohesion of parts such as connate filaments and tubules. The berries are large, often insipid, usually 5-locular.

The third clade has two distinct lineages, an American lineage that includes our blueberries (*V.* section *Cyanococcus*) as well as a SE Asian lineage of *Agapetes* and various *Vaccinia* taxa. Many of these taxa have racemose inflorescences on old wood, a pseudo-10-locular ovary and superficial phellogen. Should this three lineage hypothesis for the Vaccinieae gain support after further analyses and adding new taxa to the matrix, *Vaccinium* as we know it will surely disappear. Only the lineage containing the type *V. uliginosum* will remain in *Vaccinium*, the remainder will become segregate genera. Thus *V.* section *Cyanococus* could be transferred to the genus *Agapetes* or could become *Cyanococcus* Rydberg. The high bush blueberry would become *Agapetes corymbosa* or *Cyanococcus corymbosus*. If neither appeals, then for an expansion of generic boundaries for *Vaccinium* that would include all the features now found in the tribe would be necessary. Conversely, the genus *Vaccinium sensu stricto* could contain as few as four species: the circumboreal and holoarctic *V. uliginosum* as well as three taxa from *V.* section *Hemimyrtillus* viz., the Caucasian *V. arctostaphylos* L., *V. padifolium* Smith from Madeira, and *V. cylindraceum* Smith from the Azores.

*Vaccinium sensu lato*, was first promoted by Hooker (1854) and most recently by Camp (1940) who showed that both the flowers and the stamens of the North American *V.* section *Cyanoccocus* Asa Gray, closely resemble those found in the segregate neotropical genus *Thibaudia* R. and P., the only difference being that in *V.* section *Cyanoccocus* the filaments are free and in *Thibaudia* the filaments are somewhat connate.

Nonetheless Camp's main thesis cannot be ignored: the differences between *Vaccinium corymbosum* and *V. myrtillus* are greater than those between the genera *Vaccinium* and *Thibaudia*. This hypothesis certainly is supported by current molecular data. A sample protolog for *Vaccinium sensu lato*, would read as follows:

## TAXONOMY

*Vaccinium* Linnaeus, Gen. Plant, ed. 5:166 (1954); Sleumer, Bot. Jahrb. 71: 413 (1941); Sleumer, Fl. Malesiana 6: 746 (1972); Vander Kloet, the Genus *Vaccinium* in North America, Agriculture Canada Publication 1828 (1988); Luteyn, Oliver & Stevens in Kubizki, Genera Plantarum 5: 54 (1996).

*Oxycoccus* J. Hill, Brit. Herbal: 324 (1756)
*Ceratostema* Jussieu, Gen. Pl.: 163 (1789)
*Thibaudia* Ruiz & Pavin ex Jaume Saint-Hilaire, Expos. Fam. 1:362 (1805)
*Gaylussacia* Kunth, Nov. Gen. Sp. Pl. 3, ed. Fol. 215, ed. qu. 275, t.275 (1819)
*Symphysia* K.B. Presl, Epist. Symphysia (1827)
*Cavendishia* Lindley, Bot. Reg. 21: pl.1791 (1835)
*Sphyrospermum* Poeppig & Endlicher, Nov. Gen. Sp. Pl. 1:4 (1835)
*Agapetes* D. Don ex G Don, Gen. Hist. 3: 862 (1834)
*Macleania* W.J. Hooker, Icon. Pl. 2: pl. 109 (1837)
*Anthopterus* Hooker, Icon. Pl. 3: pl. 243 (1840)
*Oreanthes* Bentham, Pl. Hartweg.: 140 (1844)
*Satyria* Klotzsch, Linnaea 24: 14 (1851)
*Orthaea* Klotzsch, Linnaea 24:23 (1851)
*Semiramisia* Klotzsch, Linnaea 24:25 (1851)
*Themistoclesia* Klotzsch, Linnaea 24:41 (1851)
*Psammisia* Klotzsch, Linnaea 24:42 (1851)
*Paphia* Seeman, J. Bot. 2: 77 (1864)
*Notopora* J. D. Hooker, Icon. pl. 12: 53 pl. 1159 (1876)
*Dimorphanthera* F. Mueller, Wing's South. Sci Rec. N.S. 2 (1886)
*Rusbya* Britton. Bull. Torrey Bot. Club 20:68 (1893)
*Costera* J.J. Smith, Icon. Bog. 4: 77 tab 324 (1910)
*Mycerinus* A.C. Smith, Bull. Torrey Bot. Club 58:441 (1931)
*Lateropora* A.C. Smith, Contr. U.S. Nat. Herb. 28:333 (1932)

*Diogenesia* Sleumer, Not. Bot. Gart. Berlin-Dahlem 12: 121 (1934)
*Demosthenesia* A.C. Smith, Bull. Torrey Bot. Club 63: 310 (1936)
*Plutarchia* A.C. Smith, Bull. Torrey Bot. Club 63: 311 (1936)
*Polyclita* A.C. Smith, Bull. Torrey Bot. Club 63: 314 (1936)
*Anthopteropsis* A.C. Smith, Ann. Miss. Bot. Gard. 28: 441 (1941)
*Utleya* Wilbure & Luteyn, Brittonia 29: 267 (1977)

Shrubs, small trees, or woody vines, terrestrial or epiphytic; stems of epiphytes often with hypocotylar basal swelling and the roots or rhizomes sometimes enlarged into tubers. *Leaves* alternate, spirally arranged, deciduous or evergreen, coriaceous, subcoriaceous, or thin; blades glabrous, glaucous, or beset with some form of indumentum; margins entire, crenulate or serrate, whether or not clad with simple, gland-capitellate or glandular-muriculate hairs, sometimes with a distinct basal marginal glands; petioles short or absent. *Racemes* axillary or pseudo axillary (terminal), on old wood or not, sometimes reduced to a solitary flower, caducously perulate or not, rarely eperulate. Pedicels subtended by a caduceus or ± persistent (foliaceous) bract and provided with two, sometimes very early caduceus bracteoles. *Calyx* tube cupshaped, turbinate or not, limb (4) 5 (6-8) partite to various degree, rarely absent. *Corolla* various, but often campanulate, tubular or urceolate, the (4) 5 (6-8) lobes imbricate in bud. *Stamens* (4) 8-10 (12-14) isomorphic or not inserted at the outer margin of the disk or near the base of the corolla tube; anthers dorsifix, whether or not 2-horned dorsally, ending with tubules of various length or not, the pore strictly terminal and round, or introrse (extrorse) as a spore elliptical opening or narrow slit. *Disk* annular, prominent or not. *Ovary* inferior, 4-5 (or falsely 8-10) rarely 10 celled, placentas bifid, with ∞ (rarely two) ovules. *Style* usually as long as the corolla tube, sometimes longer, rarely shorter; stigma obtuse. *Fruit* a juicy and soft berry with one or many seeds per locule, sometimes crowned by the disk and persistent calyx lobes. *Seeds* small, illipsoid, albuminous; rarely pyrenes (*Gaylussacia*); testa firm and brown or scarcely developed and white (gelatinous) embryo white or green.

## LITERATURE CITED

Camp, W.H. 1940. Changing generic concepts. Bull. Torrey Bot. Club 67: 381-389.

Hooker, J.D. 1854. *Vaccinium erythrinum*. Bot. Mag. 78t: 4688.

Klotzsch, J.F. 1851. Studien über die natürliche Klasse Bicornes Linné, Linnaea 24: 1-88.

Kron, K.A., A.E. Powell, and J.L. Luteyn. 2002. Phylogenetic relationships within the blueberry tribe (Vaccinieae, Ericaceae) based on sequence data from *mat K* and nuclear ribosomal ITS regions, with comments on the placement of *Satyria*. Amer. J. Bot. 89: 327-336

Linnaeus, C. 1737. Flora Lapponica, Amsterdam, Nederland.

Sleumer, H. 1967. Ericaceae [Flora Malesiana], Van Steenis in Flora Malesiana. 5:669-914 Wolters-Noordhoff, Groningen, Nederland.

Stevens, P.F. 1972. Notes on the infrageneric classification of *Agapetes*, with four new taxa from New Guinea. Notes Royal Bot. Garden Edinburgh 32: 13-28.

Vander Kloet, S.P., J.L. Baltzer, J.H. Appleby, R.C. Evans and D.T. Stewart. 2003. A re-examination of the taxonomic boundaries of *Symphysia* (Ericaceae) Taxon 52: in press.

Vander Kloet, S.P., T.A. Dickinson, and W. Strickland. 2003. From Nepal to Formosa, a much larger footprint for *Vaccinium* § *Aëthopus*. Acta Bot. Yunnanica 24 in press.

Vander Kloet, S.P. 1988. The Genus *Vaccinium* in North America. Agriculture Canada Publication #1028, Ottawa, ON, 201 pp.

Vander Kloet, S.P. 1985. On the generic status of *Symphysia*. Taxon 34: 440-447.

Vander Kloet, S.P. 1981. On the lectotypification of *Vaccinium* L. Taxon 30: 646-648.

Wilbur, R.L. and J.L. Luteyn. 1978. Flora of Panama VIII Ericaceae. Ann. Missouri Bot. Garden 65: 27-144.

Wright, R. 1850. Icones Fl. Indica 4: 1180-1194.

# An Overview of Weed Management in the Wild Lowbush Blueberry– Past and Present

Klaus I. N. Jensen

David E. Yarborough

**SUMMARY.** The wild lowbush blueberry (*Vaccinium angustifolium* Ait.) is an important successional species of cleared woodland and abandoned farmland of northeastern North America where commercial, managed blueberry fields have been developed. Unlike other fruit crops, the weed flora is unique and consists mainly of a broad range of native herbaceous and woody perennial species that thrive under the two-year cropping system. Traditionally, weedy vegetation was controlled or suppressed by burning, cutting, and roguing, and regenerating woody and herbaceous species were the major weed problems. The introduction of phenoxyalkanoic herbicides in the late 1940s lead to the early development by innovative growers of selective roller/wiper applicators that could control the taller, weedy overstory. Several selective preemergence herbicides (terbacil and diuron) were introduced in the 1970s to control grasses and some broadleaved weeds, and hexazinone was approved in

Klaus I. N. Jensen is Weed Scientist at Agriculture and Agri-Food Canada, Atlantic Food and Horticulture Research Centre, Kentville, NS B4N 1J5, Canada. David E. Yarborough is Professor and Extension Blueberry Specialist, University of Maine, Cooperative Extension, 5722 Deering Hall, Orono, ME 04469-5722.

This work was done jointly and the part of the work was completed by Klaus I. N. Jensen, for the Department of Agriculture and Agri-Food, Government of Canada. © Minister of Public Works and Government Services Canada (2003).

[Haworth co-indexing entry note]: "An Overview of Weed Management in the Wild Lowbush Blueberry–Past and Present." Jensen, Klaus I. N., and David E. Yarborough. Co-published simultaneously in *Small Fruits Review* (Food Products Press, an imprint of The Haworth Press, Inc.) Vol. 3, No. 3/4, 2004, pp. 229-255; and: *Proceedings of the Ninth North American Blueberry Research and Extension Workers Conference* (ed: Charles F. Forney, and Leonard J. Eaton) Food Products Press, an imprint of The Haworth Press, Inc., 2004, pp. 229-255. Single or multiple copies of this article are available for a fee from The Haworth Document Delivery Service [1-800-HAWORTH, 9:00 a.m. - 5:00 p.m. (EST). E-mail address: docdelivery@haworthpress.com].

Canada in 1982 and in Maine in 1983. This soil-applied, broad spectrum herbicide has controlled many of the common woody and herbaceous weeds. Its widespread use lead rapidly to increased yields and, directly or indirectly, it has contributed to changes in other production practices, such as the further development of mechanical harvesters and increased fertilizer use. However, the almost total reliance on the repeated use of hexazinone has introduced other problems, including shifts in weed species, the development of resistance, and soil degradation on vegetation-free soils. The highly soluble nature of the herbicide has resulted in wide-spread detection of hexazinone in groundwater adjacent to managed blueberry fields. Best Management Practices have been introduced to minimize problems associated with hexazinone use and is leading to new approaches to vegetation management that employ reduced risk herbicides, lower rates, mulches and ground covers. *[Article copies available for a fee from The Haworth Document Delivery Service: 1-800-HAWORTH. E-mail address: <docdelivery@haworthpress.com> Website: <http://www. HaworthPress.com>]*

**KEYWORDS.** *Vaccinium angustifolium*, weed control, hexazinone, sulfonylurea herbicides, selective weed control, herbicide tolerance, Best Management Practices

## *INTRODUCTION*

The wild lowbush blueberry is the most important fruit crop of Maine, Quebec and the four Atlantic Canadian provinces with over 53,000 ha (132,000 acres) under management with a mean annual production of 65,000 tonnes (Yarborough, 2003), although some recent crops have exceeded 88,000 tonnes. Unlike other fruit crops, commercial fields have been developed from native stands on deforested or abandoned agricultural land (Hall, 1959; Hall et al., 1979), and commercial fields are comprised of numerous distinct and variable clones (Hepler and Yarborough, 1991; Nams, 1994). Crop cover and yield have been increased over time by various management practices that include land improvements, pruning, fertility management, pollination, and control of insects, diseases and weeds (Barker et al., 1964; Blatt et al., 1989; Yarborough, 1996a). Also unique to lowbush blueberry production is the 2-year crop cycle, where the plants are pruned to near ground level by either burning or flail mowing to stimulate new shoot production in the first or vegetative year. The plants bloom and produce fruit in the second year. Weedy vegetation has traditionally been (Belzile, 1951; Chandler and Mason, 1946; Kinsman, 1986; 1993), and continues to be, a major problem affecting blueberry productivity and management (Bouchard, 1986; Jensen, 1985; Trevett, 1972), despite the introduction of some effective herbicides during the last fifty years. In this review we will examine the origin and nature of vegetation in lowbush blueberry fields, including some aspects of the crop, and how vegetation is changing in response to changes in production practices. We will also discuss weed control practices, particularly the introduction and use of herbicides, and their effects and consequences.

## VEGETATION OF WILD LOWBUSH BLUEBERRY FIELDS

*The lowbush blueberry.* The lowbush blueberry, principally *Vaccinium angustifolium* Ait.,[1] is a low-growing, subarctic, rhizomatous calcifuge shrub native to northeastern North America that is distributed from eastern Newfoundland west to eastern Manitoba and Minnesota and from Labrador south to Virginia (Hall et al., 1979; Vander Kloet, 1978). With the exception of the Lac-Saint-Jean region of Quebec, commercial production is largely confined to coastal areas. Although *V. angustifolium* is found on a wide range of soils, including organic ones, it is most common on coarse textured, well-drained, infertile podzols with low pH of glacial or alluvial origin (Bouchard, 1986; Hall et al., 1979; Lavoie, 1968; Vander Kloet, 1978). *V. angustifolium* is common on headlands, bogs, moors, and is an important member of the herb-dwarf shrub stratum of open and semi-open forests (Hall, 1959; Hall et al., 1979). It is particularly common on 'barrens' with frequent fire cycles (Hall and Aalders, 1968; Kinsman, 1986) and in the early successional stages following forest fires or clear cutting (Flinn and Wein, 1977; Lavoie, 1968; Whittle et al., 1997). *V. angustifolium* is a cryptophyte with underground rhizomes that endures fire and other disturbances (Flinn and Wein, 1977; Rowe, 1983). These are usually found in the organic duff layer on previously uncultivated soils but can also form in the mineral zones when the organic layer has been degraded by fire or lost by cultivation (Flinn and Wein, 1977; Trevett, 1956). New shoots arise from and clones spread by these rhizomes (Hall et al., 1979; Trevett, 1956), and shoot production can also be stimulated by severing or cutting them (Belzile, 1951; Hall, 1963). Established plants have deep taproots (Hall, 1957) that account, in part, for a moderate drought tolerance (Hicklenton et al., 2000). *V. angustifolium* also produces abundant viable and non-dormant seed (Hall et al., 1979; Wesley et al., 1986) but its seedbank is nonpersistent. Seedlings are rare, usually less than 1 $m^{-1}$ (Hall, 1959; Hall and Aalders, 1968; Wesley et al., 1986). In commercial fields, it is unlikely that seedlings would survive a thermal prune (Kender and Eggert, 1966) or the widely used residual herbicides; hence, stem density and clone expansion is entirely dependent on rhizome growth and vigor (Trevett, 1956; 1972). Seed however, especially in bird and animal droppings, may be an important means of dissemination (Vander Kloet and Hall, 1981; Wesley et al., 1986) and this is likely the origin of clones in abandoned farmland. The origin of blueberry clones in commercial fields originating from woodland are those shade tolerant, non-flowering plants that are likely surviving remnants of larger clones formed after earlier forest fires or clearing (Hall, 1955). The lowbush blueberry will grow and persist under a wide range of light conditions (Hall, 1958; Hall and Ludwig,1961; Hoefs and Shay, 1981). Plants can persist under a conifer canopy at > 0.5% full sunlight, but at least 50% is required for flower bud formation and fruiting (Hall, 1955; Hall et al.,1971). Shade from taller weeds reduces yield when sunlight is reduced below 80% (Chandler and Mason, 1946).

A second low-growing species, the velvet-leaf blueberry (*V. myrtilloides* Michaux) is often associated with *V. angustifolium*. In Maine and New Brunswick up to 30% of the clones in commercial blueberry fields may be *V. myrtilloides* (Vander Kloet, 1978). It was most common in fields originating from woodland but it has decreased over time (Hall, 1959). Vander Kloet (1994) showed both were equally fire tolerant and suggested that the diploid (2n = 24) *V. myrtillloides* may be less adaptable and competitive compared to the tetraploid *V. angustifolium*. The relative decrease of *V. myrtilloides* in blueberry fields is also due, in part, to its greater sensitivity to the widely used herbicide hexazinone (Velpar) (Jensen, 1985). At full maturity there is little difference between the fruit of the two species, but *V. myrtilloides* matures later (Lavoie, 1968) giving the fruit a more tart flavor, which may account for its alternate name, the 'sour-top blueberry.'

*Weed flora of lowbush blueberry fields.* Several factors have influenced the weedy flora of blueberry fields: (1) their origin, (2) the traditional method of pruning by burning, and more recently, (3) the widespread change to pruning by mowing, and (4) the use of selective herbicides in the 1980s. Hall (1955; 1959) reported that weeds occurring in fields originating from woodland were principally surviving woodland species, e.g., some ferns, bunchberry (*Cornus canadensis* L.), black chokeberry or barrenberry [*Aronia melanocarpa* (Michx.) Elliot = *Pyrus melanocarpa*] (Hall et al., 1978), sheep-laurel (*Kalmia angustifolia* L.), and *Prunus* spp.; whereas, those from abandoned farmland were predominately grasses, composite species, and narrow-leaved meadowsweet [*Spiraea alba* DuRoi var. *latifolia* (Aiton) Dippel], or those species associated with open sites. However, without aggressive weed control, these farmland sites became invaded with woody species that greatly reduced their productivity (Belzile, 1951; Chandler and Mason, 1946; Hall, 1959). Hall and his colleagues have described the biology of some of the most important of these weeds (Hall, 1975; Hall and Sibley, 1976; Hall et al., 1973; 1974; 1976; 1978). In contrast to other crops, this traditional weed flora is unique because it consists almost entirely of native species that are typically not considered to be weeds elsewhere.

Burn pruning has also been a determining factor in selecting the weed flora (Belzile, 1951; Lavoie, 1968). Generally, lists or descriptions of the weedy flora in blueberry fields are remarkably similar in Quebec (Bouchard, 1986; Lavoie, 1968), New Brunswick (Hall, 1955; 1959), Maine (Chandler and Mason, 1946), and Nova Scotia (Hall and Aalders, 1968; Jensen, 1985) prior to the widespread use of selective herbicides. This vegetation, in turn, is similar to that of early successional stages following forest fires in northeastern North America (Flinn and Wein, 1977; Lynham et al., 1998; Martin, 1954; Whittle et al., 1997). However, fire intensity and frequency, surrounding vegetation, climatic and topographical factors may influence species mix on a local level (Lavoie, 1968; Rowe, 1983). For example, Lavoie (1968) demonstrated that plant associations in Quebec blueberry fields reflected interactions between soil type, moisture, and fire history. Managed blueberry fields can be considered as 'proclimaxes' (Rowe, 1983) where succession is arrested at the 'herb-dwarf shrub stage' by pruning and herbicide use. Rowe (1983) has categorized post-fire successional species according to their means of regeneration as invaders, evaders, avoiders, resisters and endurers. Prior to the 1980s, many of the common woody weeds were 'endurers,' that is, cryptophytes like the lowbush blueberry capable of regenerating from vegetative structures in either the organic or mineral soil zone (Flinn and Wein, 1977; Lavoie, 1968; Rowe, 1983). Among the most common were sheep-laurel, sweet-fern [*Comptonia perigrina* (L.) Coult], and others on dry sites and Canada rhododendron [*Rhododendron canadense* (L.) Torr.] and Labrador tea (*Ledum groenlandicum* Oeder) on wet ones (Lavoie, 1968). Some rhizomatous perennial broadleaved weeds, such as goldenrods (*Solidago* spp.), bunchberry, spreading dogbane (*Apocynum androsaemifolium* L.), violets (*Viola* spp.), and eastern bracken [*Pteridium aquilinum* (L.) Kuhn var. *latiusculum* Desv.] are also adapted to survive low intensity pruning fires (Hall, 1959; Lavoie, 1968; Lynham et al., 1998; Rowe, 1983). These species often 'peak' in the second or third year after burning (Swan, 1970) which coincides with the crop cycle. Burning also destroys many weed seeds but it creates open space for 'invaders' such as birches (*Betula* spp.), aspen poplar (*Populus tremuloides* Michx.), wild red raspberry [*Rubus idaeus* var. *strigosus* (Michx.) Maxim.] and such herbaceous species as fireweed (*Epilobium angustifolium* L.), goldenrods, and other compositae (Hall, 1955; Lynham et al., 1998; Martin, 1954; Rowe, 1983; Swan, 1970). Some tree species are 'resisters' that regenerate from basal buds following low intensity fires, e.g., maples (*Acer*) and birches (Rowe, 1983). Burning also eliminates many shallow-rooted woodland species and most conifers and tends to increase herbaceous species at the expense of grasses and sedges (Chandler and Mason, 1946; Hall, 1959; Rowe, 1983; Swan, 1970).

The introduction and widespread use of the broad spectrum, selective herbicide hexazinone (Velpar) in the early 1980s drastically changed the traditional weed flora (Jensen, 1985; Yarborough and Bhowmik, 1989a). Prior to that time, the principal herbicide use involved grass control with terbacil (Sinbar) (Ismail, 1974b) and selectively spot spraying weed patches with phenoxyalkanoic herbicides or treating species that grew taller than blueberry with roller/wiper applicators, which provided partial control of some herbaceous and woody broadleaved species (Abdalla, 1967; Trevett, 1952). In contrast, initial hexazinone applications controlled most grasses and sedges, herbaceous broadleaved weeds and such important woody species as sheep-laurel, sweetfern, meadow-sweet, and others (Jensen, 1985; Yarborough and Bhowmik, 1989a). The latter woody species were considered the most serious weed problems, compared to grasses or herbaceous broadleaves, because of their competitive ability to 'crowd out' the lowbush blueberry (Chandler and Mason, 1946; Hall, 1959; Trevett, 1952).

Early hexazinone use, as expected, also resulted in an increasing predominance of species with inherent hexazinone tolerance, e.g., bunchberry, spreading dogbane, black chokeberry, tufted vetch (*Vicia cracca* L.), and others (Yarborough, 1991; K.I.N. Jensen, unpubl. data). In Maine, hexazinone-induced bare areas are often colonized by several annual grasses, witch grass (*Panicum capillare* L.) and fall panicum (*Panicum dichotomiflorum* Michx.) and the native colonial bent grass (*Agrostis capillaris* L.) and many fields now require treatment with selective graminicides (Yarborough and Hess, 1999b) or rotation with other preemergence applications of terbacil (Sinbar) or diuron (Karmex). A comparison of weed surveys conducted in Nova Scotia in 1984 and 1985 (McCully and Sampson, 1991) and again in 2001 and 2002 (K.I.N. Jensen and M.G. Sampson, unpubl. data) show an increasing diversity of the weed flora that has resulted from herbicide use (primarily hexazinone) and other changes in production practices over time (Table 1). There has been a 2-fold increase in the number of species in every category and eight species of annual grasses have been recorded for the first time. Many of these annuals are common arable weeds, e.g., crabgrasses (*Digitaria* spp.), witch grass (*Panicum capillare* L.), redroot pigweed (*Amaranthus retroflexus* L.), hemp-nettle (*Galeopsis tetrahit* L.), lambsquarters (*Chenopodium album* L.), and annual fleabane (*Erigeron annuus* (L.) Pers.). Biennials such as goat's-beard (*Trapopogon patensis* L.) and wild carrot (*Daucus carota* L.) are also becoming more common. These are potentially serious problems because they are invasive, rank, produce copious amounts of seed, and will necessitate herbicide use in both the prune and fruiting year. The trend is towards species that are spread by seed and invade and colonize bare ground in contrast to the traditional fire-tolerant, perennial weeds that spread slowly by vegetative means. No doubt prune mowing has also contributed to the survival

TABLE 1. Comparison of the number of species recorded in surveys of lowbush blueberry fields conducted in Nova Scotia in 1984-1985 and in 2001-2002.

| Weed classification | Total number of species[z] | |
|---|---|---|
| | 1984-1985 | 2001-2002 |
| Trees | 9 | 19 |
| Shrubs | 16 | 26 |
| Biennial/perennial broadleaves | 40 | 82 |
| Annual broadleaves | 18 | 35 |
| Perennial grasses | 12 | 22 |
| Annual grasses | 0 | 8 |
| Sedges and Rushes | 5 | 8 |
| Other monocots | 5 | 8 |
| Cryptogramma | 4 | 10 |
| Total species | 109 | 218 |

[z] The 1984-1985 weed survey included 115 fruiting year fields (McCully and Sampson, 1991); the one in 2001-2002 included 128 (Jensen and Sampson, unpubl. data).

of many of these recent species and has resulted in greater seed deposition and seed banks. Dissemination of seed by mowers, harvesters, and other equipment by contractors is also an important, but undocumented, factor. Finally, there is also evidence in Atlantic Canada that some native perennial grasses that were originally controlled by hexazinone, e.g., poverty oat grass (*Danthonia spicata* (L.) P. Beauv ex Roem. & Schult.), Mexican muhly (*Muhlenbergia mexicana* (L.) Trin.), and several *Festuca* and *Agrostis* spp., have now developed resistance to the herbicide, and often occur in large single-species stands. On the other hand, the increasing appearance of such herbaceous perennials as sheep sorrel (*Rumex acetocella* L.) and narrow-leaved goldenrod [*Euthamia graminifolia* (L.) Nutt. = *Solidago graminifolia*] is more likely the result of reduced hexazinone rates used by producers rather than to development of herbicide resistance (K. I. N. Jensen, unpubl. data).

Surveys conducted in the mid-1990s in Quebec blueberry fields (LaPointe and Rochefort, 2001) showed a decrease in weed frequency and abundance compared to one conducted prior to hexazinone use (Bouchard, 1986), and there was no indication that major species shifts or development of tolerance was occurring at that time. However, it is likely that in all areas of lowbush blueberry production the weed flora will continue to change and evolve in response to herbicide use and other changes in production practices.

## VEGETATION MANAGEMENT

*Herbicides.* Herbicides have been used as the primary means of weed control in lowbush blueberry for more than 50 years. Here we will discuss the development of chemical weed control within three categories of herbicide use: (1) non-selective postemergence treatments applied only to the weed topgrowth, (2) pre-emergence soil-applied treatments that are principally active via root uptake, and (3) selective postemergence treatments applied broadcast to foliage (Table 2). We conclude by discussing the limitations of chemical and nonchemical weed control and the growing importance of integrated vegetation management in this crop.

*Non-selective, postemergence herbicide treatments.* Chandler and Mason (1946) give early weed control recommendations for Maine using a 4,6-dinitrophenol and a number of inorganic compounds, e.g., $H_2SO_4$, ammonium sulfamate, and calcium chlorate. These contact herbicides were generally applied as 10% to 15% aqueous solutions to

**TABLE 2.** Herbicides approved for use in lowbush blueberry in Canada and the United States.

| Herbicide[z] | | Herbicide | |
|---|---|---|---|
| Common name | Trade name(s) | Common name | Trade name(s) |
| Non-selective postemergence | | Selective postemergence | |
| 2,4-D ester*[y] | various | clethodim** | Select 240 EC |
| dicamba* | Banvel II 480 | clopyralid‡ | Lontrel 360/Stinger |
| glyphosate | Roundup/Transorb | fluazifop-P-butyl | Venture L |
| triclopyr* | Garlon 4E | nicosulfuron/rimsulfuron* | Ultim DF |
| | | sethoxydim** | Poast |
| Selective preemergence | | tribenuron* | Spartan 75 DF |
| atrazine* | Aatrex Nine-0 | | |
| diuron** | Karmex 80 WP | | |
| hexazinone | Velpar 75 DF/Pronone 10G | | |
| pronamide* | Kerb 50 WP | | |
| simazine | Princep Nine-T/Simazine 80W | | |
| terbacil | Sinbar 80 W | | |

[z] Herbicides with a single asterisk (*) are approved only in Canada; those with a double asterisk (**) are only approved in the USA; and those with ‡ are pending minor use registration.
[y] 2,4-D esters are only approved for use in combination with dicamba.

topgrowth of herbaceous and woody species and provided a temporary chemical burn. These compounds usually required repeated applications and were corrosive, hazardous to the applicator, and contact with the crop caused damage. Ferns and grasses could be controlled for one year with organic solvents, e.g., varsol and kerosene, but these too could cause crop injury (Hall, 1954). In many cases, these treatments were not cost effective compared to manual or mechanical weed control at that time (Chandler and Mason, 1946).

The introduction of the hormonal phenoxyalkanoic herbicides 2,4-D and 2,4,5-T into agriculture after World War II provided effective control of many woody and herbaceous broadleaved weeds. By the early 1950s, cutting and roguing were no longer recommended practices except in the control of escapes and bracken fern (Trevett, 1952). In Maine, Trevett (1950; 1952) developed susceptibility tables and classified weeds according to their sensitivity to foliar applications of iso-octyl esters of 2,4-D and 2,4,5-T: (1) sensitive and could be controlled by a single application of 2,4-D [e.g., alders (*Alnus* spp.), birches, willows (*Salix* spp.), sweet-fern]; (2) moderately tolerant and required either several applications of 2,4-D or addition of 2,4,5-T for control [e.g., chokecherry (*Prunus* spp.), Canada rhododendron, *Spiraea* spp., and sheep-laurel)]; and (3) resistant and could not be controlled even with repeated applications [e.g., conifers, eastern bracken, maples (*Acer* spp.), and brambles (*Rubus* spp.)]. Ammonium sulfamate (AMS) was also widely used against woody weeds, including a number of phenoxy-tolerant ones (e.g., brambles, conifers, and maples). However, AMS and 2,4,5-T were removed from food crop use in the 1970s.

Lack of crop tolerance to phenoxyalkanoic herbicides requires selective means of applying them so that crop contact and injury can be minimized. Broadcast sprays can only be applied safely in two situations. One is where a taller weed canopy intercepts the spray as in dense stands of bayberry (*Myrica pensylvania* Mirbel) and sweet-fern (Abdalla, 1967; Trevett, 1952). The other is to control evergreen species, e.g., sheep-laurel and teaberry (*Gaultheria procumbens* L.), with post-harvest, mid-fall applications after blueberry foliage has senesced (Abdalla, 1967; Everett et al., 1968). Here herbicide absorbed into the blueberry stems does not translocate to the rhizomes or cause damage after the crop is pruned (Jensen and North, 1987). The early senescence of foliage of harvested blueberries can also be exploited to safely control some late-senescing woody species with directed spot applications of 2,4-D or dicamba (Banvel), e.g., alders (Jensen and North, 1987) and bayberry (Abdalla, 1967). Dicamba is approved for blueberry use only in Canada and is superior to 2,4-D in controlling some broadleaved weeds. Mid-fall applications are also safer on the crop than 2,4-D (Jensen and North, 1987; Trevett and Durgin, 1972). Other attempts to increase crop tolerance to foliarly applied herbicides have involved using endothall, a contact herbicide, in mid-summer to selectively defoliate blueberry plants prior to applications of 2,4-D (Trevett and Durgin, 1973) and glyphosate (Yarborough and Ismail, 1979a; 1980) to black chokeberry (or barrenberry), a serious weed and fruit contaminant in Maine. Although herbicide treatments reduced black chokeberry by about 90% and had no long-term effect on blueberry productivity, treatments failed to reduce fruit contamination of processed blueberries below minimum grade standards (Yarborough and Ismail, 1979b; 1980). This contamination could also be reduced with foliar applications of the growth regulator ethephon which selectively induces abortion of black chokeberry fruit (Ismail, 1974a).

A number of techniques were developed in the 1950s to apply nonselective herbicides to weeds with minimal crop contact, such as manually wiping using brushes, gloves or wipers made with absorbent fabric. Tractor mounted mechanical roller/wipers were designed and manufactured, often on-farm, to treat woody species that grew above the crop canopy (Blatt et al., 1989; Kinsman, 1993; Trevett, 1950; 1952). Also used to control some woody weeds were solutions of 8% to 20% (w/v) iso-octyl esters of phenoxyalkanoic herbicides in diesel or fuel oil that were selectively applied to cut stumps and stems with shielded nozzles (Abdalla,1967; Trevett, 1950; 1952). These continue to be used in Canada in new developing fields (Atlantic Comm. Fruit Crops, 1999), but this use has been discontinued in the United States. In Canada, triclopyr (Garlon 4) is also approved for several oil-based stump and basal bark treatments that will also control a number of 2,4-D resistant species, e.g., maples, ash (*Fraxinus* spp.), and beech (*Fagus* spp.) (Atlantic Comm. Fruit Crops, 1999).

Today, glyphosate (Roundup, Transorb, Touchdown and others) are widely used either as directed foliar sprays of 1% to 2% aqueous solutions or as wiping treatments using 20% to 33% solutions with a variety of implements that range from hand-held 'hockey-stick' applicators to vehicle driven wick wipers or rollers (Atlantic Comm. Fruit Crops, 1999; Yarborough, 1996b; Yarborough and Ismail, 1979b). In addition to controlling a broad range of woody species, glyphosate also controls many herbaceous species, including ferns, sedges, and grasses. Here, the addition of ammonium sulfate will improve control of hardwood species, but not herbaceous ones (Yarborough and Hess, 2000b). There is also the potential of applying glyphosate broadcast after harvest to late-senescing or evergreen species, e.g., sheep-laurel (Ismail and Yarborough, 1981).

*Selective, pre-emergence herbicide treatments.* By the 1970s, Trevett (1972) postulated that it was necessary to increase yields 2- to 3-fold and to develop mechanical harvesting if the lowbush blueberry industry was to survive competition from the higher yielding, cultivated highbush blueberry (*Vaccinium corymbosum* L.). These developments were dependent on: (1) broad spectrum weed control, (2) higher fertility management, (3) mowing vs. burning as a pruning practice, (4) managing soil pH, and (5) land leveling, that is, Trevett's 'Big 5 Plan.' Selective herbicides were required to control weeds under increased fertility regimes because weeds, especially grasses, were more responsive to fertilizer than the crop. Low fertility management using ~ 20 kg ha$^{-1}$ N, or less, was practiced only on new fields developed from woodland or others with few weeds, but even here repeated fertilizer use aggravated weed problems (Trevett, 1972; Trevett and Durgin, 1972). The results of a long-term screening program identified terbacil (Sinbar) as a potential herbicide that controlled native grasses [redtop (*Agrostis gigantea* Roth), poverty oat grass (*Danthonia spicata* [L.] P. Beauv. Ex Roem. & Schult.), rice-grass (*Oryzopsis asperfolia* Michx.), feather grass (*Stipa canadensis* L.)], and sedges (Trevett and Durgin, 1972). Terbacil also controlled some perennial broadleaved weeds, e.g., yarrow (*Achillea millefolium* L.), hawkweeds (*Hieraceum* spp.), cinquefoils (*Potentilla* spp.), fireweed, and several others (Trevett and Durgin, 1972). Lowbush blueberry tolerated up to 7 kg/ha terbacil without injury when applied preemergence after the pruning operation (Ismail, 1974b). When weeds were controlled, terbacil, either alone and in combination with fertilizer, increased both vegetative growth and yield by up to 2-fold due to increases in stem density and fruit buds (Ismail et al., 1981). However, its usefulness was limited because of its poor activity on some common broadleaved weed species, especially *Solidago* spp., *Aster* spp., sheep sorrel, and others, which increase dramatically with terbacil use (Smagula and Ismail, 1981; Yarborough and Ismail, 1985). In Maine, diuron (Karmex) may be tankmixed with terbacil to improve broadleaved weed control (Yarborough, 2002a), but even this combination gives poor control of many common weeds, including all woody ones. Terbacil at 1 kg in 500 L of water can also be applied as a spot spray to control hay-scented fern [*Dennstaedtia punctilobula* (Michx.) T. Moore] (Jensen, 1985).

Reports on broad spectrum, non-agricultural herbicides in the late 1970s revealed that lowbush blueberry was not controlled on hexazinone-treated railroad and forestry plots in New Brunswick. Initial trials in lowbush blueberry demonstrated that pruning offered a 'window' prior to emergence in which hexazinone could be safely and effectively applied. Later post emergence applications caused severe injury, whereas fall applications were safe after blueberry plant dormancy, but weed control was poor (Jensen and Kimball, 1985). Broadcast, pre-emergence applications of 2 or more kg ha[-1] hexazinone provided excellent control of most grasses and sedges, as well as, many herbaceous perennials (including many terbacil-tolerant ones) and some important woody weeds, such as sheep-laurel, *Rubus* spp., meadow-sweet, Canada rhododendron and others. Lists of hexazinone-tolerant and susceptible species are given in Jensen (1985), Jensen et al. (1981), Sampson et al. (1990), and Yarborough and Bhowmik (1989a). The long-term effect of repeated hexazinone use on changes to the weedy flora was described above. A comparison of weed and crop response in these early trials shows that hexazinone was more effective at lower rates and against a wider range of species (and crop injury was correspondingly increased) on the heavier, upland soils in Nova Scotia than on the sandy barrens in Maine. In Maine, for example, control of yarrow, sweet-fern, sheep sorrel, and poplar with approximately 2 kg ha$^{-1}$ hexazinone was rated poor (Yarborough and Bhowmik, 1989a), whereas control was rated good in Nova Scotia (Jensen, 1985). In multi-site experiments, differences in hexazinone performance between sites was attributed to vegetation type and density, soil type, organic matter, and drainage (Jensen, 1986; Yarborough and Ismail, 1985). This reflects the importance of adjusting hexazinone rate to site characteristics (Atlantic Comm. Fruit Crops, 1999; Yarborough, 2002b; Yarborough and Jemison, 1997).

Hexazinone provided broad spectrum, selective weed control in lowbush blueberry that was superior to previous residual herbicides or their combinations (Hoelper and Yarborough, 1985; Jensen et al., 1981) and the effect on blueberry was significant. Blueberry plants respond positively to weed control at rates that did not cause significant injury (usually < 4.5 kg ha$^{-1}$ hexazinone) with linear increases in number of branches and fruit buds per stem, yield, and sometimes stem density. Increasing injury at higher rates resulted in yield decreases. Although highly variable among locations, yield of plots typically doubled with the first hexazinone application that eliminated weed competition. The greatest relative increase on those with an initially low crop-to-weed density (Jensen, 1986; Yarborough and Ismail, 1985; Yarborough et al., 1986). Injury occurred at higher rates resulting in yield losses, but in most cases, new foliar growth was produced from axillary buds following defoliation of plants that were in the vegetative growth stage and plants recovered (Jensen, 1986). Crop injury was also more pronounced on plots where crop vigor was reduced by heavy weed competition, on poorly drained sites or those with low organic matter soils (Jensen, 1986; Yarborough, 1995; 2002b; Yarborough and Ismail, 1985). A series of trials was also conducted in the early 1990s to examine the cause of crop injury that was being reported by growing numbers of producers. These trials consistently demonstrated that crop injury resulted from late application (late May or later) of hexazinone and not from changes that had occurred since the herbicide was introduced, such as changes in pruning method (mow vs. burn), formulation (dry flowable, granular, or fertilizer impregnated vs. liquid), or the degradation of the organic layer (Jensen et al., 1993).

There also appears to be differential tolerance among clones (Jensen, 1985), as there is amongst highbush blueberry cultivars (Barron and Monaco, 1986; Jensen, 1981). The reduction of *Vaccinium myrtilloides* in fields is likely also related to its greater sensitivity to hexazinone

compared to *V. angustifolium* (Yarborough, 2002a). Differential tolerance among *Vaccinium* and weed taxa results from differences in the
rates of uptake and translocation of hexazinone and its degradation to
non-toxic hydroxylated and demethylated metabolites (Barron and Monaco, 1986; Jensen and Kimball, 1990). Metabolism of hexazinone in
lowbush blueberry is rapid and similar to that reported in other plant
species and results in negligible or undetectable residues in fruit, even
following fruiting-year applications (Jensen and Kimball, 1985). Similar results have been reported among lowbush blueberry clones for
terbacil (Jensen et al., 1985).

Hexazinone was registered for use in Canada and Maine in 1982 and
1983, respectively, and it was immediately adopted by the industry. Applications of 1.0 kg ha$^{-1}$ hexazinone are also approved to control sensitive
herbaceous weeds in the spring of the fruiting year. These postemergence applications are safe provided they are applied before the majority of floral buds separate and show the white corolla tube (Jensen and
Specht, 2002). This treatment does not affect yield parameters, e.g.,
stem density or fruit bud numbers, and is intended to improve harvest
efficiency in weedy fields. It can also be used for weed control in fields
managed on a 3-year cycle with two consecutive crop years.

In addition to liquid formulations, hexazinone is also available in
solid form as a large pellet (Velpar Gridball), as 10% granules (Pronone
MG or 10G), and impregnated onto fertilizer granules. Only the 10%
granules are used in Canada. Solid formulations do not adhere to dry
blueberry foliage and can therefore be applied later in the spring or summer until vegetative growth ceases at the tip die-back stage (Yarborough,
2002a; 2002b). Effectiveness is dependent on rainfall to leach the herbicide from the product into the soil. Yarborough and Ismail (1981) obtained control of some tree species by placing several pellets at the base
of the cut stump. 'Patchy' crop injury is common due to uneven distribution of granules or redistribution after application (Jensen et al.,
1993; Yarborough and Hess, 2000a). Air-assist applicators are superior
to mechanical spreaders in providing even granule distribution (Yarborough
and Hess, 2000a). Solid formulations are more commonly used on newly
developed land where crop injury is less critical and late applications
generally improve control of woody species. However, aqueous applications out performed granular and fertilizer impregnated applications
in weed control and crop tolerance when applied preemergence after
pruning (Jensen et al., 1993).

Unlike earlier experience with terbacil, weed control (at least initially) was not affected by high fertility regimes, nor were there detri-

mental hexazinone × N interactions (Yarborough et al., 1986). Trends in incremental yield increases over time were attributed almost entirely to weed control and not to increased fertility (Eaton, 1994; Penney and McRae, 2000). For example, Eaton (1994) reported hexazinone use alone resulted in a 2.5-fold increase in average annual yield over a 10-year period compared to check plots, while the application of 56 kg ha$^{-1}$ N (in 1:1:1 NPK) to herbicide-treated plots only accounted for an additional 16% yield increase. Yield increases were due to increases in stem density and number of flower buds per stem. Perhaps reduced competition and release of nutrients from organic sources provides sufficient nutrients for good blueberry growth for at least four or five cycles after the initial hexazinone application. The widespread use of hexazinone and improved weed control was a major factor in the large increase in lowbush blueberry production that increased from about 18,700 tonnes in 1980 to about 41,600 in 1990 and 66,500 tonnes in 2000 (Yarborough, 2003). Other factors were increased use of honeybees, new land entering production, mechanical harvesting, land leveling and other improvements and inputs (Smagula and Yarborough, 1990; Yarborough 1997b, 2003).

Hexazinone's effectiveness depends, in part, on its residual properties. Its half-life in typical sandy loam blueberry field soils is about 4 or 5 weeks and about 5% persists to the following year (Jensen and Kimball, 1987; Yarborough and Jensen, 1993). This is sufficient to provide long-term control of sensitive species. Following application, residue levels remain highest near the soil surface but the herbicide readily leaches to lower levels too. Hexazinone is microbiologically degraded to a number of metabolities (Kubilius and Bushway, 1998) and the partially detoxified *N*-demethylated derivative, metabolite B, is the most common metabolite in both blueberry field soils (Jensen and Kimball, 1987) and groundwater (Keizer er al., 2001; Kubilius and Bushway, 1998). Hexazinone's extreme water solubility (33 g L$^{-1}$) and low sorption results in high soil mobility which, in turn, contributes to both its effectiveness against deep-rooted perennials and to its potential for surface and ground water contamination. Concerns about the latter have lead to an unsuccessful attempt to ban hexazinone in Maine and to studies monitoring levels of hexazinone and its metabolites in ground water in Maine (Kubilius and Bushway, 1998; Yarborough and Jemison, 1997), New Brunswick (Keizer et al., 2001), and Nova Scotia (Maddison, 1997). Maximum levels of hexazinone typically range from 5 to 15 ppb, well below the US EPA Health Advisory Limit of 400 ppb, even in areas of widespread hexazinone use on light textured soils with high infiltration rates. There is evidence that hexazinone degrades in ground water under both anaerobic and anoxic conditions (Keizer et al., 2001). Nevertheless, Best Management Practices have been developed to help growers manage hexazinone use to minimize potential surface and groundwater contamination (Yarborough, 2002b; Yarborough and Jemison, 1997). These include using the lowest economically effective rate and alternative weed control strategies to minimize hexazinone use, using later applications of granular formulations that are less likely to leach, adapting application to match site conditions, and using only properly calibrated and operated equipment. In practice, these measures appear to reduce hexazinone levels in water sources (Yarborough, 1997a; Yarborough and Jemison, 1997).

It is critical for the industry to retain essential uses of hexazinone. Finding a more environmentally acceptable, broad spectrum alternative is unlikely. In Canada, atrazine at 4 kg ha$^{-1}$ can be used after the prune to control a number of annual and perennial herbaceous weeds, including hawkweeds (*Hieracium* spp.), goldenrods (*Solidago* spp.), and others (Atlantic Comm. Fruit Crops, 1999; Jensen, 1985; Sampson et al., 1990), but generally the use of atrazine is discouraged. Blueberry plants are very tolerant to atrazine and their response when weeds are controlled is similar to that of hexazinone (Jensen, 1986). Fall applications of propyzamide (Kerb) are also permitted in Canada for grass control (Atlantic Comm. Fruit Crops 1999), and although rarely used in the past, it may find use against certain hexazinone and fluazifop-P-butyl tolerant (Fusilade, Venture) biotypes (e.g., *Festuca* spp.) that have recently appeared (Jensen, unpublished data). Herbicide screening of preemergence herbicides in Maine have demonstrated that azafenidin (Milestone) and pendimethalin (Prowl) provide selective control of some annual grasses and broadleaved weeds during the prune year (Yarborough and Hess, 1999a; 2001). Imadazolinone herbicides were generally too phytotoxic to be used in lowbush blueberry (Yarborough and Bhowmik, 1989b) and most sulfonylurea herbicides tested are ineffective when applied preemergence (Howatt, 1991; Yarborough and Bhowmik, 1989b). The increasing prevalence of annual weeds and others arising from seed will result in an increasing need to introduce new herbicides that control weeds in the seedling stage in both the prune and crop year, such as those that are predominately used in field and horticultural crops.

*Selective, postemergence treatments.* The repeated use of hexazinone has resulted in dramatic, but not unexpected, changes in weed species of blueberry fields (Jensen, 1985; Yarborough and Bhowmik, 1989a).

Consequently, finding selective postemergence herbicides to control these new species has been given much attention, but to date this has been only partially successful. Advantages of postemergence treatments are that susceptible species can be specifically targeted which reduces herbicide load and potential crop injury can be minimized by timing and careful directed spot applications.

Bunchberry is one of the most common hexazinone-tolerant weeds that rapidly expanded in blueberry fields when other weeds were controlled (McCully and Sampson, 1991; Yarborough and Bhowmik, 1993). Some sulfonylurea herbicides, particularly tribenuron [Express (USA) or Spartan (Canada)], appeared promising against it (Howatt, 1991; Yarborough and Hess, 1995). To be safe and effective, 40 g ha$^{-1}$ tribenuron is applied broadcast after the pruning operation when the first flowers of the bunchberry plants open and the blueberry shoots are less than 2.5 cm tall. Later applications result in stunting of the blueberry plants that may reduce yields (McCully et al., 1995). Late summer applications drastically reduce flower bud numbers and yield. This effect may be a characteristic of other sulfonylureas (Jensen and Specht, 2003). Directed spot applications of tribenuron at 0.25 g L$^{-1}$ water during mid summer also provides excellent control of several perennial weeds, including: wild rose (*Rosa virginiana* Mill.), eastern bracken, yellow loosestrife [*Lysimachia terrestris* (L.) BSP], and speckled alder [*Alnus incana* (L.) Moench] (Howatt, 1991; Jensen and Specht, 2003). Eastern bracken was previously controlled with asulam (Jackson, 1981) but this product was discontinued. Blueberry plants contacted by tribenuron may show some stunting and discoloration, but there is negligible long-term effect. Broadcast and directed applications of tribenuron are currently only approved for use in Canada. Recent screening trials have shown that chlorimuron (Classic) at 0.02 g L$^{-1}$ water selectively controlled trailing blackberries (*Rubus* spp.) and black chokeberry (K.I.N. Jensen, unpubl. data). However, more work is required to identify selective herbicides with activity against herbaceous and woody broadleaved weeds.

The increasing problem of annual and perennial grasses in hexazinone-treated blueberry fields was discussed above. The cyclohexenone graminicides clethodim (Select) and sethoxydim (Poast) and the arloxyphenoxy propionate fluazifop-P-butyl are very safe on the crop and can be applied at any time with negligible effect on blueberry plants. They are very effective against annual panicoid grasses, e.g., witch grass and fall panicum (Yarborough and Hess 1999b). However, graminicide activity against perennial native grasses is variable and depends on species and time of application. The effects range from good control [rough hair grass (*Agrostis scabra* Willd. = *A. hyemalis*)], to short-term control/ suppression [little bluestem (Yarborough, 1988); poverty oat grass, quack grass (*Elytrigia repens* (L.) Desv. ex D. B. Jacks) (Atlantic Comm. Fruit Crops, 1999), *Poa* spp., *Muhlenbergia* spp., and others (K.I.N. Jensen, unpubl. data)], to resistant (*Festuca* spp.). In Maine, multiple spot and broadcast applications of these herbicides can be made in the prune year (Yarborough, 1999), but in Canada only a single application of fluazifop-P-butyl is registered (Atlantic Comm. Friut Crops, 1999). In Canada, a recent approval was obtained for a fruiting-year application of fluazifop-P-butyl to suppress sensitive grasses to improve harvest efficiency.

In addition to grasses, other monocot species are increasing, e.g., rushes and sedges. In Canada, black bulrush (*Scirpus atrovirens* Willd.), a tall, tussock-forming sedge has become a major problem. The commercial 1:1 mixture of nicosulfuron/rimsulfuron (Ultim) at 0.031 g L$^{-1}$ water is approved for use as a mid-summer spot application against black bulrush (Jensen and Specht, 2003). This herbicide may cause unacceptable crop injury and yield losses when applied broadcast, except perhaps after harvest. However, nicosulfuron/rimsulfuron has not proven effective against other problem monocots, e.g., *Carex* or *Juncus* spp. Weed diversity of lowbush blueberry fields is increasing more rapidly than the potential of finding effective herbicide controls. Since many of these weeds arise from seed, or have a life cycle with a sensitive seedling stage, there may be a number of available herbicides currently used in conventional agromonic and horticultural crops that merit evaluation in lowbush blueberry.

*Nonchemical weed control and integrated weed management.* Other means of controlling weeds, besides herbicides, include biological, mechanical, preventative, and cultural control, but the applicability of each is strongly influenced by the nature of the weeds and crop. Biological weed control, for example, has traditionally been most successful against single, introduced, non-native species growing in relatively undisturbed habitats. Hence, most weed species of blueberry fields would not be suitable candidates for this approach. However, in Atlantic Canada, high levels of control of both seedlings and mature plants of St. John's wort (*Hypericum perforatum* L.), a non-native species, are obtained with naturally recurring epidemics of *Colletotrichum gloeosporioides* f. spp. *hyperici* (Morrison et al., 1998). Although other insects and pathogens (e.g., *Cryptomycena pteridii* on eastern bracken or *Puccinia* rust on narrow-leaved goldenrod) may occasionally impact individual species, this has not been documented. Mechanical weed control is largely impractical because of cost (Yarborough and Marra, 1997), except for a few species like eastern bracken and conifers. However, newer cutting implements that apply herbicides (usually glyphosate) to the cut surface would likely be useful against many perennial weeds but need further assessment (Yarborough and Hess, 2002). Preventative weed control that aims to reduce the introduction and spread of weed seed should be more widely practiced since many of the recent herbaceous weed problems propagate and spread by seed. The increasing mechanization of lowbush blueberry production, particularly mowers and harvesters, has likely contributed to these problems. Cleaning equipment and preventing seed production, particularly of rapidly spreading invasive species such as goat's-beard, lamb's-quarters, or black bulrush, by mowing or cutting seed heads should be practiced. In the past, prune burning helped reduced weed seed, seedlings and some shallow-rooted or fire intolerant species (Rowe, 1983). Although burning is no longer cost effective in fields that can be mowed, prescriptive burning may be recommended as part of an integrated approach to manage certain insect pests (e.g., blueberry leaf-tier [*Coesia cuevalana* (Kearfott)] or the blueberry stem gall wasp [*Hemadas nubilipennis* (Ashmead)]), diseases (e.g., *Godronia* and *Phomopsis* stem canker), and weeds, e.g., hair-cap moss [(*Polytrichum commune* L.] and some grasses).

A recent vegetation survey in Nova Scotia revealed that the mean coverage of blueberry plants was ~75% (range ~40% to 98% of 128 fields surveyed; Jensen and Sampson, unpubl. data). Weeds often occur in areas not occupied by the crop or near the clone margins. Practices that encourage a vigorous competitive crop that spreads and occupies available space will directly favor weed control. The long-term reduction of the organic layer (Trevett, 1956) and erosion and degradation on herbicide-treated vegetation-free areas (Eaton and Jensen, 1997) is detrimental to blueberry rhizome growth and expansion of the clone. Organic mulches have been used to control weeds and they promote rhizome expansion and establishment of blueberry plants in bare or eroded soils (Kender and Eggert, 1966; Sanderson and Cutcliffe, 1991; Hicklenton et al., 2000). Mulches in combination with interplanting either rooted cuttings or blueberry seedlings have also been used to increase crop cover (Hicklenton et al., 2000; DeGomez and Smagula, 1990). Other soil management practices that favor blueberry growth while suppressing some competing species include optimizing fertilizer use and lowering soil pH to < 5 with gypsum (Tevett, 1972).

Growers must also recognize that not all species are detrimental to blueberry production. Some may be beneficial in stabilizing and improving soil conditions that favor blueberry expansion, encouraging pollinators or other benefical insects, trapping and holding a snow cover, recycling nutrients and others. The lowbush blueberry has co-evolved with other early successional species and monoculture (through herbicide use) may not be sustainable. Currently, biotypes of certain hexazinone-tolerant grasses, namely poverty oat grass (Burgess, 2003) and rough hair grass (K.I.N. Jensen, unpubl. data) are being studied as promising ground covers or companion species that will reduce soil degradation and encourage blueberry expansion in herbicide-treated fields (K.I.N. Jensen, unpubl. data). Clearly the aim of vegetation management in the future will have to be different than in the past. To be successful a number of techniques must be used in addition to herbicides and these must be integrated within a crop management system.

## CONCLUSIONS

For over 50 years, selective and non selective herbicides have been introduced into lowbush blueberry culture and have largely replaced the traditional methods of weed control, burning, and cutting. Improved weed control has been either directly or indirectly involved in other recent changes in production practices, e.g., fertility management and mechanical harvesting, and has contributed to the large increases in yields since the 1980s. The traditional weed flora has changed in response to changes in weed control and other production practices. Long-term weed control with repeated hexazinone use has helped to select a new weed flora and has raised concerns about environmental residues and soil degradation. Although some new, narrow-spectrum herbicides are potentially available to control some individual weed species, herbicides must be integrated with other weed control practices in order to maintain a manageable level of vegetation within blueberry fields. A managed complex of associated plant species within blueberry fields may have long-term advantages.

## GROWER BENEFITS

Lowbush blueberry production without weed control is unprofitable. Weed control has contributed to large increases in yields and has facilitated other changes in production practices. Growers need to understand the limitations of available herbicides and to identify and understand the biology of their weeds in order to be able to use the herbicides judiciously and to prioritize their weed management efforts. American producers would benefit from approval and use of some herbicides registered in Canada, e.g., tribenuron (Express), nicosulfuron/rimsulfuron (Ultim), and triclopyr (Garlon). Likewise, Canadian producers could benefit from increased use of grass-specific herbicides and from the registration of diuron and its tankmix with terbacil.

## NOTE

1. Nomenclature follows Darbyshire et al. (2000), if given. Otherwise, common names of some native species follow Zinck (1998).

## LITERATURE CITED

Abdalla, D.A. 1967. Weed control in lowbush blueberry. Blueberry Info. Guide No. 5. Coop. Ext. Serv., Univ. Maine, Orono.

Atlantic Committee on Fruit Crops. 1999. Guide to weed control for lowbush blueberry production in Atlantic Canada. Factsheet ACC 1014, Agdex No. 235/641.

Barker, W.G., I.V. Hall, and G.W. Wood. 1964. The lowbush blueberry industry in eastern Canada. Econ. Bot. 18:357-365.

Barron, J.J. and T.J. Monaco. 1986. Uptake, translocation and metabolism of hexazinone in blueberry (*Vaccinium* spp.) and hollow goldenrod (*Solidago fistula*). Weed Sci. 34: 824-829.

Belzile, A. 1951. The problem of weeds in blueberry barrens, pp. 120-124. Proc. 5th Mtg. Eastern Section Natl. Weed Committes. Ottawa. pp. 120-124.

Blatt, C.R., I.V. Hall, K.I N. Jensen, W.T.A. Neilsen, P.D. Hildebrand, N.L. Nickerson, R.K. Prange, P.D. Lidster, L.Crozier, and J.D. Sibley. 1989. Lowbush blueberry production. Publ.1477/E. Agr. Canada, Ottawa.

Bouchard, A.R. 1986. La végétation, les sols et la productivité fruitière de *Vaccinium angustifolium* et *V. myrtilloïdes* dans les bleuetières du Saguenay-Lac-Saint-Jean. Naturaliste-can. 113:125-133.

Burgess, P.M. 2003. Vegetation management in lowbush blueberry (*Vaccinium angustifolium* Ait.): the influences of sub-lethal doses of herbicides on *Danthonia spicata* (L.) Beauv. when used as a living mulch. Dalhousie Univ., Halifax, N.S. MSc Thesis.

Chandler, F.B. and I.C. Mason. 1946. Blueberry weeds in Maine and their control. Bul. 443. Maine Agr. Expt. Sta., Orono, ME.

Darbyshire, S.J., F. Fareau, and M. Murray. 2000. Common and scientific names of weeds in Canada. Publ. 1398/B. Agr.and Agri-Food Canada, Ottawa.

DeGomez, T. and J. Smagula. 1990. Filling bare spots in blueberry fields. Fact sheet 221. Univ. Maine Coop. Ext., Orono, ME.

Eaton, L.J. 1994. Long-term effects of herbicide and fertilizers on lowbush blueberry growth and production. Can. J. Plant Sci. 74:341-345.

Eaton, L.J. and K.I.N. Jensen. 1997. Protection of lowbush blueberry soils from erosion. Lowbush blueberry fact sheet, Nova Scotia Dept. Agr. Mktg. Truro, N.S.

Everett, C.F., J.C. Dunphy, and F.A. Stewart. 1968. Timing of application of herbicides for control of lambkill in lowbush blueberry, p. 194. Natl. Weed Committee. Res. Rpt. (Eastern Canada).

Flinn, M.A. and R.W. Wein. 1977. Depth of underground plant organs and theoretical survival during fire. Can. J. Bot. 55:2550-2554.

Hall. I.V. 1954. Ecological studies. Pp. 18-23. In: Progress Report 1949-1953. Can. Dept. Agr., Dominion Blueberry Substation., Tower Hill, N.B.

Hall, I.V. 1955. Floristic changes following the cutting and burning of a woodlot for blueberry production. Can. J. Agr. Sci. 35:142-152.

Hall, I.V. 1958. Some effects of light on native lowbush blueberries. J. Am. Soc. Hort. Sci. 72:216-218.

Hall, I.V. 1957. The taproot of the lowbush blueberry. Can. J. Bot. 35:933-935.

Hall, I.V. 1959. Plant populations in blueberry stands developed from abandoned hayfields and woodlots. Ecology 40:742-743.

Hall. I.V. 1963. Note on the effect of a single intensive cultivation on the composition of an old blueberry stand. Can. J. Plant Sci. 43:417-419.

Hall, I.V. 1975. The biology of Canadian weeds. 7. *Myrica pensylvanica* Loisel. Can. J. Plant Sci. 55:163-169.

Hall, I.V. and L.E. Aalders. 1968. The botanical composition of two barrens in Nova Scotia. Naturaliste-can. 95:393-396.

Hall, I.V., L.E. Aalders, and C.F. Everett. 1976. The biology of Canadian weeds. 16. *Comptonia peregrina* (L.) Coult. Can. J. Plant Sci. 56:147-156.

Hall, I.V., L.E. Aalders, N.L. Nickerson, and S. P. Vander Kloet. 1979. The biological flora of Canada. 1. *Vaccinium angustifolium* Ait., Sweet lowbush blueberry. Can. Field-Naturalist 93:415-427.

Hall, I.V., F.R. Forsythe, L.E. Aalders, and L.P. Jackson. 1971. Physiology of the lowbush blueberry. Econ. Bot. 26:68-73.

Hall. I.V., L.P. Jackson, and C.F. Everett. 1973. The biology of Canadian weeds. 1. *Kalmia angustifolia* L. Can. J. Plant Sci. 53s:865-873.

Hall, I.V. and R.A. Ludwig. 1961. The effects of photoperiod, temperature, and light intensity on the growth of the lowbush blueberry (*Vaccinium angustifolium* Ait.). Can. J. Bot. 39: 1733-1739.

Hall, I.V., R.A. Murray, and L.P. Jackson. 1974. The biology of Canadian weeds. *Spiraea latifolia* L. Can. J. Plant Sci. 54:141-147.

Hall, I.V. and J.D. Sibley. 1976. The biology of Canadian weeds. 20. *Cornus canadensis* L. Can. J. Plant Sci. 56:885-892.

Hall, I.V., G.W. Wood, and L.P. Jackson. 1978. The biology of Canadian weeds. 30. *Pyrus melanocarpa* (Michx.) Willd. Can. J. Plant Sci. 57:499-504.

Hepler, P.R. and D.E. Yarborough. 1991. Natural variability in yield of lowbush blueberry. HortScience 26:245-246.

Hicklenton, P.R., J.Y. Reekie, and R.J. Gordon. 2000. Physiological and morphological traits of lowbush blueberry (*Vaccinium angustifolium* Ait.) plants in relation to post-transplant conditions and water availability. Can. J. Plant Sci. 80:861-867.

Hoefs, M.E.G. and J.M. Shay. 1981. The effects of shade on shoot growth of *Vaccinium angustifolium* Ait. after fire pruning in southeastern Manitoba. Can. J. Bot. 59: 166-167.

Hoepler, A.L. and D.E. Yarborough. 1985. Hexazinone and terbacil to suppress weeds in a commercial lowbush blueberry field. Proc. Northeastern Weed Sci. Soc. 39:151-156.

Howatt, S.M. 1991. Control of hexazinone tolerant weeds in lowbush blueberries. McGill Univ., Montreal, PQ. M.Sc. Thesis.

Ismail, A.A. 1974a. Selective thinning of black barrenberry fruit in lowbush blueberry fields with ethephon. HortScience 9:346-347.

Ismail, A.A. 1974b. Terbacil and fertility effects on yield of lowbush blueberry. HortScience 9:457.

Ismail, A.A., J.M. Smagula, and D.E. Yarborough. 1981. Influence of pruning method, fertilizer and terbacil on the growth and yield of the lowbush blueberry. Can. J. Plant Sci. 61:61-71.

Ismail, A.A. and D.E. Yarborough. 1981. Lambkill control in lowbush blueberry fields with glyphosate and 2,4-D. J. Amer. Soc. Hort. Sci. 106:393-396.

Jackson, L.P. 1981. Asulam for control of eastern bracken fern in lowbush blueberry fields. Can. J. Plant Sci. 61:475-477.

Jensen, K.I.N. 1981. Hexazinone, a promising herbicide for highbush blueberry. HortScience 16: 315-316.

Jensen, K.I.N. 1985. Weed control in lowbush blueberries in eastern Canada. Acta Hort. 165:259-265.

Jensen, K.I.N. 1986. Response of lowbush blueberry to weed control with atrazine and hexazinone. HortScience 2:1143-1144.

Jensen, K.I.N., S.M. Campbell, and E.G. Specht. 1993. Results of 1992-1993 field trials with hexazinone in lowbush blueberry. Expert Committee Weeds (Eastern Canada Section) Res. Rpt. 2: 8-12.

Jensen, K.I.N., D.J. Doohan, and J.Thompson. 1981. Weed control in lowbush blueberries with hexazinone. Proc. Northeastern Weed Sci. Soc. 35:147.

Jensen, K.I.N. and E.R. Kimball. 1985. Tolerance and residues of hexazinone in lowbush blueberries. Can. J. Plant Sci. 65:223-227.

Jensen, K.I.N. and E.R. Kimball. 1987. Persistence and degradation of the herbicide hexazinone in soils of lowbush blueberry fields in Nova Scotia, Canada. Bul. Environ. Contamination Toxicology 38:232-239.

Jensen, K.I.N. and E.R. Kimball. 1990. Uptake and metabolism of hexazinone in *Rubus hispidus* L. and *Pyrus melanocarpa* (Michx) Willd. Weed Res. 30:35-41.

Jensen, K.I.N., E.R. Kimball, and I.V. Hall. 1985. Tolerance of select clone lowbush blueberries to terbacil and its residues in fruit. Pp. 12-14. Agr. Agri-Food Canada, Kentville Res. Sta. Annu. Rpt., Kentville, N.S.

Jensen, K.I.N. and L H. North. 1987. Control of speckled alder in lowbush blueberry with selective fall herbicide treatments. Can. J. Plant Sci. 67:369-372.

Jensen, K.I.N. and E.G. Specht. 2002. Response of lowbush blueberry (*Vaccinium angustifolium*) to hexazinone applied early in the fruiting year. Can. J. Plant Sci. 82:781-783.

Jensen, K.I.N. and E.G. Specht. 2004. Use of two sulfonylurea herbicides in lowbush blueberry. Small Fruits Review 3(3/4):257-272.

Keizer, J.P., P.H. MacQuarrie, P.H. Milburn, K.V. McCully, R.R. King, and E.J. Embleton. 2001. Long-term ground water quality impacts from the use of hexazinone for the commercial production of lowbush blueberries. Ground Water Monitoring Remediation 21:128-135.

Kender W.J. and F.P. Eggert. 1966. Several soil management practices influencing the growth and rhizome development of the lowbush blueberry. Can. J. Plant Sci. 46:141-149.

Kinsman, G. 1986. The history of the lowbush blueberry industry in Nova Scotia 1880-1950. Nova Scotia Dept. Agr. Mktg. Truro, N.S.

Kinsman, G. 1993. The history of the lowbush blueberry industry of Nova Scotia 1950-1990. Blueberry Producers' Assoc. N.S., Debert, N.S.

Kubilius, D.T. and R.J. Bushway. 1998. Determination of hexazinone and its metabolites in groundwater by capillary electrophoresis. J. Chromatography A 793:349-355.

Lapointe, L. and L. Rochefort. 2001. Weed survey of lowbush blueberry fields in Saguenay-Lac-Saint-Jean, Québec following eight years of herbicide application. Can. J. Plant Sci. 81:471-478.

Lavoie, V. 1968. La phytosociologie et l'aménagement des bleuetières. Naturaliste-can. 95:397-412.

Lynham, T.J., G.M. Wickware, and J.A. Mason. 1998.Soil chemical changes and plant succession following experimental burning in immature jack pine. Can. J. Soil Sci. 78:93-104.

Maddison, L. 1997. A framework for hexazinone-related risk assessment, Cumberland County, Nova Scotia, Technical Univ. N.S., Halifax. B.Sc. Thesis.

Martin, J.L. 1954. The vegetational characteristics of certain forest burn communities on the southern upland of Nova Scotia with observations on secondary growth. Dalhousie Univ., Halifax, N.S. MSc Thesis.

McCully, K.V., K. Jensen, G. Sampson, and D. Doohan. 1995. Bunchberry control in wild blueberries with Spartan 75 DF. New Brunswick/Canada Coop. Agreement on Econ. Diversification, Wild Blueberry Factsheet C4.3.0.

McCully, K.V. and M.G. Sampson. 1991.Weed survey of Nova Scotia lowbush blueberry (*Vaccinium angustifolium* Ait.) fields. Weed Sci. 39:180-185.

Morrison, K.D., E.G. Reekie, and K.I.N. Jensen. 1998. Biocontrol of common St. Johnswort (*Hypericum perforatum*) with *Chrysolina hyperici* and a host-specific *Colletotrichum gloeosporioides*. Weed Technol. 12:426-435.

Nams, V.O. 1994. Increasing sampling efficiency of lowbush blueberry. Can. J. Plant Sci. 74: 573-576.

Penney, B.G. and K.B. McRae. 2000. Herbicidal weed control and crop-year NPK fertilization improves lowbush blueberry (*Vaccinium angustifolium* Ait.) production. Can. J. Plant Sci. 80:351-361.

Rowe, J.S. 1983. Concepts of fire effects on plant individuals and species. Pp. 135-154. In R.W. Wein and D.A. MacLean (eds.). The role of fire in northern circumpolar ecosystems. John Wiley and Sons Ltd., Toronto.

Sampson, M.G., K.V. McCully, and D.L. Sampson. 1990. Weeds of eastern Canadian blueberry fields. Nova Scotia Agr. College Bookstore, Truro, N.S.

Sanderson, K.R. and J.A. Cutcliffe. 1991. Effect of sawdust mulch on yield of select clones of lowbush blueberry. Can. J. Plant Sci. 71:1263-1266.

Smagula, J.M. and A.A. Ismail. 1981. Effects of fertilizer application, preceded by terbacil, on growth, leaf nutrient concentration, and yield of lowbush blueberry. Can. J. Plant Sci. 61:961-964.

Smagula, J.M. and D.E. Yarborough. 1990. Changes in the lowbush blueberry industry. Fruit Varieties J. 44:72-77.

Swan, F.R. 1970. Post-fire response of four plant communities in south-central New York state. Ecology 51:1074-1082.

Trevett, M.F. 1950. Weed control. Pp. 32-42. In: Producing blueberries in Maine. Bul. 479. Maine Agr. Exp. Sta., Univ. Maine, Orono, Maine.

Trevett, M.F. 1952. Control of woody weeds in lowbush blueberry fields. Bul. 499. Maine Agr. Exp. Sta., Univ. Maine, Orono, Maine.

Trevett, M.F. 1956. Observations on the decline and rehabilitation of lowbush blueberry fields. Bul. 626. Maine Agr. Exp. Sta., Univ. Maine, Orono, Maine.

Trevett, M.F. 1972. The integrated management of lowbush blueberry fields–a review and forecast. Bul. 699. Maine Life Sci. and Agr. Exp. Sta., Univ. Maine, Orono, Maine.

Trevett, M.F. and R.E. Durgin. 1972. Terbacil: a promising herbicide for the control of perennial grass and sedge in unplowed lowbush blueberry fields. Univ. Maine Agr. Exp. Sta., Res. Life Sci. 19(15):1-13.

Trevett, M.F. and R.E. Durgin. 1973. Leaf dessicants and the selective eradication of barren berry (*Pyrus floribunda* and *Pyrus melanocarpa*) in lowbush blueberry fields. Univ. Maine Agr. Exp. Sta., Res. Life Sci. 20(9):1-6.

Vander Kloet, S.P. 1978. Systematics, distribution and nomenclature of the polymorphic *Vaccinium angustifolium* Ait. Rhodora 80:538-376.

Vander Kloet, S.P. 1994. The burning tolerance of *Vaccinium myrtilloides* Michaux. Can. J. Plant Sci. 74:577-579.

Vander Kloet, S.P. and I.V. Hall. 1981. The biological flora of Canada 2. *Vaccinium myrtilloides* Michx., Velvet-leaf blueberry. Can. Field-Naturalist 95:329-345.

Wesley, S.L., N.M. Hill, and S.P. Vander Kloet. 1986. Seed banks of *Vaccinium angustifolium* Aiton on managed and unmanaged barrens in Nova Scotia. Naturaliste-can. 113:309-316.

Whittle, C.A., L.C. Duchesne, and T. Needham. 1997. The impact of broadcast burning and fire severity on species composition and abundance of surface vegetation in a jack pine (*Pinus banksiana*) clearcut. Forest Ecol. Mgt. 94:141-148.

Yarborough, D.E. 1988. Effect of sethoxydim on little bluestem in lowbush blueberry. Proc. Northeastern Weed Sci. Soc. 42:214-215.

Yarborough, D.E. 1991. Effect of hexazinone on species distributions and weed competition in lowbush blueberry fields in Maine. Univ. Massachusetts, Amherst, MA. PhD Thesis.

Yarborough, D.E. 1995. Velpar for weed control in lowbush blueberries. Fact Sheet 238. Univ. Maine Coop. Ext., Orono, Maine.

Yarborough, D.E. 1996a. Wild blueberry culture in Maine. Chronica Hort. 36: 8-10.

Yarborough, D.E. 1996b. Roundup for weed control in lowbush blueberry. Fact Sheet 237. Univ. Maine Coop. Ext., Orono, Maine.

Yarborough, D.E. 1997a. Best management practices to reduce hexazinone in groundwater in wild blueberry fields. 1997 Brighton Crop Protection Conf., Weeds 3:1091-1098.

Yarborough, D.E. 1997b. Production trends in the wild blueberry industry of North America. Acta Hort. 446:33-36.

Yarborough, D.E. 1999. Postemergence grass control for wild blueberry. Fact Sheet 221. Univ. Maine Coop. Ext., Orono, Maine.

Yarborough, D.E. 2002a. Weed control guide for wild blueberries. Fact Sheet 239. Univ. Maine Coop. Ext. Ororno, Maine.

Yarborough, D.E. 2002b. Hexazinone best management system for wild blueberry fields. Fact Sheet 250. Univ. Maine Coop. Ext., Orono, Maine.

Yarborough, D.E. 2004. Factors contributing to the increase in productivity in the wild blueberry industry. Small Fruit Rev. 3(1/2):33-43.

Yarborough, D.E. and P.C. Bhowmik. 1989a. Effect of hexazinone on weed populations and on lowbush blueberries in Maine. Acta Hort. 241:344-349.

Yarborough, D.E. and P.C. Bhowmik. 1989b. Evaluation of sulfonyl urea and imidazoline compounds for bunchberry control in lowbush blueberry fields. Proc. Northwestern Weed Sci. Soc. 43 :142-145.

Yarborough, D.E. and P.C. Bhowmik. 1993. Lowbush blueberry-bunchberry competition. J. Amer. Soc. Hort. Sci. 118:54-62.

Yarborough, D.E., J.J. Hanchar, S.P. Skinner, and A.A. Ismail. 1986. Weed response, yield, and economics of hexazinone and nitrogen use in lowbush blueberry production. Weed Sci. 34: 723-729.

Yarborough, D.E. and T.M. Hess. 1995. Control of bunchberry in wild blueberry fields. J. Small Fruit Viticult. 3:125-132.

Yarborough, D.E. and T.M. Hess. 1999a. Effect of azafenidin and rimsulfuron on weeds in wild blueberries. Proc. Northeastern Weed Sci. Soc. 53:2-3.

Yarborough, D.E. and T.M. Hess. 1999b. Grass control alternatives for wild blueberries. Proc. Northeastern Weed Sci. Soc. 53:90-91.

Yarborough, D.E. and T.M. Hess. 2000a. Effect of rate, formulation and application method on efficiency and phytotoxicity of granular hexazinone in wild blueberry fields. Proc. Northeastern Weed Sci. Soc. 54:4-5.

Yarborough, D.E. and T.M. Hess. 2000b. Comparison of sulfosate and glyphosate for weed control in wild blueberries. Proc. Northeastern Weed Sci. Soc. 54:6-7.

Yarborough, D.E. and T.M. Hess. 2001. Environmental factors and timing affect efficacy of azafenidin, rimsulfuron and pendimethalin on weeds in wild blueberries. Proc. Northeastern Weed Sci. Soc. 55: 78-79.

Yarborough, D.E. and Y.M. Hess. 2002. Comparison of the sprout-less weeder with cutting and wiping for hardwood control in wild blueberry fields. Proc. Northeastern Weed Sci. Soc. 56:92-93.

Yarborough, D.E. and A.A. Ismail. 1979a. Barrenberry control in lowbush blueberry fields through selective application of 2,4-D and glyphosate. J. Amer. Soc. Hort. Sci. 104: 786-789.

Yarborough, D.E. and A.A. Ismail. 1979b. Effect of endothall and glyphosate on a native barrenberry and lowbush blueberry stand. Can. J. Plant Sci. 59:737-740.

Yarborough, D.E. and A.A. Ismail. 1980. Effect of endothall and glyphosate on blueberry and barrenberry yield. Can. J. Plant Sci. 60:891-894.

Yarborough, D.E. and A.A. Ismail. 1981. Hexazinone pellets for spot treatments of woody weeds in lowbush blueberry fields. Proc. Northeastern Weed Sci. Soc. 35:148-151.

Yarborough, D.E. and A.A. Ismail. 1985. Hexazinone on weeds and on lowbush blueberry growth and yield. HortScience 20:406-407.

Yarborough, D.E. and J.M. Jemison. 1997. Developing best management practices to reduce hexazinone in ground water in wild blueberry fields. Acta Hort. 446:303-307.

Yarborough, D.E. and K.I.N. Jensen. 1993. Hexazinone movement in blueberry soils in North America. Acta Hort. 346:278-283.

Yarborough, D.E. and M.C. Marra. 1997. Economic thresholds for weeds in wild blueberry fields. Acta Hort. 446:293-301.

Zinck, M. 1998. Roland's flora of Nova Scotia. Nimbus Publishing and Nova Scotia Museum, Halifax, N.S.

# Use of Two Sulfonyl Urea Herbicides in Lowbush Blueberry

Klaus I. N. Jensen

Eric G. Specht

**SUMMARY.** The continuous use of the broad spectrum herbicide hexazinone since 1981 has resulted in many changes to the weedy flora of lowbush blueberry fields, including shifts to hexazinone-tolerant species. Many of these occur in patches and could best be controlled by selective, foliar herbicide treatments. Preliminary assessments of several sufonyl urea (SU) herbicides indicated that tribenuron (Spartan or Express 75% DF) was effective against some important weed species with little risk to the crop. Long-term control (> 95%) of eastern bracken [*Pteridium aquilinum* (L.) Kuhn.], yellow loosestrife [*Lysimachia terrestris* (L.) BSP], speckled alder [*Alnus incana* (L.) Moench], common wild rose (*Rosa virginiana* Mill.) and several others was obtained with mid summer, prune-year foliar applications of tribenuron at 0.2 g L$^{-1}$ water with 0.2% Agral 90. Screening trials targeting black bulrush (*Scirpus atrovirens* Willd.), currently the most serious weed problem in Nova Scotia blueberry fields, indicated that SU's with activity against grasses were also effective against this weed, and a 1:1 commercial mixture of

Klaus I. N. Jensen is Weed Scientist and Eric G. Specht is Research Technician, Agriculture and Agri-Food Canada, Atlantic Food and Horticulture Research Centre, Kentville, NS, B4N 1J5, Canada.

This study was funded in part by Health Canada's Minor Use of Pesticides Program Funding Initiative and the Wild Blueberry Producers' Association of Nova Scotia.

For the Department of Agriculture and Agri-Food, Government of Canada. © Minister of Public Works and Government Services Canada (2002).

[Haworth co-indexing entry note]: "Use of Two Sulfonyl Urea Herbicides in Lowbush Blueberry." Jensen, Klaus I. N., and Eric G. Specht. Co-published simultaneously in *Small Fruits Review* (Food Products Press, an imprint of The Haworth Press, Inc.) Vol. 3, No. 3/4, 2004, pp. 257-272; and: *Proceedings of the Ninth North American Blueberry Research and Extension Workers Conference* (ed: Charles F. Forney, and Leonard J. Eaton) Food Products Press, an imprint of The Haworth Press, Inc., 2004, pp. 257-272. Single or multiple copies of this article are available for a fee from The Haworth Document Delivery Service [1-800-HAWORTH, 9:00 a.m. - 5:00 p.m. (EST). E-mail address: docdelivery@haworthpress.com].

http://www.haworthpress.com/web/SFR
Digital Object Identifer: 10.1300/J301v03n03_03

nicosulfuron/rimsulfuron (Ultim DF) was selected for further work. Summer and fall prune-year applications of nicosulfuron/rimsulfuron at 0.031 g $L^{-1}$ water with 0.2% Agral 90 generally provided > 90% control, except when plants were stressed by drought. Although blueberry plants were injured (< 20% injury) when sprayed directly with these SU herbicides, there was little effect on plants in the understory following applications to weeds. Both herbicides have become registered for use in lowbush blueberry in Canada for control of the above species. *[Article copies available for a fee from The Haworth Document Delivery Service: 1-800-HAWORTH. E-mail address: <docdelivery@haworthpress.com> Website: <http://www.HaworthPress.com>]*

**KEYWORDS.** Weed control, tribenuron, nicosulfuron/rimsulfuron, *Vaccinium angustifolium*

## INTRODUCTION

Weed control is important in the development and long-term management of commercial stands of the native lowbush blueberry (*Vaccinium angustifolium* Ait.). Unlike most fruit crops, the principle weeds include a wide range of woody and herbaceous native species (Hall, 1959; Jensen, 1985). These weeds are mostly perennials that also survive and thrive under the two-year prune management system. Prior to the 1980s, weed control depended on cutting and roguing and on the limited use of several nonselective herbicides, e.g., 2,4-D and dicamba, that were applied as directed spot sprays or by roller/wiper applicators (Trevett, 1950). The introduction in the early 1980s of hexazinone, a selective, broad-spectrum and residual herbicide that was applied broadcast in newly-pruned blueberry fields, provided control of most of these native species (Jensen, 1985; Yarborough and Bhowmik, 1989). Improved weed control resulted in rapidly increasing yields. The repeated use of this single herbicide has since resulted in changes in the weed flora to species that, like the lowbush blueberry, are also tolerant to the herbicide, e.g., bunchberry and black chokeberry (see Table 1 for all common and Latin names). Some species have developed tolerant populations after repeated exposure (e.g., some native *Agrostis* and *Festuca* spp.), whereas others are injured but eventually recover from labeled rates (e.g., wild rose and black bulrush) or re-establish from seed after soil residues have dissipated (e.g., lamb's quarters and some goldenrods) (authors' unpublished data). These species often occur in patches in hexazinone-treated blueberry fields and could best be controlled by spot or directed foliar applications of selective herbicides. Currently in Canada there are only nonselective herbicides approved for use in lowbush blueberry that can be used as foliar sprays, e.g., 2,4-D/dicamba mixtures, triclopyr, and glyphosate. These cause severe injury when the crop is contacted by spray or spray drift (Atlantic Committee on Fruit Crops, 1999). Hence, they are used primarily in early stages of field development. There is a need to find selective herbicides that can be used within producing fields to control escaping weed species. This article will describe the selective control of some important hexazonone-tolerant weed species with spot applications of two sulfonylurea herbicides.

Sulfonylurea (SU) herbicides are used in many crops and, depending on the herbicide, have activity on monocotyledonous or dicotyledonous plants, or both. They are mostly foliarly applied and translocate in the phloem to the meristems where they inhibit growth of sensitive species by blocking acetolactate synthase, a key enzyme in amino acid synthesis (Brown, 1990). Preliminary screening trials (authors' unpublished data; Howatt, 1992) indicated some SU herbicides were too phytotoxic to be applied broadcast in lowbush blueberry (chlorsulfuron, metsulfuron) but tribenuron only caused minor injury and appeared promising. Bunchberry, one of the most common hexazinone-tolerant weeds, was effectively controlled by a broadcast, postemergence application of 30 g ha$^{-1}$ tribenuron applied after the spring pruning operation when bunchberry flowers were opening and new blueberry shoot growth did not exceed 2.5 cm (McCully et al., 1995). This use was approved under Health Canada's User Requested Minor Use Label Expansion (URMULE) program in 1994. Other studies were underway at that time to find other species that could be controlled with directed tribenuron applications that were safer on the crop. This paper summarizes some of the results of these studies.

By the late 1990s, black bulrush had become one of the most serious weed problems in lowbush blueberry in eastern Canada. This native, 1-m tall, tussock-forming, perennial sedge survives hexazinone applications, rapidly colonizes fields, competes aggressively with the crop, and impedes mechanical harvesting. It has not previously been described as a weed although it is a common wetland plant in most of eastern North America. However, several annual weedy *Scirpus* species were controlled in rice by sulfonylurea herbicides that also control annual grasses (Tanaka et al., 1998). Preliminary greenhouse and field screening of approved sulfonylurea herbicides with annual grass activity by the authors identified rimsulfuron and a commercial 1:1 nicosulfuron/rimsulfuron mixture as potential candidates for selective black bulrush control. This paper also summarizes results of trials that supported the 2002 URMULE registration of the nicosulfuron/rimsulfuron mixture for black bulrush control in Canada.

## MATERIALS AND METHODS

*Herbicide treatments and application.* Herbicide trials assessing directed foliar sprays of tribenuron and nicosulfuron/rimsulfuron were conducted in commercial lowbush blueberry fields in Nova Scotia, Canada from 1991 to 2001. All fields were in the vegetative or prune year of the 2-year production cycle. When applied as a directed spot spray, the concentration of tribenuron was 0.2 g $L^{-1}$ water and that of nicosulfuron/rimsulfuron was 0.031 g $L^{-1}$ water. All herbicide solutions contained 0.2% (v/v) of the wetting agent Agral 90. The rate of tribenuron had been determined from earlier trials that assessed rates of 0.05 to 0.4 g $L^{-1}$ water (authors' unpublished data) and that of nicosulfuron/rimsulfuron was equivalent to the labeled broadcast rate of 25 g $ha^{-1}$ applied in 800 L $ha^{-1}$ (Anonymous, 2002). The latter was based on the assumptions that (i) spot applications deliver about 4 times the volume of spray to plants as broadcast ones applied in 200 L $ha^{-1}$ water, and that (ii) generally volumes of spot sprays would not exceed 800 L $ha^{-1}$. Hence, for regulatory purposes, the amount of herbicide applied per hectare as a spot spray would not exceed the maximum labeled broadcast rate. Earlier greenhouse and field screening trials had also indicated this rate of nicosulfuron/rimsulfuron was effective in controlling some target weeds.

Both herbicides are products of DuPont Canada. Tribenuron is the active ingredient of Spartan 75 DF. An equivalent product in the USA is Express 75% DF. The rate of 0.2 g $L^{-1}$ tibenuron is equivalent to 0.36 oz Spartan 75 DF in 10 US gal. The 1:1 nicosulfuron/rimsulfuron formulation used was the commercial premix Ultim 37.4% DF but a new formulation contains 75% active ingredient. This premix is only available in Canada.

Herbicide solutions were applied using a hand-held, $CO_2$-pressurized plot sprayer operated at 175 kPa. The sprayer used a 2-m wide boom fitted with four 8003 XR TeeJet nozzles for broadcast applications and a handgun with an adjustable 5500 ConeJet nozzle for directed foliar sprays. The latter were applied to thoroughly cover and wet the foliage to the point of run-off. Preweighed herbicide samples were mixed with water and surfactant in the field immediately before application.

*Preliminary assessment of tribenuron directed foliar sprays.* Observational trials were conducted from 1991 to 1994 to determine which weed species could be controlled with the above spot application of tribenuron. The experimental unit (plot) differed with target species. For larger woody species, e.g., some trees, individuals were labeled and treated; for shrubs (e.g., wild rose and sweet-fern) patches with at least 4 or 5 individuals were labeled and treated; for herbaceous species with high stem densities (e.g., goldenrods and yellow loosestrife), all individuals within a 0.5- to 1.5-m$^2$ area were treated. The treatment was applied to at least 3 (but usually 6) plots of each species at any site and there were at least 2 sites for each species tested. Blueberry was also included. In tribenuron trials, herbicide injury to either the crop or the weed was assessed using a linear 0 to 9 scale where 0 is no effect and 9 is complete death of the tissue or topgrowth. If only minor discoloration was observed on some, but not all, treated plants the treatments were scored < 0.5 (trace). In this series of trials, herbicide injury was scored on both the shoot meristems and on the remaining shoots. Because the herbicide is translocated to meristems, symtoms are often most obvious there and include reduction and cessation of growth and die-back; foliar symptoms are generally discoloration, often chlorosis or reddening of the foliage changing to eventual necrosis and abscission when injury is severe. Adjacent untreated plants served as controls when scoring treatment effects. Treatments were applied during July and August and assessments were made 4 to 6 weeks later. Plots were generally revisited the following year to observe recovery. Assessments were pooled for each species over years and locations and data is presented as means ± standard deviation.

*Effect of timing of tribenuron application on crop tolerance.* An experiment was established in a weed-free stand of prune-year lowbush blueberry to determine if timing of application affected crop response to tribenuron. Plots measuring 2 m × 4 m were treated with a single broadcast application of 30 g ha$^{-1}$ tribenuron in 200 L ha$^{-1}$ with 0.2% Agral 90 at approximately 2 week intervals from 17 June to 29 September 1991. Treated plots were paired with adjacent untreated plots in a randomized complete block design with treatments replicated 3 times at two locations. In addition to visual assessments 3 to 4 weeks after each application and again June 1992, the plots at one location were harvested at maturity.

*Assessment of tribenuron spot applications on four sensitive species.* Additional trials were established in 1995 and 1996 on eastern bracken, yellow loosestrife, speckled alder, and common wild rose to corroborate the promising results in the preliminary trials for the purpose of supporting an URMULE registration. The experimental units remained the same for speckled alder and wild rose but treatments were replicated at least 6 times per site and the number of sites ranged from 2 to 7 per year. Treatments were applied in mid summer (7 July to 18 August). Because these two species retain their foliage into the fall, early fall applications from 29 September to 14 October were included in 1996 to determine if later applications were effective. The experimental unit for eastern bracken was an area that had at least 25 fronds. These and the fronds outside the area were treated. For yellow loosestrife, an area of about 2 m² was treated but only the center m² was included in assessments. In addition to visual injury assessments, shoots of these two species were also recorded before application and about one year later. Where possible, assessments were made of injury to blueberry plants in the understory of weed control plots.

*Effect of rate and timing of rimsulfuron and nicosulfuron/rimsulfuron directed spot applications on black bulrush control.* Earlier trials had indicated rimsulfuron alone may also control black bulrush and that control with both herbicides products could be improved by increasing the rate. Hence, a factorial experiment was established with three rates of herbicide (1×, 2×, and 3×, where 1× is 0.018 g L⁻¹ rimsulfuron and 0.031 g L⁻¹ nicosulfuron/rimsuluron) and three times of application (19 June, 4 August, and 2 October 1998). Rimsulfuron is available as the following DuPont Inc. products: Elim or Prism 25% (Canada) and Matrix 25% (USA). The 1× rate is the spot application dilution equivalent to 15 g ha⁻¹ rimsulfuron or 60 g ha⁻¹ product (or 0.85 oz acre⁻¹). Treatments were replicated 6 times and the experiment was repeated in two adjacent prune-year blueberry fields. In black bulrush trials, the experimental unit was a separate patch of 10 to 20 mature, flowering plants. Bulrush control was rated on a 0 to 9 scale where 0 is no control and 9 is complete control. Mortality (%) was determined by examining each plant in each plot in May 1999 to determine if all tillers in the tussock were dead or if some had survived. Where possible, injury on blueberry plants growing on the plots was also scored. Only ratings taken in late May of the year after application are included here. The data from both sites were combined and analyzed as a factorial with ANOVA.

*Additional trials with nicosulfuron/rimsulfuron on black bulrush.* Trials were established in three locations in 1999 to confirm efficacy of the nicosulfuron/rimsulfuron treatment using the same methods as above to meet URMULE requirements. Treatments were applied at two times (early vs. late). Trials were repeated in nearby fields in 2000 because of erratic control the previous year. The same methods as above were employed.

## RESULTS AND DISCUSSION

Directed foliar applications of 0.2 g $L^{-1}$ tribenuron were promising in preliminary trials in controlling a number of important weed species of lowbush blueberry fields (Table 1). Assessments taken 4 to 6 weeks after mid-summer applications showed severe injury and necrosis to meristems and foliage of some herbaceous (bunchberry, yellow loosestrife, lion's-paw, tufted vetch and several Compositae species) and woody species (speckled alder, trembling aspen and common wild rose). Observations in the year after application showed long-term control of these species, and also of eastern bracken which showed few symptoms after application. Control of red maple and several *Rubus* species had appeared promising, but these recovered in the year after treatment with new apical growth from the base of dead shoot tips. It may be possible to control these species with higher rates or with split or repeated applications. Several ericaceous species (sheep laurel and black huckleberry), including lowbush blueberry, were semi-tolerant and showed only minor symptoms from this treatment. Symptoms on these species included mostly shoot stunting and chlorosis of the foliage, but not meristem die-back. Conifers and some other woody weed species were highly tolerant and showed few, if any, symptoms (Table 1).

Eastern bracken, yellow loosestrife, speckled alder, and wild rose were selected for further studies leading to minor use registration (URMULE) based on these preliminary trials. Tribenuron spot applications gave similar control of eastern bracken as the older, discontinued herbicide asulam (Table 2). Control was consistent between sites and the tribenuron treatment reduced the stem density of eastern bracken and yellow loosestrife > 95% in the year after application. Some shoots emerging in treated plots may have arisen from rhizomes of plants outside the plots. These species senesce in late summer so treatments should be applied early. Tribenuron treatments also gave a high level of control of both speckled alder and wild rose in both the year of and the year after application (Table 3). Applications in early fall were also effective against these later-senescing woody species.

TABLE 1. Response of some perennial herbaceous and woody weeds to mid-summer spot applications of tribenuron[z] in observational trials conducted in prune-year lowbush blueberry fields.

| Species[y] | Common name | Injury (0-9)[x] Meristem | Foliage | |
|---|---|---|---|---|
| **Herbaceous species** | | | | |
| *Apocynum androsaemifolium* L. | Spreading dogbane | 4.6 ± 1.72 | 3.2 ± 1.03 | (n = 12) |
| *Aster umbellatus* P. Mill. | Tall white aster | 8.0 ± 0.81 | 4.3 ± 1.11 | (n = 9) |
| *Centaurea nigra* L. | Black knapweed | 6.0 ± 1.00 | 5.0 ± 1.22 | (n = 12) |
| *Cornus canadensis* L. | Bunchberry | 8.4 ± 0.42 | 7.2 ± 0.36 | (n = 18) |
| *Euthamia graminifolia* (L.) Nutt. | Narrow-leaved goldenrod | 4.0 ± 0.97 | 3.5 ± 0.74 | (n = 18) |
| *Lysimachia terrestris* (L.) BSP | Yellow loosestrife | 9.0 | 8.1 ± 0.19 | (n = 18) |
| *Oenothera biennis* L. | Evening primrose | 9.0 | 9.0 | (n = 9) |
| *Prenanthes trifoliata* (Cass.) Fern. | Lion's-paw | 8.0 ± 0.22 | 6.2 ± 1.33 | (n = 9) |
| *Pteridium aquilinum* (L.) Kuhn. | Eastern bracken | 0 | 1.2 ± 0.15 | (n = 12) |
| *Scirpus atrovirens* Willd. | Black bulrush | 3.2 ± 0.54 | 2.6 ± 0.32 | (n = 12) |
| *Solidago canadensis* L. | Canada goldenrod | 9.0 | 7.9 ± 0.40 | (n = 15) |
| *Solidago rugosa* Ait. | Rough goldenrod | 8.5 ± 0.31 | 2.2 ± 1.80 | (n = 15) |
| *Vicia cracca* L. | Tufted vetch | 8.8 ± 0.14 | 8.0 ± 0.36 | (n = 18) |
| **Woody weeds** | | | | |
| *Abies balsamea* (L.) Mill. | Balsam fir | 0 | 0 | (n = 12) |
| *Acer rubrum* L. | Red maple | 7.2 ± 0.56 | 5.5 ± 1.09 | (n = 15) |
| *Alnus incana* (L.) Moench | Speckled alder | 8.1 ± 0.57 | 8.0 ± 0.49 | (n = 12) |

| Species | Common name | | | |
|---|---|---|---|---|
| Aronia melanocarpa (Michx.) Ell. | Black chokeberry | 0 | < 0.5 | (n = 12) |
| Betula populifolia Marshall | Grey birch | < 0.5 | 1.2 ± 1.15 | (n = 15) |
| Crataegus spp. | Hawthorn | 9.0 | 8.0 ± 0.55 | (n = 9) |
| Gaylussacia baccata (Wang.) K. Koch | Black huckleberry | 0 | < 0.5 | (n = 12) |
| Ilex glabra (L.) Gray | Inkberry | < 0.5 | < 0.5 | (n = 12) |
| Juniperus communis L. | Ground juniper | 0 | 0 | (n = 12) |
| Kalmia angustifolia L. | Sheep laurel | 2.5 ± 1.55 | 3.5 ± 1.04 | (n = 15) |
| Myrica pensylvanica Mirbel | Bayberry | 0 | 0 | (n = 15) |
| Populus tremuloides Michx. | Trembling aspen | 9.0 | 8.2 ± 0.33 | (n = 18) |
| Picea spp. | Spruce | 0 | 0 | (n = 15) |
| Pyrus malus L. | Wild apple | 9.0 | 9.0 | (n = 9) |
| Rosa virginiana Mill. | Common wild rose | 8.6 ± 0.10 | 7.7 ± 0.41 | (n = 18) |
| Rubus allegheniensis Porter | Allegheny blackberry | 8.5 ± 0.22 | 2.2 ± 0.78 | (n = 9) |
| Rubus hispidus L. | Trailing blackberry | 8.1 ± 0.24 | 4.0 ± 0.88 | (n = 18) |
| Rubus idaeus L. var. stigosus (Michx.) Maxim | Wild raspberry | 7.0 ± 1.08 | 6.1 ± 1.22 | (n = 12) |
| Spirea latifolia (Ait.) Borkh. | Meadowsweet | 2.2 ± 1.77 | 2.7 ± 1.01 | (n = 15) |
| **Vaccinium angustifolium Ait.** | **Lowbush blueberry** | **< 0.5** | **1.4 ± 0.99** | **(n = 24)** |

[z] Tribenuron was applied to run-off at 0.2 g $L^{-1}$ water containing 0.2% (v/v) Agral 90.

[x] Nomenclature follows Dartyshire et al., 2000

[y] Injury to the terminal meristems and foliage was rated 4 to 6 weeks after application where 0 is no effect and 9 is complete death of topgrowth. Less than 0.5 indicates some minor sypmtoms (usually discoloration) were occasionally observed on some, but not all, treated plants.

TABLE 2. Effect of tribenuron applied as a directed foliar application to eastern bracken and yellow loosestrife from mid-July to early August.

| Species | Year | Sites (no.) | Herbicide[z] | Visual injury (0-9)[y] | | Stem density (no plot$^{-1}$)[x] | |
|---|---|---|---|---|---|---|---|
| | | | | Blueberry | Weed | Pretreatment | Year after |
| Eastern bracken | 1995 | 3 | Tribenuron | < 0.5 | < 0.5 | 37 ± 9.2 | 0.8 ± 0.92 |
| | | | Asulam | 0 | < 0.5 | 39 ± 7.6 | 0.6 ± 0.54 |
| | 1996 | 3 | Tribenuron | 0 | < 0.5 | 41 ± 9.5 | 0 |
| Yellow loosestrife | 1995 | 3 | Tribenuron | < 0.5 | 8.3 ± 0.42 | 163 ± 39 | 5 ± 5.2 |
| | 1996 | 5 | Tribenuron | < 0.5 | 8.1 ± 0.37 | 181 ± 30 | 7 ± 5.6 |

[z] Tribenuron was applied to run-off at 0.2 g L$^{-1}$ water plus 0.2% Agral 90. Asulam was applied as a 0.75% (v/v) solution of Asulox F (now discontinued in North America) and was included as a standard treatment in 1995 only.
[y] An eastern bracken plot was an area that had at least 25 mature fronds; a loosestrife plot measured 1 m$^2$.
[x] Injury was rated 3 to 4 weeks after application where 0 was no effect and 9 was complete death of the topgrowth. The blueberry plants in the understory were also assessed for injury.

TABLE 3. Effect of tribenuron applied as a direct foliar application to speckled alder and common wild rose at two times of application.

| Species | Time of application[z] | Sites (no) | Injury rating (0-9)[y] | | | Year after control rating (0-9) |
|---|---|---|---|---|---|---|
| | | | Blueberry | Weed meristem | Weed foliage | |
| Speckled alder | mid-summer 1995 | 2 | 0 | 8.2 ± 0.38 | 8.0 ± 0.58 | 8.5 ± 0.26 |
| | mid-summer 1996 | 4 | 0 | 7.1 ± 0.81 | 6.3 ± 0.44 | 8.6 ± 0.17 |
| | early fall 1996 | 4 | - | - | - | 8.9 ± 0.09 |
| Common wild rose | mid-summer 1995 | 3 | < 0.5 | 9.0 | 7.2 ± 0.82 | 8.0 ± 0.29 |
| | mid-summer 1996 | 7 | 0.8 | 8.8 ± 0.22 | 6.0 ± 0.60 | 8.5 ± 0.15 |
| | early fall 1996 | 3 | - | - | - | 8.4 ± 0.25 |

[z] Mid summer applications were made from late July to mid-August; early fall applications from late September to mid-October.
[y] Injury was rated 3 to 4 weeks after application where 0 was no effect and 9 was complete death of topgrowth. No ratings were made after early fall applications because plants had begun to senesce.

Overall, there was little effect of tribenuron sprays on the blueberry plants growing in the understory (Tables 1, 2, and 3), but injury was inconsistent and injury scores ranged from 0 to > 2, i.e., approximately 20% to 25% stunting and chlorosis. As growth inhibitors, SU herbicides can be expected to have their greatest effect on young, expanding vegetative or meristematic tissue. We observed that mature blueberry plants showed little effect of this treatment apart from some minor chlorsis, whereas, new, elongating shoot growth in May and June was sensitive (McCully et al., 1998; Yarborough and Hess, 1995). Broadcast applications to weed-free blueberry plots also showed that August and early September applications after the 'black tip' stage of crop development drastically reduced yields (Figure 1). The 'black tip' stage is when the vegetative growth in the prune year stops and fruit bud initiation begins. Yield reductions were due to reductions in bloom which were estimated to be > 75% on some plots, whereas, injury scores on vegetative shoots did not exceed 2 in this experiment (from the 17 June application, data not shown).

The commercial mixture of nicosulfuron/rimsulfuron gave better control of black bulrush than rimsulfuron alone (Table 4), although pre-

FIGURE 1. The effect of timing on blueberry yield of a broadcast application of 30 g ha$^{-1}$ tribenuron made during the prune-year of the 2-year crop cycle. The asterisks indicate means that are significantly (P < 0.05) lower than the untreated control by a t-test.

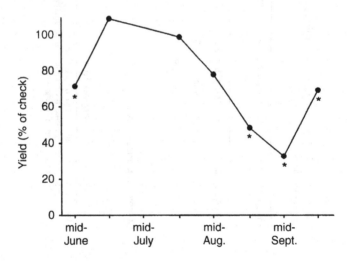

TABLE 4. The effect of timing and rate[z] of application of two sulfonylurea herbicides applied as spot applications to black bulrush and the lowbush blueberry plants occurring in the plots. Values are means of two sites and assessments are those from the year after application (May 19, 1999).

| Herbicide | Application date 1998) | Black bulrush | | | | | | Lowbush blueberry | | |
|---|---|---|---|---|---|---|---|---|---|---|
| | | Control (0-9)[y] | | | Mortality (%)[x] | | | Crop injury (0-9)[w] | | |
| | | 1× | 2× | 3× | 1× | 2× | 3× | 1× | 2× | 3× |
| Rimsulfuron | 19 June | 5.1 | 5.6 | 6.3 | 17 | 29 | 22 | 0 | 0 | 0 |
| | 4 Aug. | 7.6 | 8.4 | 8.4 | 30 | 60 | 72 | 0 | 0 | < 0.5 |
| | 2 Oct. | 8.1 | 8.2 | 8.6 | 48 | 47 | 74 | 0 | 0 | < 0.5 |
| Nicosulfuron/rimsulfuron | 19 June | 8.2 | 9.0 | 9.0 | 62 | 94 | 100 | 1.7 | 2.7 | 1.9 |
| | 4 Aug. | 8.5 | 8.9 | 9.0 | 90 | 93 | 100 | 1.4 | 1.4 | 2.2 |
| | 2 Oct. | 8.3 | 8.9 | 8.9 | 74 | 95 | 95 | 0 | 0.6 | 1.6 |

[z] The 1× rates of rimsulfuron and nicosulfuron/rimsulfuron were 0.018 and 0.031 g L$^{-1}$, respectively.
[y] The LSD (0.05) for comparing any two means of bulrush injury is 1.0, and for comparing values within a given application date or rate it is 0.6 and 0.7, respectively.
[x] The LSD (0.05) for comparing any two means of bulrush mortality is 19%, and for comparing values within a given application date or rate it is 8% and 10%, respectively.
[w] Crop injury was assessed only on plots with sufficient blueberry plants to do so. Values are means of 3 to 8 observations.

liminary screening had suggested the latter was promising. Mortality in the year after application with the 1 × rate (0.031 g L$^{-1}$) of the mixture ranged from 62% to 90% and was influenced by timing of application. Higher rates gave consistently excellent control. Again, earlier applications appeared to have a greater effect on the crop, but there appeared to be no difference between rates on visible crop injury. However, when weed-free blueberry clones were treated directly with the 1 × rate to obtain fruit samples for residue analysis, yields were reduced 20% to 48% over 6 sites compared to untreated clones (data not shown). There were also differences in effectiveness of nicosulfuron/rimsulfuron treatments applied in 1999 and 2000 (Table 5) that may be related to rainfall. Below average rainfall occurred in 1999 from May until mid-August and the black bulrush, a wetland species, showed symptoms of drought stress by mid summer. In these trials and others, early applications in June and early July were generally more effective than mid-summer ones due, perhaps, to generally better moisture conditions and rapid vegetative growth.

Sulfonylurea herbicides have shown promise in controlling both woody and herbaceous broadleaved weeds (Bowes and Spurr, 1996; Meyer and Bovey, 1990), as well, as sedges and grasses (Tanaka et al., 1998) and their usefulness in lowbush blueberry merits evaluation. Tol-

TABLE 5. Effect of timing of directed foliar applications of 0.031 g L$^{-1}$ nicosulfuron/rimsulfuron on black bulrush control and mortality in a season with below normal (1999) and normal (2000) rainfall.[z] Assessments were taken in the year after application and are means of 6 observations.

| Site | 1999 applications | | | 2000 applications | | |
|------|--------|-------------|----------------|--------|-------------|----------------|
|      | Timing | Control (0-9) | Mortality (%) | Timing | Control (0-9) | Mortality (%) |
| 1 | 11 June | 7.9 ± 1.07 | 56 ± 25 | 4 July | 8.0 ± 0.61 | 71 ± 18 |
|   | 9 July | 2.3 ± 0.92 | 0 | 23 Aug. | 6.3 ± 1.77 | 47 ± 12 |
| 2 | 16 June | 4.9 ± 3.21 | 24 ± 20 | 4 July | 8.7 ± 0.24 | 78 ± 16 |
|   | 3 Sept. | 3.6 ± 2.11 | 9 ± 14 | 23 Aug. | 7.0 ± 0.82 | 73 ± 16 |
| 3 | 11 June | 7.6 ± 0.97 | 63 ± 22 | 4 July | 8.9 ± 0.11 | 84 ± 21 |
|   | 8 Aug. | 3.2 ± 1.16 | 12 ± 11 | 23 Aug. | 7.7 ± 1.01 | 68 ± 23 |

[z] Monthly precipitation recorded at Kentville, Nova Scotia during the growing season of the dry (1999) vs. normal (2000) year were: 26 May vs. 58 cm; 24 June vs. 45 mm; 25 July vs. 72 mm; 11 Aug. vs. 37 mm; 20 Sept. vs. 69 mm. In Aug. 1999, the first significant rainfall (65 mm) occurred on 15 Aug.

erance to marginally selective herbicides can be enhanced by careful, directed applications and the 'patchiness' of some weed infestations lend themselves to these applications. Directed sprays of tribenuron and nicosulfuron/rimsulfon were effective against some important weed species. Weed control may be enhanced by applying the herbicides to young, rapidly growing weeds, which was also observed by Bowes and Spurr (1996). Despite high levels of control there was generally some survival and regrowth from rhizomes or tillers. To maintain long-term weed control, follow-up applications to regrowth would be beneficial. These herbicides generally produced only minor symptoms on the vegetative growth of blueberry plants in the understory. However, the effect on fruit buds and yield may be considerable and this needs to be studied further, especially if these herbicides were to be broadcast against larger infestations of sensitive weeds.

## CONCLUSIONS

Selective, long-term control of eastern bracken, yellow loosestrife, wild rose, and speckled alder can be obtained with directed foliar applications of 0.2 g L$^{-1}$ tribenuron. Black bulrush can be controlled with 0.031 g L$^{-1}$ nicosulfuron/rimsulfuron. Applying the treatments under conditions when the weeds are actively growing favors good control. Although crop tolerance may be marginal under some conditions, crop tolerance can be enhanced by careful application. These two herbicide treatments have been registered in Canada and would likely be good candidates for minor use registration in the USA. Additional species could likely be added to the labels with further testing. New SU herbicides should also be evaluated in lowbush blueberry to determine if they could provide selective control of weeds that are not currently being controlled satisfactorily.

## GROWER BENEFITS

In Canada, lowbush blueberry growers can now use directed post-emergence applications of tribenuron and nicosulfuron/rimsulfuron to control a number of important weed species. Unlike glyphosate (Roundup) or 2,4-D that have been used for spot application, these SU herbicides are much safer and crop injury can be minimized by careful application. Heavy infestations of weeds like wild rose and black bulrush make blueberry plants unproductive and difficult to harvest, and these can now be efficiently controlled with minimal effect on the crop.

## LITERATURE CITED

Anonymous. 2002. Guide to weed control. Ont. Min. Agr., Food and Rural Affairs. Publ. 75. Toronto, Ont.

Atlantic Committee on Fruit Crops. 1999. Guide to weed control for lowbush blueberry production in Atlantic Canada. Factsheet ACC 1014, Agdex No. 235/641.

Bowes, G.G. and D.T. Spurr. 1996. Control of aspen poplar, balsam poplar, prickly rose and western snowberry with metsulfuron-methyl and 2,4-D. Can. J. Plant Sci. 76:885-889.

Brown, H.M. 1990. Mode of action, crop selectivity and soil relations of the sulfonylurea herbicides. Pestic. Sci. 29:263-281.

Darbyshire, S.J., M. Favreau, and M. Murray. 2000. Common and scientific names of weeds in Canada. Agriculture and Agri-Food Canada, Publ. 1397/B, Ottawa, Ont.

Hall, I.V. 1959. Plant populations in blueberry stands developed from abandoned hayfields and woodlots. Ecology 40:742-743.

Howatt, S.M. 1992. Control of hexazinone tolerant weeds in lowbush blueberries. McGill Univ., Montreal, MSc Thesis.

Jensen, K.I.N. 1985. Weed control in lowbush blueberries in eastern Canada. Acta Hort. 165:259-265.

McCully, K., K. Jensen, D. Doohan, and G. Sampson. 1995. Bunchberry control in wild blueberries with Spartan 75 DF. Factsheet C 4.3.0. In: J. Argall and G. Chaisson (eds.). Wild blueberry production guide. New Brunswick Dept. Agr., Fredericton, N.B.

Meyer, R.E. and R.W. Bovey. 1990. Influence of sulfonylurea and other herbicides on selected woody and herbaceous species. Weed Sci. 38:249-255.

Tanaka, Y., T. Yamawaki, and H. Yoshikawa. 1998. Factors affecting herbicidal activity of imazosulfuron against weeds in paddy fields. J. Weed Sci. Technol. 43:195-202.

Trevett, M.F. 1950. Weed control–recommended practices for the control of woody weeds. Maine Agr. Exp. Sta. Bul. 479. pp. 32-42.

Yarborough, D.E. and P.C. Bhomik. 1989. Effect of hexazinone on weed populations and on lowbush blueberries in Maine. Acta Hort. 241:344-349.

Yarborough, D.E. and T.M. Hess. 1995. Control of bunchberry in wild blueberry fields. J. Small Fruit Viticult. 3:125-132.

# Determination of Evapotranspiration and Drainage in Lowbush Blueberries (*Vaccinium angustifolium*) Using Weighing Lysimeters

Gordon Starr
Rose Mary Seymour
Fred Olday
David E. Yarborough

**SUMMARY.** Lowbush blueberry growers need recommendations for irrigation scheduling. This study was conducted to determine irrigation water use potential. Initial data were collected from the end of April through early August 2002 using weighing lysimeters to measure evapotranspiration and drainage. High rainfall and drainage rates near the end of April indicated a period of rapid leaching when chemical applications could be beneficially postponed. Nighttime increases in lysimeter weight

---

Gordon Starr is Soil Scientist, USDA-ARS, New England Plant, Soil, and Water Laboratory, University of Maine, Orono, ME 04469.

Rose Mary Seymour is Public Service Assistant, University of Georgia, at the Griffin, GA Campus.

Fred Olday is Director of Farm Research, Jasper Wyman and Son, Deblois, ME.

David E. Yarborough is Professor of Horticulture, University of Maine, Orono, ME.

Research was supported by the U.S. Department of Agriculture, University of Maine, and the Maine Wild Blueberry Commission.

[Haworth co-indexing entry note]: "Determination of Evapotranspiration and Drainage in Lowbush Blueberries (*Vaccinium angustifolium*) Using Weighing Lysimeters." Starr, Gordon et al. Co-published simultaneously in *Small Fruits Review* (Food Products Press, an imprint of The Haworth Press, Inc.) Vol. 3, No. 3/4, 2004, pp. 273-283; and: *Proceedings of the Ninth North American Blueberry Research and Extension Workers Conference* (ed: Charles F. Forney, and Leonard J. Eaton) Food Products Press, an imprint of The Haworth Press, Inc., 2004, pp. 273-283. Single or multiple copies of this article are available for a fee from The Haworth Document Delivery Service [1-800-HAWORTH, 9:00 a.m. - 5:00 p.m. (EST). E-mail address: docdelivery@haworthpress.com].

http://www.haworthpress.com/web/SFR
Digital Object Identifer: 10.1300/J301v03n03_04

were observed indicating that direct vapor deposition is a significant factor in the water balance. Results from this ongoing study will be used for scheduling irrigation water applications to improve water use and production efficiency. *[Article copies available for a fee from The Haworth Document Delivery Service: 1-800-HAWORTH. E-mail address: <docdelivery@ haworthpress.com> Website: <http://www.HaworthPress.com> © 2004 by The Haworth Press, Inc. All rights reserved.]*

**KEYWORDS.** Lysimeter, irrigation, blueberry, Maine, evapotranspiration

## INTRODUCTION

The lowbush blueberry growers of Maine need technical support and recommendations for scheduling irrigation to improve their water use efficiency. A common grower irrigation scheduling practice is to supplement rainfall to ensure that roughly one inch (2.5 cm) per week of water reaches the plants during the growing season. Another common belief is that less irrigation water is required near the Atlantic coast where temperatures are cooler and dense fog and dew occur frequently. The fog is thought to relieve plant water stress even when there is no rainfall. More water is thought to be needed at inland locations where temperatures are higher and humidity lower.

Several studies point to the substantial yield benefits of irrigating lowbush blueberries (Benoit et al., 1984; Seymour et al., 2003; Struchtemeyer, 1956). Irrigation is also an important factor in root development of both lowbush (Hicklenton et al., 2000) and highbush (*Vaccinium corymbosum* L.) blueberries (Spiers, 2000). The observed yield benefits have led to large increases in irrigated acreage in recent years. However, continued expansion of irrigated production is currently being hindered by competing interests in water supplies and government licensing requirements covering water withdrawals from surface waters and wetlands.

Tuning an irrigation system to supply water in response to crop needs saves water, reduces leaching, and improves production efficiency. Studies of plant water use in highbush blueberries show that highbush evapotranspiration (ET) varies widely with location, stage of plant development, and temporally throughout the growing season (Byers and Moore, 1987; Haman et al., 1997; Storlie and Eck, 1996). Basic information on plant water use for lowbush blueberries as functions of time and proximity to the coast would help establish supplemental water requirements for optimal yield. Kosmas et al. (1998) found large amounts of water were adsorbed directly onto soil from atmospheric water vapor at night along the Mediterranean coast. If significant amounts of water are being supplied to the lowbush blueberry crop through condensation or adsorption, then this needs to be quantified and further understood.

To address these issues, weighing lysimeters are being used to quantify the water balance by measuring ET and drainage in lowbush blueberries. These lysimeters are installed at three field locations where conditions of fog, humidity, temperature, and distance inland from the Maine coast differ to such an extent that blueberry irrigation requirements may be affected. This study has a long term objective of improving irrigation scheduling and water use efficiency for lowbush blueberry production in Maine. Initial results reported here are being used to focus future studies for maximum benefit to growers and the environment.

## MATERIALS AND METHODS

The lysimeter design used in this study was taken from Storlie and Eck (1996) with slight modifications intended to improve the systems' accuracy. The design incorporates an inner chamber containing the plants, supported by a weighing load cell in an outer chamber with a drainage collection system. The inner chamber (Figure 1a) has a 1.25 in (3.2 cm) treated plywood frame supported by 1/4 in (6.4 mm) steel plate on the bottom. The inner chamber has inside dimensions of 18 in $\times$ 18 in (46 cm $\times$ 46 cm) for a basal area of 2.25 ft$^2$ (0.21 m$^2$) and a depth of 17.5 in (44.5 cm). A single temperature compensated load cell and a ball bearing support the inner chamber on a 1/4 in (0.64 cm) steel base in the outer chamber (Figure 1b). The only substantive modifications made to the Storlie and Eck (1996) design were the roller bearings mounted on the inside walls of the outer chamber (Figure 1b). These represent the point of contact between inner and outer chambers and were intended to reduce weighing errors caused by springs used in the Storlie and Eck (1996) design. Repeated tests of weighing accuracy showed that the roller bearings could give comparable error (about 3% of changes in weight) to the previous design, but did not justify the added expense as compared with springs.

After thorough wetting of a chosen area, a 2.25 ft$^2$ (0.21 m$^2$) square plug of blueberry sod was extracted using shovels and sod pullers in the spring of the bearing year during 2002 (Figure 1c). The depth of sod

FIGURE 1. Construction and installation of weighting lysimeter showing inner chamber (a), outer chamber (b), blueberry sod (c), partial installation (d), and completed installation (e).

a.

b.

c.

d.

e.

ranged from 4-8 in (10-20 cm). Soil was then carefully excavated from the 2.25 ft$^2$ (0.21 m$^2$) area to a depth of 15 in (38 cm). The depth range that each bucket of soil came from was recorded so that the soil profile could be reconstructed. Perforated drainage pipe was placed in the bottom of the inner chamber followed by two inches (5 cm) of 1/4-1/2 in (6.4-13 mm) gravel and covered by horticultural cloth. The excavated subsoil was then wet packed into the chamber to approximately the original density using the ordering and depths recorded for buckets that were removed during the excavation (Figure 1d). The final layer of subsoil was loosely placed in the inner chamber and the sod placed on top. The reconstituted column (Figure 1e) was then completely saturated and allowed to drain.

Four lysimeters were installed in a cluster (several meters apart) at the Blueberry Hill experimental farm operated by the University of Maine near Jonesboro, Maine. Several clusters are planned for each of at least three sites. Initial data are reported for only one lysimeter installed at a coastal site owned by Sanford Kelley on Kelley Point adjacent to Jonesport, Maine and one lysimeter at an inland site farmed by Jasper Wyman and Son in Deblois, Maine. The Kelley Point site is around 1/2 mile (0.8 km) inland, whereas Blueberry Hill is about four miles (6 km), and Wyman's farm is roughly 10 miles (16 km) inland from the jagged Maine coast. Precipitation, relative humidity, and temperature data were obtained from a weather station located at Blueberry Hill. Irrigation was applied at Blueberry Hill using moveable piping with sprinkler heads set on risers (Seymour et al., 2003). The Wyman and Son's farm uses buried piping with sprinkler heads set on risers. Sanford Kelley's farm was not irrigated. The soils at all sites had an organic mat overlying coarse textured soil (sandy loam or loamy sand) with varying amounts of coarse fragments.

## RESULTS AND DISCUSSION

Evapotranspiration was determined by measuring the change in lysimeter weight per day for a 24 h period from midnight to midnight on days having no rain (Storlie and Eck, 1996) and expressing this as an equivalent depth of water per week. These units (1 in/wk = 2.54 cm/wk) are used because of their convenience for supplemental irrigation scheduling. This is illustrated in Figure 2 which shows daily rainfall and average lysimeter weight versus time from 6 June through 25 June at the Blueberry Hill site. The ET for d 159 and 160 averaged 0.48 in/wk (1.2

FIGURE 2. Lysimeter weight and rainfall equivalent weight versus time at Blueberry Hill farm

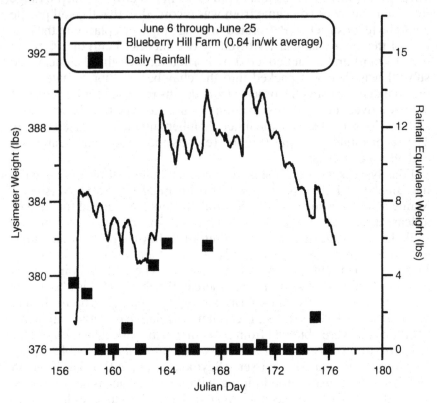

cm/wk) whereas d 164 and 165 averaged only 0.10 in/week (0.25 cm/wk). On all four of these days, strong increases in nighttime lysimeter weight were evident. By contrast, the nighttime rise in lysimeter weight was not as pronounced for d 172 through 174 and ET averaged 1.0 in/wk (2.5 cm/wk).

The nighttime increases in lysimeter weight were a persistent feature seen in the data, particularly at the two sites nearest the coast. A comparison of Blueberry Hill and Wyman's farm (Figure 3) contrasts two sites with different nighttime deposition rates. For d 194 and 195, it is the nighttime rise in weight that appears to make the difference between the 0.99 in/wk (2.5 cm/wk) recorded at Blueberry Hill and the 1.25 in/wk (3.2 cm/wk) recorded at Wyman's farm. The difference in ET between Blueberry Hill (1.0 in/wk or 2.5 cm/wk) and Kelley Point (0.61 in/wk or

FIGURE 3. Comparison of lysimeter weight over time at Blueberry Hill and Wyman's farms

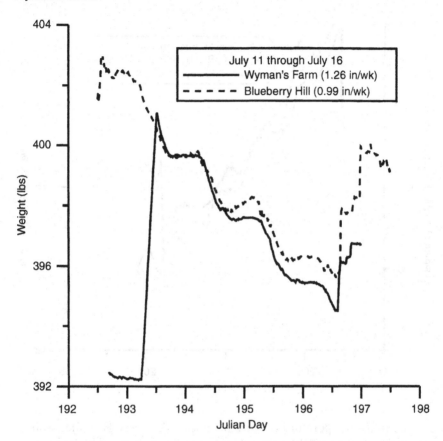

1.5 cm/wk) could not be entirely explained by nighttime rises in lysimeter weight (Figure 4). The nighttime rises were evident at both sites yet Blueberry Hill still had much higher ET (Figure 4). The daytime temperature has a strong effect on ET and the Kelley Point site is persistently much cooler than either of the other sites located further inland.

The nighttime rise in weight is clearly a significant flux of water and should be studied further. Kosmas et al. (1998) saw similar effects in their weighing lysimeters containing bare soil near the Mediteranean coast and attributed them to influxes of cool, moist air from the sea. The water vapor from the air was thought to adsorb directly onto the soil. Increases in relative humidity at night characteristically accompanied de-

FIGURE 4. Comparison of lysimeter weight over time at Blueberry Hill and Kelley Point farms

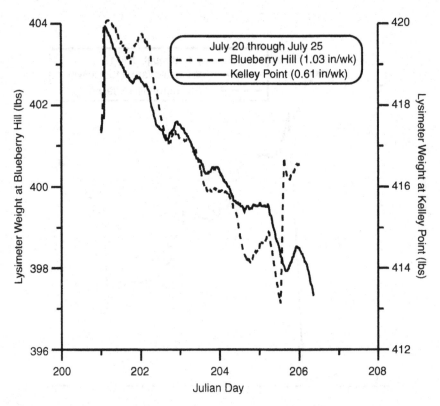

creases in air temperature (Figure 5) at the Blueberry Hill site, so it is reasonable to suspect the same phenomena are at work. The lysimeters in this study contain lowbush blueberry plants that will frequently collect heavy dew as moist evening air condenses on leaves and stems. It is not clear how much of the water deposited on the lysimeters at night comes from dew and how much (if any) is directly adsorbed into the soil. In an attempt to resolve this question in the future, soil moisture tension sensors (heat dissipation sensors) have been installed just below the soil surface to measure an expected decrease in tension if water is being adsorbed from atmospheric vapor or drip from the plants.

A projected water balance was constructed for the Blueberry Hill site (Table 1) using rainfall data recorded daily and averaged over each period, cumulative drainage data and a sampling of ET data from the

FIGURE 5. Diurnal fluctuations of relative humidity and temperature at Blueberry Hill farm

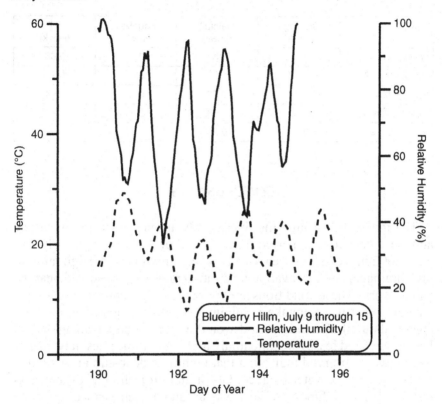

lysimeters. Irrigation was the average amount applied over the time period to keep the soil water tension below around 20 centibars as measured by several tensiometers buried at a depth of 4-6 in (10-15 cm) within and around the cluster of lysimeters. The period from 24 April through 1 May had high rainfall and drainage, but low ET. These conditions are common for this time of year and are favorable for ground water recharge. Applications of agricultural chemicals during this period would be subject to a high leaching hazard and should be postponed if possible to reduce chemical losses. Drainage decreased through May and June and was zero from 10 July through 9 August. The ET remained low until the end of June and remained constant at 1 in/wk (2.5 cm/wk) from 19 June-9 August. Additional site-years and replications of the experiments are needed to further substantiate these initial results.

TABLE 1. Projected water balance for Blueberry Hill Farm including rainfall, drainage, irrigation, and evapotranspiration (ET).

| Dates | Drainage (in/wk) | Rainfall (in/wk) | Irrigation (in/wk) | ET (in/wk) |
|---|---|---|---|---|
| April 24-May 1 | 1.1 | 1.8 | 0 | 0.68 |
| May 2-May 22 | 0.23 | 0.99 | 0 | N.D. |
| May 23-June 18 | 0.24 | 0.58 | 0 | 0.64 |
| June 19-July 9 | 0.16 | 0.88 | 0 | 1.0 |
| July 9-August 9 | 0 | 0.25 | 0.73 | 1.0 |

## CONCLUSIONS

The initial data from a multiyear study are presented in this article and these data cover a period from the end of April through early August, 2002 in the harvest year of a two year growth cycle. High rainfall, high drainage, and low evapotranspiration rates were observed near the end of April. These conditions are favorable for groundwater recharge, whereas plant water uptake is minimal. Waiting to apply agricultural chemicals until after this period would appear to be advantageous as leaching related losses are potentially high. Nighttime rises in lysimeter weight were observed indicating that nighttime deposition is a significant factor in the water balance. Our initial projection of evapotranspiration from 19 June-9 August was around 1 inch per week for the bearing year at the Blueberry Hill site near Jonesboro, Maine. Several years of additional data are needed to adequately address the complex interactions of soil, crop, and coastal atmospheric phenomena that have a bearing on irrigation scheduling in the region. However, it is clear from the nighttime increases in lysimeter weight that condensation and/ or adsorption of atmospheric water vapor cannot be ignored as a substantial flux of water, particularly near the coast.

## GROWER BENEFITS

Although the data presented here need to be supplemented with additional replications in future years, they do generally agree with grower observations that coastal fog reduces irrigation requirements for blue-berries. These data also imply that waiting to apply agricultural chemicals until after the high drainage period around the end of April should reduce leaching related losses of chemicals and improve efficacy.

## LITERATURE CITED

Benoit, G.R., W.J. Grant, A. Ismail, and D.E. Yarborough. 1984. Effect of soil moisture and fertilizer on the potential and actual yield of lowbush blueberries. Can. J. Plant Sci. 64:683-689.

Byers, L.P. and J.N. Moore. 1987. Irrigation scheduling for young highbush blueberry plants in Arkansas. HortScience 22:52-54.

Haman, D.Z., R.T. Pritchard, A.G. Smajstrla, F.S. Zazueta, and P.M. Lyrene. 1997. Evapotranspiration and crop coefficients for young blueberries in Florida. Appl. Eng. Agr. 13:209-216.

Hicklenton, P.R., J.Y. Reekie, and R.J. Gordon. 2000. Physiological and morphological traits of lowbush blueberry (*Vaccinium augustifolium* Ait.) plants in relation to post-transplant conditions and water availability. Can. J. Plant Sci. 80:861-867.

Kosmas, C., N.G. Danalatos, J. Poesen, and B. van Wesemael. 1998. The effect of water vapor adsorption on soil moisture content under Mediterranean climatic conditions. Agr. Water Mat. 36:157-168.

Seymour, R.M., G.C. Starr, and D. Yarborough. 2004. Lowbush blueberry (*Vaccinium angustifolium*) with irrigated and rain-fed conditions. Small Fruits Rev. 3(1/2): 45-56.

Spiers, J.M. 2000. Influence of cultural practices on root distribution of 'Gulfcoast' blueberry. Acta Hort. 513:247-252.

Storlie, C.A. and P. Eck. 1996. Lysimeter-based crop coefficients for young highbush blueberries. HortScience 31:819-822.

Struchtemeyer, R.A. 1956. For larger yields, irrigate lowbush blueberries. Univ. of Maine Farm Res. 4(2):17-18.

# Field-Edge Based Management Tactics for Blueberry Maggot in Lowbush Blueberry

Judith A. Collins

Francis A. Drummond

**SUMMARY.** Blueberry maggot flies were heavily aggregated within the first 30 m into fields. Pupae were present in the soil of pruned fields and in wooded areas with *Vaccinium angustifolium* as an under-story plant adjacent to fields. Perimeter fencing was ineffective as a non-insecticidal approach to control blueberry maggot fly (BMF). A perimeter application of phosmet (Imidan) 70 WP resulted in a significant reduction in the number of BMF captured on AM traps between the treated compared to untreated check areas. Results with Nu-Lure Insect Bait were inconclusive. *[Article copies available for a fee from The Haworth Document Delivery Service: 1-800-HAWORTH. E-mail address: <docdelivery@haworthpress. com> Website: <http://www.HaworthPress.com> © 2004 by The Haworth Press, Inc. All rights reserved.]*

Judith A. Collins is Assistant Scientist, Maine Agricultural and Forest Experiment Station, University of Maine, 5722 Deering Hall, Orono, ME 04469.

Francis A. Drummond is Professor of Insect Ecology/Entomology, Department of Biological Sciences, University of Maine, 5722 Deering Hall, Orono, ME 04469.

This work was supported by the Maine Wild Blueberry Commission.

Mention of any product or material in this report does not constitute endorsement by the University of Maine, Maine Cooperative Extension, or the Main Agricultural and Forest Experiment Station.

[Haworth co-indexing entry note]: "Field-Edge Based Management Tactics for Blueberry Maggot in Lowbush Blueberry." Collins, Judith A., and Francis A. Drummond. Co-published simultaneously in *Small Fruits Review* (Food Products Press, an imprint of The Haworth Press, Inc.) Vol. 3, No. 3/4, 2004, pp. 285-293; and: *Proceedings of the Ninth North American Blueberry Research and Extension Workers Conference* (ed: Charles F. Forney, and Leonard J. Eaton) Food Products Press, an imprint of The Haworth Press, Inc., 2004, pp. 285-293. Single or multiple copies of this article are available for a fee from The Haworth Document Delivery Service [1-800-HAWORTH, 9:00 a.m. - 5:00 p.m. (EST). E-mail address: docdelivery@haworthpress.com].

http://www.haworthpress.com/web/SFR
© 2004 by The Haworth Press, Inc. All rights reserved.
Digital Object Identifer: 10.1300/J301v03n03_05

**KEYWORDS.** Lowbush blueberry, blueberry maggot, *Rhagoletis mendax*, *Vaccinium angustifolium*

## INTRODUCTION

The blueberry maggot fly (BMF), *Rhagoletis mendax* (Curran), is considered the most important pest of commercially grown lowbush blueberries (*Vaccinium angustifolium* Aiton 'lowbush') in Maine. This fly is found infesting blueberry in all production areas in northeastern North America, except Quebec and Newfoundland (Estabrooks, 1995). Maggots in fruit are particularly damaging since the markets for fresh, canned, and frozen fruit have a "near zero to zero" tolerance for infested fruit (Brown and Ismail, 1981). In Maine, approximately 25% of the 60,000 acres of blueberry are treated with an insecticide each year (Yarborough, personal communication). The current IPM program for BMF has been very successful, in Maine and in Canada, by reducing unnecessary insecticide applications and producing a crop with little maggot infestation (Gaul et al., 1995). Increasing public awareness and concern about the use of pesticides, and the passage and implementation of the Food Quality Protection Act of 1996 mandates continuing research into strategies that are not only effective, but also socially and environmentally acceptable.

The objective of this research was to evaluate exclusion barriers, perimeter insecticide sprays and baits for their potential to enhance control of blueberry maggot fly in lowbush blueberry. Work was also completed to evaluate the colonization pattern of the flies into fields and to confirm the importance of pruned fields and unmanaged areas as important sources of infestation.

## MATERIALS AND METHODS

*Colonization of lowbush blueberry fields by blueberry maggot flies (BMF).* In 2000, emergence traps, each covering a basal area of 0.36 m², were placed in three pruned lowbush blueberry fields in Washington Co., Maine to identify the habitats in which BMF emerge in the spring. Eight traps were set at each site, four in the field and four in nearby wooded areas with unmanaged blueberries. In 2001 the study was expanded and 15 traps were set in each of two commercially managed lowbush blueberry fields. Ten of the traps were placed along the field edge; five in a fruit-bearing section of each field and five in a pruned section. The remaining five traps were placed in an adjacent wooded area or an area with low shrubs and unmanaged blueberries in the forest under story. Nine additional traps were set at Blueberry Hill Farm Experiment Station, Jonesboro, Maine; three in the fruit-bearing section, three in the pruned section, and three in the forest adjacent to the fruit-bearing blueberry field. In both years, a baited, yellow Pherocon® AM trap (Great Lakes IPM, Vestaburg, Michigan) was placed in each area to monitor the presence of flies. Traps were checked one or two times per week and any BMF were counted and removed from the traps.

In conjunction with the habitat trials, we also completed work to evaluate the colonization pattern of BMF into lowbush blueberry fields. In each of three years (1999-2001), AM traps were placed in three fields. The traps were distributed in line transects. For each transect, one trap was set in the woods outside the field edge. The next trap was at the field edge; subsequent traps were set 3, 6, 15, 30, 61, 91, and 152 m (10, 20, 50, 100, 200, 300, and 500 ft) along a line running into the field. Traps were checked and any captured BMF counted and removed one or two times per week.

*Exclusion of blueberry maggot from field plots using mesh fencing.* In 2000 and again in 2001, 3-sided, u-shaped, plots were set in three, fruit-bearing wild blueberry fields. Each plot measured $21 \times 46 \times 21$ m$^3$ ($70 \times 150 \times 70$ ft$^3$) and was enclosed with black fiberglass window screening, 1 m (4 ft) high attached to wooden stakes. Effectiveness was evaluated based on seasonal density of BMF captured on baited, yellow Pherocon AM traps and on the number of maggots found in the fruit at harvest.

*Control of blueberry maggot with perimeter application of phosmet (Imidan) 70 WP.* In 2000 and again in 2001, baited, yellow Pherocon AM traps were distributed in transects in each of three fields (4 transects/field). For each transect, one trap was set 3 m (10 ft) into the field from the edge. Subsequent traps were set 15, 30, and 46 m (50, 100, and 150 ft) into the field. Phosmet (Imidan) 70 WP 1.5 L/ha (21.3 oz/acre) was applied with an air blast sprayer in a 24 m (80 ft) swath along the perimeter of each field and in such a way as to incorporate two of the four trap transects in the treated area. Efficacy was evaluated on numbers of BMF captured on the AM traps before and after the application.

*Attractiveness of Nu-Lure Insect Bait to blueberry maggot flies.* Six, $24 \times 122$ m$^2$ ($80 \times 400$ ft$^2$) plots were established in a fruit-bearing blueberry field. Two baited, yellow Pherocon AM traps were placed in each plot. One trap was in the center of the plot; the second was 15 m

(50 ft) in from one edge. When trap captures indicated a suitable BMF population was present, an air blast sprayer was used to apply Nu-Lure Insect Bait at a rate of 3.4 L/ha (48 oz/acre) to three of the plots. An untreated check plot was left between each treated plot. The AM traps were checked 4 and 7 d after the application date. Any BMF were counted and removed from the traps.

## RESULTS AND DISCUSSION

*Colonization of lowbush blueberry fields by blueberry maggot flies (BMF).* Only a small number of BMF were captured in emergence traps over the duration of the trial in either year. Similar numbers of flies emerged in wooded and pruned areas (Figure 1). As expected, no flies emerged from fruit-bearing fields in 2001. A large number of flies were captured on AM traps at all the sites in both years. Captures on AM traps in the fruit-bearing areas did lag slightly behind those in pruned fields and wooded areas (the habitats of emerging BMF). This may indicate that flies are moving into fruit-bearing fields from these areas.

*Exclusion of blueberry maggot from field plots using mesh fencing.* Trends in infestation in 2000 (Table 1) did suggest that there was less maggot infestation within the mesh barriers. Consistently fewer flies were captured inside the enclosures, and there was a significant difference for all sites combined when data was analyzed using a RCBD ($P = 0.02$); the overall reduction in seasonal fly density was only 23%. In 2001, the overall reduction was 58.3%, but the difference was not significant ($P = 0.39$). There was no significant difference when data from both years was combined ($P = 0.16$). There were also no significant differences at $P < 0.05$) in numbers of maggots found in fruit at harvest in

FIGURE 1. Emergence of BMF from different habitats.

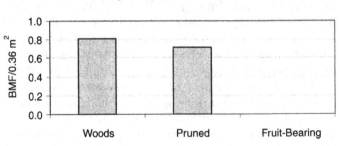

TABLE 1. Density of BMF in open vs. enclosed plots.

| Treatment | BMF Seasonal density | Maggots/qt |
|---|---|---|
| | 2000 | |
| Enclosed | 6.7 * | 1.5 ns |
| Open | 8.7 | 2.7 |
| | 2001 | |
| Enclosed | 1.5 ns | 0.5 ** |
| Open | 3.6 | 1.3 |
| | Both years, combined | |
| Enclosed | 4.1 ns | 1.0 ** |
| Open | 6.1 | 2.0 |

* Significant at $P \leq 0.05$, ** significant at $P \leq 0.10$), ns = nonsignificant.

either year (2000 $P = 0.32$, 2001 $P = 0.08$) or for both years, combined ($P = 0.09$).

*Control of blueberry maggot with perimeter application of phosmet (Imidan) 70 WP.* In 2000 ($P = 0.0001$) and 2001 ($P = 0.01$), the perimeter application of phosmet (Imidan) 70 WP resulted in a significant reduction in the number of BMF captured on AM traps between the treated compared to untreated check areas. When data from both years were combined, there was a significant treatment × year interaction ($P = 0.01$) (Figure 3). Better control was obtained in 2000. There was also a treatment × distance interaction ($P = 0.005$) (Figure 4). In 2001, captures in the treated areas were low near the field edge, but increased sharply further into the field. Captures in the control area were high near the edge and remained essentially unchanged further into the field.

*Attractiveness of Nu-Lure Insect Bait to blueberry maggot flies.* The attractiveness of Nu-Lure to the BMF extended well beyond the edge of the treated areas. The number of flies in the edge of the untreated check areas was much higher than the fly numbers in the middle of the check areas. A comparison between the middle of the check areas and the middle of the treated areas showed that the density of flies was twice as high in the treated areas at 4 and 7 d. However, the difference was not significant (ANOVA, d 4 $P = 0.24$; d 7 $P = 0.61$) (Figure 5) due to high variation between replicates.

FIGURE 2. Distribution of BMF captures within a blueberry field.

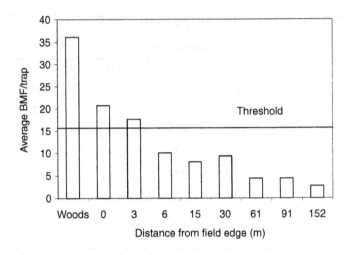

FIGURE 3. BMF captures in check and treated fields following a perimeter application of phosmet (Imidan) 70 WP.

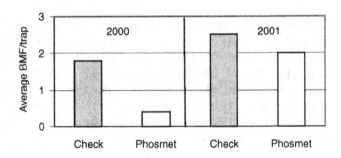

## CONCLUSIONS

*Colonization of lowbush blueberry fields by blueberry maggot flies (BMF).* The focus of this habitat study was to determine if wooded areas with unmanaged wild blueberries are an important source of infestation of fruit-bearing fields. Despite the low numbers of BMF collected in the emergence traps, this study does confirm the presence of pupae in the soil of pruned fields and in wooded areas with *V. angustifolium* as an under-story plant adjacent to fields.

FIGURE 4. Treatment × distance interaction following perimeter application of phosmet (Imidan) 70 WP.

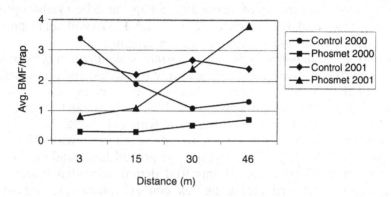

FIGURE 5. Average trap capture of BMF in untreated check areas vs. areas treated with Nu-Lure Insect Bait.

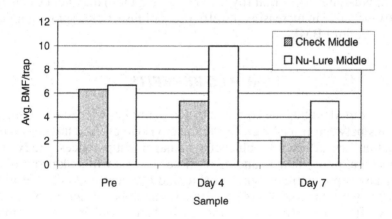

Research between 1999 and 2001 focused on colonization patterns of BMF into lowbush blueberry fields. Work in 1999 and 2000 suggested that most of the BMF population in a field is aggregated within the first 30 m (100 ft) into the field (Figure 2). Flies colonize fruit-bearing fields as a wave starting from the border (Drummond and Collins, unpublished data). Similar behavior has been observed with the closely related apple maggot fly and is the basis for a perimeter tactic in apples in Massachusetts (Prokopy and Mason, 1996).

*Exclusion of blueberry maggot from field plots using mesh fencing.* Despite the consistent trend of lower fly numbers and less fruit infestation by maggots, perimeter fencing does not appear to be a viable option as a non-insecticidal approach to control of BMF. We will not be pursuing research aimed at the use of perimeter fencing.

*Control of blueberry maggot with perimeter application of phosmet (Imidan) 70 WP.* The purpose of this study was to test and implement a tactic of field perimeter spraying to reduce the amount of insecticide sprayed for control of BMF. Lowbush blueberry production involves the pruning of the crop by burning or mowing every other year. Typically, within a region, blueberry fields represent a patchwork of bearing and pruned fields. The flies emerge from pruned fields and migrate to bearing fields. The results of this trial demonstrate that insecticide sprays around the perimeter of the field could eliminate spraying entire fields if only a small portion of the field (around the edge) is actually inhabited by BMF.

*Attractiveness of Nu-Lure Insect Bait to blueberry maggot flies.* The use of Nu-Lure Insect Bait (hydrolyzed corn gluten) may not be consistently effective in increasing the efficiency of insecticide sprays applied for the control BMF.

## GROWER BENEFITS

It is important that we continue to add to our knowledge of the interactions between insect species and the environment and how these interactions are affected by blueberry management practices. Study and refinement of our understanding of these systems will make control of pest insect populations more reliable and efficient and enhance wild blueberry production. The goal of this research is to improve on an already effective BMF control strategy in Maine which is based upon knowledge of its behavior and ecology.

## LITERATURE CITED

Brown. H.L. and A.A. Ismail. 1981. 1980 Blueberry integrated pest management program. Bull. 639, Univ. Maine Cooperative Extension Service Publ. Orono, Maine.

Estabrooks, E.N. 1995. Biology and management of the blueberry maggot, pp. 41-45. In: Management of blueberry insect pests. Pest Management Alternatives Office, Agr. Agri-Food Canada, Ottawa.

Gaul, S.O., W.T.A. Neilson, E.N. Estabrooks, L.C. Crozier, and M. Fuller. 1995. Deployment and utility of traps for management of *Rhagoletis mendax* (Diptera: Tephritidae). J. Econ. Entomol. 88: 134-139.

Prokopy, R.J. and J. Mason. 1996. Behavioral control of apple maggot flies, pp. 267-289, In: Fruit fly pests: a world assessment of their biology and management. B.A. McPherson and G.J. Steck, (eds.), St. Lucie Press, Delray Beach, FL.

# An Overview of RAPD Analysis
# to Estimate Genetic Relationships
# in Lowbush Blueberry

Karen Burgher-MacLellan
Kenna MacKenzie

**SUMMARY.** Randomly amplified polymorphic DNA (RAPD) analysis, a simple dominant molecular marker technique, has been used extensively for cultivar identification and relatedness studies in many perennial woody species. Thus, this technique should provide genetic information for lowbush blueberry (*Vaccinium angustifolium* Ait.). Young leaves of lowbush blueberry from field clones with varying phenotype were collected for DNA extraction. Pre-screening of RAPD primers resulted in 11 polymorphic primers and 140 consistent RAPD fragments. Eight primers were selected as useful for our study, the fragments scored and the data analyzed with Genstat5 to calculate similarity, produce dendrograms and perform a principal coordinate analysis. The RAPD analysis was able to identify distinct field clones. Average genetic similarity among field clones was 68% reflecting expected genetic variation.

Karen Burgher-MacLellan is Research Technician and Kenna MacKenzie is Research Scientist, Atlantic Food and Horticulture Research Center, Agriculture and Agri-Food Canada, 32 Main Street, Kentville, NS B4N 1J5, Canada (E-mail: burgherk@agr.gc.ca).

The authors thank Walter Wojtas and Katherine Williams for technical assistance and Ken Mcrae and Brad Walker for statistical assistance.

[Haworth co-indexing entry note]: "An Overview of RAPD Analysis to Estimate Genetic Relationships in Lowbush Blueberry." Burgher-MacLellan, Karen, and Kenna MacKenzie. Co-published simultaneously in *Small Fruits Review* (Food Products Press, an imprint of The Haworth Press, Inc.) Vol. 3, No. 3/4, 2004, pp. 295-305; and: *Proceedings of the Ninth North American Blueberry Research and Extension Workers Conference* (ed: Charles F. Forney, and Leonard J. Eaton) Food Products Press, an imprint of The Haworth Press, Inc., 2004, pp. 295-305. Single or multiple copies of this article are available for a fee from The Haworth Document Delivery Service [1-800-HAWORTH, 9:00 a.m. - 5:00 p.m. (EST). E-mail address: docdelivery@haworthpress.com].

Approximately 15% of the field clones were not related. RAPD analysis is a useful tool for genetic relationship studies in lowbush blueberry and may provide similarity information for future pollination/productivity research. *[Article copies available for a fee from The Haworth Document Delivery Service: 1-800-HAWORTH. E-mail address: <docdelivery@haworthpress. com> Website: <http://www.HaworthPress.com>* © *2004 by The Haworth Press, Inc. All rights reserved.]*

**KEYWORDS.** Lowbush blueberry, Genetics, RAPD

## INTRODUCTION

Lowbush blueberry (*Vaccinium angustifolium* Ait.), a valuable commercial crop in Maine, Quebec, and Atlantic Canada, is typically harvested from managed perennial fields. New fields are usually created by deforestation of marginal lands and allowing the native stands (established plants and/or new seedlings) to grow and fill in. This growth pattern results in a population that is a composite of many highly variable plants. Plants that spread and form uniform clusters of vegetation within a field are termed "clones." Such clonal populations continue further stand development through clonal spreading by rhizomes exclusively or new seedlings plus clonal spreading (Eriksson, 1993). Due to the conditions in which lowbush blueberry are managed and based on the natural dispersal of seeds due to animal behavior, it appears that the former scenario prevails as new seedlings are rare and mainly found along the edges of commercial fields (Crossland and Vander Kloet, 1996). The high level of variability found in lowbush blueberry can easily be observed as differences in leaf color and growth habit when viewing an established field. As well, different fields and clones within a field can show large variations in berry color and size and, more importantly, yield. Maintaining optimal berry yield is critical for commercial fields, thus an understanding of how population structure and genetic diversity affect yield is important. To date few studies on how genetic variation relates to field performance and/or clonal genetic relatedness have been reported for lowbush blueberry.

Molecular markers using the polymerase cycle reaction (PCR) have become very popular and are useful tools to study variability, relatedness, and diversity issues in many plant species (Hodgkin et al., 2001; Joshi et al., 1999). RAPD analysis is a dominant molecular marker technique that can be used to estimate genetic similarity. The technique is relatively easy but reveals limited genetic information as dominant markers can not distinguish between the heterozygote and homozygote for a dominant allele (Williams et al., 1990). Thus when scoring RAPD markers, they are interpreted for one of two conditions: (1) the presence of a band is considered the heterozygote or dominant homozygote and (2) the absence of a band is regarded as the recessive homozygote. Regardless of this limitation, when scoring dominant markers, RAPD analysis remains a very popular technique and has recently been reported widely for diversity studies in many plant species ranging from agronomically important annuals such as barley (*Hordeum vulgare*) (Fernandez et al., 2002), soybean (*Glycine max* L.) (Li and Nelson, 2001), and flax (*Linum usitatissimum* L.) (Fu et al., 2002), to perennial woody plants such as the cultivated chestnut (*Castanea sativa* Mill.) (Goulao et al., 2001). As well, RAPD analysis has been successfully used in *Vaccinium* spp., mainly for cultivar identification and relatedness studies (Arce-Johnson et al., 2002; Aruna et al., 1995; Levi and Rowland, 1997; Polashock and Vorsa, 1997).

In a recent study, Burgher et al. (2002) used RAPD analysis with lowbush blueberry selections and native accessions that had been collected from various geographic regions in Atlantic Canada and Maine. This analysis successfully distinguished all the clones and correctly identified unknown DNA samples. An average similarity of 56% was determined among all the samples and similarity values greater than 70% were found for individuals known to be related. Given the inherent variability that exists within commercial lowbush blueberry fields, it is likely that RAPD analysis will prove useful for examining diversity within field populations. Thus the objectives of this study were to apply RAPD analysis in a commercial field situation to (1) establish if RAPD analysis could distinguish genetic differences among phenotypic clones situated in a typical established field, (2) determine how much variability exist and estimate the relatedness (similarity) among these clones, and (3) compare these results with our previous study.

## MATERIALS AND METHODS

A 100 m × 13 m plot was randomly selected and marked in a lowbush blueberry field near Londonderry, N.S., Canada in late May of 1999. Twenty-six visually distinct clones, identified by growth type and leaf color, were measured, mapped, and numbered by location in the plot. Three samples (a, b, and c) of 6-8 young terminal leaf buds were randomly collected from each clone for a total of 78 sub-samples. The leaf buds were stored on ice and returned to the Kentville Agriculture Centre for total DNA extraction in liquid nitrogen within 6 to 8 h after collection. The DNA extraction protocol had been modified for lowbush blueberry as reported by Burgher et al. (2002). The DNA quantity was determined by spectrophotometry and then each samples was diluted to 10 ng·μL$^{-1}$ for PCR. A set of RAPD primers (Biotechnology Laboratory, Univ. B.C., Vancouver, B. C., Canada) was pre-screened and 11 were found to be polymorphic. Eight of these primers were found to be highly polymorphic in lowbush blueberry and were used in this study. The RAPD procedure was followed as previously reported by Burgher et al. (2002) and was repeated three times for each sub-sample. The PCR DNA products were separated by electrophoresis on 1.4% agarose gels, visualized with ethidium bromide (0.5 μg·mL$^{-1}$) and documented under ultraviolet light with the Gel Doc 2000 system. The DNA fragments were scored for their presence or absence and fragments that were not reproducible or weak were not scored. The data set was analyzed by Genstat 5 to calculate the Nei and Li (1979) coefficient of similarity (*s*). A dendrogram was constructed with average linkage cluster and nearest neighbor analysis. Principal coordinate ordination (3-D PCO) was used to demonstrate spatial relationships.

### RESULTS AND DISCUSSION

The eight polymorphic RAPD primers resulted in 77 consistent fragments for the 26 lowbush blueberry clones. The size of the RAPD fragments scored ranged from 180 to 2300 base pairs, with 6-14 fragments per primer and an average 8.4 fragments per primer (data not shown). A unique RAPD fragment was not found for any of the clones which would have indicated a marker for that clone, however one 350 base pair fragment from primer UBC-292 was found in only 4 of the 33 samples. The number of unique patterns for the primers ranged from 7 to 12 (Table 1), and when all primer-clone combinations were combined a total of 33 distinct clones were identified from the 26 original clones, indicating variation within the a, b, and c sub-samples of 5 clones.

The selected UBC primers proved to be highly polymorphic as 87% of the RAPD fragments showed variation among the lowbush blueberry clones. A high percentage of polymorphic markers usually indicates a higher level of genetic variability. This high level of polymorphism is

TABLE 1. Name, sequence, percentage of polymorphic bands, and number of unique band profiles for eight highly polymorphic RAPD 10-mer primers used for estimating similarity in lowbush blueberry.

| UBC-primer | Sequence 5'-3' | Polymorphic Bands/primer (%) | Unique band profiles (no.) |
|---|---|---|---|
| 203 | CACGGCGAGT | 75 | 7 |
| 222 | AAGCCTCCCC | 89 | 7 |
| 239 | CTGAAGCGGA | 100 | 12 |
| 244 | CAGCCAACCG | 75 | 8 |
| 268 | AGGCCGCTTA | 100 | 11 |
| 280 | CTGGGAGTGG | 75 | 9 |
| 287 | CGAACGGCGG | 88 | 10 |
| 292 | AAACAGCCCG | 75 | 7 |

larger than the 53% polymorphic fragments that were found in our previous study of 26 geographically diverse clones (Table 2). This means there is more variability present within the small population of 26 closely situated clones than the 26 geographically separated clones. However, the average percent genetic similarity also differed for the two studies, with 68% for the field clones and 56% for the diverse study. Average similarity is calculated from all the similarity coefficients among the clones and is derived from the relationship of shared RAPD fragments between clones, thus the more shared fragments the higher the similarity. The difference in average similarity between the two studies is not surprising as the diverse clones were collected from different geographic areas and thus, had a lower probability of being closely related. These results show that the narrow field population of related clones displayed a higher level of variability and similarity than the broad collection of geographically diverse clones. This finding is congruent with another study (Schoen and Brown, 1991), where highly out-crossing species were shown to display more genetic variation within populations than among populations, subsequently in-breeding populations display more variation among populations. As lowbush blueberry has a high level of out-crossing, this may explain the higher variation found within the narrow population than within the diverse clones.

TABLE 2. Comparison of two studies using RAPD analysis for lowbush blueberry clones. One study contained diverse clones from different geographical
areas and the other looked at clones that were closely situated in one commercial field.

| Study | RAPD primers (No.) | Total bands (No.) | Polymorphic bands (%) | Average *s* (%) |
|---|---|---|---|---|
| 26 diverse clones (Burgher et al., 2002) | 11 | 138 | 53 | 56 |
| 26 closely situated clones (this study) | 8 | 77 | 87 | 68 |

A dendrogram was calculated from the nearest neighbor analysis and
showed the clustering of the 33 unique clones (Figure 1). All the clones
started to cluster together at 60% or greater similarity, except for clone
10 which appears to be the most diverse clone as it grouped away from
all the other clones at only 50% similarity.

In our previous study it was found that *s* values among clones were
greater than 70% when the known relatedness of clones was 50% or
greater (i.e., parent/progeny association), and this was in accord with
findings by Lynch and Milligan (1994). Therefore using the *s* value of
70% as a cut off for estimating 50% relatedness, the 33 clones are clustered into 11 groups of less than 50% related clones. Five of the 33
clones (15%) appear to cluster in groups by themselves and are probably distantly related as all *s* values are less than 70%. The largest cluster (group 4) representing 39% of the clones appears to have all but
one of its clones found within the first 46 m of the plot (data not
shown). An interesting exception in group 4 is clone 25 which appears to be most closely related to clone 1 (*s* = 79%) and they are situated at either end of the plot being 96 m apart. In contrast, three other
groups (2, 3, and 5) appear to cluster in relation to clone order and
proximity within the mapped area, the most apparent being group 3
(clones 15, 16,17, and 18). Clone 23 clustered as one distinct group
although each of the sub-samples appeared to be separate individual
plants. Figure 2 demonstrates the high level of RAPD fragment polymorphism among the 3 sub-samples of clone 23 found from just one
RAPD primer.

It is interesting to note that most of the clustering in the dendrogram
is between 70% and 80% similarity and only four clusters are > 85%

FIGURE 1. Dendrogram of 33 lowbush blueberry clones calculated from average linkage cluster analysis based on Nei and Li's (1979) similarity coefficient obtained from 8 RAPD primers.

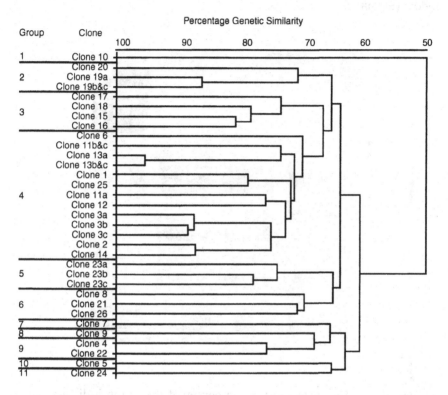

(clones 19a and 19b&c, clones 13a and 13b&c, clones 2 and 14, and clones 3 a, b, &c). In an out-crossing species, seeds from the same berry can be related as full-sibs or half-sibs depending if fertilization is from the same pollen source or from more than one pollen source (Handel, 1983). In the former situation, similarity values among the seedlings would be expected to be higher. This might explain the high similarity values found for the 4 clusters listed above, as they may represent seedlings from the same parent cross seeds.

Three-D PCO plots are useful to show the spatial separations of closely related individuals as calculated latent vectors are plotted with 3 axes. Figure 3 shows the placement of all 33 clones and demonstrates that most of the separation is due to the first two axes, as well 66% of the

FIGURE 2. RAPD amplification fragments obtained from the three sub-samples of clone 23 with primer UBC-280. The arrows indicate four polymorphic RAPD fragments. M = the molecular weight marker was a 123 base pair DNA ladder (Sigma).

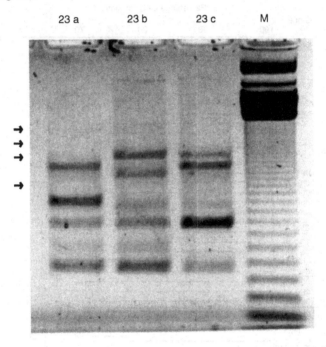

clones appear to cluster within the center of the graph indicative of a close association.

The PCO also reveals that some clones are diverse, the first grouping to the left comprises clones 5, 7, 9, 10, and 24, these are the 5 clones that clustered as five separate groups with the dendrogram. This confirms that these 5 clones are probably the most diverse in this study and clone 10 is the most unrelated as it has a further separation with the third axes. The second group to the right contains the three sub-samples of clone 3 and clones 14 and 2. These five clones are part of the large group #4 and show very little variation with the third axes. It is interesting to note that the third axes did reveal another level of separation for clone 15 and this was not apparent from the dendrogram, thus showing the benefit of using more than one type of analysis.

FIGURE 3. Principal coordinate ordination (PCO) of 33 lowbush blueberry clones based on similarity coefficients from 77 RAPD fragments.

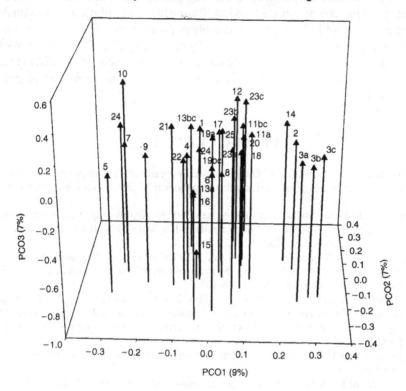

## CONCLUSIONS AND GROWER BENEFITS

The RAPD analysis found 33 distinct patterns from the initial 26 clones. The analysis found variation among potentially related clones that had distinct visual differences and was able to detect variation within 5 clones that appeared to be similar for growth type and leaf color. The average similarity was 68% and only 5 clones had average similarity values less than 70%, indicating that 85% of the clones were probably originally derived from related seed. This would be expected in a small section of a naturalized field where plant establishment would be initially from seedlings scattered due to animals and then further growth mostly through the spreading of rhizomes. This study showed that RAPD analysis can detect variation and estimate similarity within a 1300 m² plot in an established lowbush blueberry field.

Lowbush blueberry is noted as having high variability within and among fields in many aspects including growth habit and berry production. Knowledge of genetic variability within a population of lowbush blueberry could help explain these observations and allow a better understanding of field performance. In future studies, RAPD analysis will be applied as a tool to investigate correlations of genetic variability (i.e., relatedness) and clone productivity to estimate clone and/or field performance.

## LITERATURE CITED

Arce-Johnson, P., M. Rios, M. Zuniga, and E. Vergara. 2002. Identification of blueberry varieties using random amplified polymorphic DNA markers. Acta Hort. 574:221-224.

Aruna, M., M.E. Austin, and P. Ozias-Akins. 1995. Randomly amplified polymorphic DNA fingerprints for identifying rabbiteye blueberry (*Vaccinium ashei* Reade) cultivars. J. Amer. Soc. Hort. Sci.120:710-713.

Burgher, K.L., A.R. Jamieson, and X. Lu. 2002. Genetic relationships among lowbush blueberry genotypes as determined by randomly amplified polymorphic DNA analysis. J. Amer. Soc. Hort. Sci. 127:98-103.

Crossland, D.R. and S.P. Vander Kloet. 1996. Berry consumption by the American Robin, *Turdus migratorius*, and the subsequent effect on seed germination, plant vigour, and dispersal of the lowbush blueberry, *Vaccinium angustifolium*. Can. Field Naturalist. 110:303-309.

Eriksson, O. 1993. Dynamics of genets in clonal plants. Trends Ecol. Evolution 8: 313-116.

Fernandez, M.E., A.M. Figueiras, and C. Benito. 2002. The use of ISSR and RAPD markers for detecting DNA polymorphism genotype identification and genetic diversity among barley cultivars with known origin. Theoretical Appl. Genet. 104: 845-851.

Fu, Y-B., A. Diederichsen, K.W. Richards, and G. Peterson. 2002. Genetic diversity within a range of cultivars and landraces of flax (*Linum usitatissimum* L.) as revealed by RAPDs. Genet. Res. Crop Evolution 49:167-174.

Goulao, L., T. Valdiviesso, C. Santana, and C.M. Oliveira. 2001. Comparison between phenetic characterization using RAPD and ISSR markers and phenotypic data of cultivated chestnut (*Castanea sativa* Mill.) Genet. Res. Crop Evolution 48:329-338.

Handel, S.N. 1983. Pollination ecology, plant population structure, and gene flow, pp. 163-221. In: L. Real (ed.) Pollination Biology. Academic Press, Orlando, FL USA.

Hodgkin, T., R. Roviglioni, M.C. De Vicente, and N. Dudnik. 2001. Molecular methods in the conservation and use of plant genetic resources. Acta Hort. 546:107-118.

Joshi, S.P., P.K. Ranjekar, and V.S. Gupta. 1999. Molecular markers in plant genome analysis. Curr. Sci. 77:230-239.

Levi A. and L.J. Rowland. 1997. Identifying blueberry cultivars and evaluating their genetic relationships using randomly amplified polymorphic DNA (RAPD) and simple sequence repeat-(SSR-) anchored primers. J. Amer. Soc. Hort. Sci. 122: 74-78.

Li, Z. and R.L. Nelson. 2001. Genetic diversity among soybean accessions from three countries measured by RAPDs. Crop Sci. 41:1337-1347.

Lynch, M. and B.G. Milligan. 1994. Analysis of population genetic structure with RAPD markers. Mol. Ecol.3:91-99.

Nei, M. and W. Li. 1979. Mathematical model for studying genetic variation in terms of restriction endonucleases. Proc. Natl. Acad. Sci. USA 76:5269-5273.

Polashock, J.J. and N. Vorsa. 1997. Evaluation of fingerprinting techniques for differentiation of cranberry and blueberry varieties. Acta Hort. 446:239-242.

Schoen, D.J. and A.H.D. Brown. 1991. Intraspecific variation in population gene diversity and effective population size correlates with the mating system in plants. Proc. Natl. Acad. Sci. 88:4494-4497.

Williams, J.G.K., A.R. Kubelik, K.J. Livak, J.A. Rafalaki, and S.V. Tingey. 1990. DNA polymorphisms amplified by arbitrary primers are useful as genetic markers. Nucleic Acids Res.18:6531-6535.

# Identification of Host Volatile Compounds for Monitoring Blueberry Maggot Fly

Oscar E. Liburd

**SUMMARY.** In choice tests, blueberry maggot *Rhagoletis mendax* Curran flies were exposed to blueberries or marbles (surrogate berries) in a highbush blueberry *Vaccinium corymbosum* L. planting. Significantly more *R. mendax* flies were attracted to cages baited with blueberries compared with marbles. In additional experiments, five volatile compounds consisting of butyl butanoate, cis-3-hexen-1-ol, alpha-terpiniol, geraniol, and trans-2-hexenal were extracted and identified from ripening blueberries using gas chromatography and mass spectroscopy (GC-MS) techniques. Late in the season, butyl butanoate and cis-3-hexen-1-ol were as attractive to *R. mendax* as ammonium acetate (1BEEM® capsule). Increasing the load-rates of butyl butanoate and cis-3-hexen-1-ol did not significantly increase trap captures *[Article copies available for a fee from The Haworth Document Delivery Service: 1-800-HAWORTH. E-mail address: <docdelivery@haworthpress.com> Website: <http://www.HaworthPress.com> © 2004 by The Haworth Press, Inc. All rights reserved.]*

Oscar E. Liburd is Assistant Professor, Fruit and Vegetable Entomology and Nematology Department, University of Florida, Gainesville, FL 32611 USA.

The author would like to thank his technician Gisette Seferina and graduate students Erin Finn and Lukasz Stelinski for their help in preparing the figures and text for this manuscript. John Hamill and Dr. Steven Alm also contributed significantly to the data presented in this paper.

[Haworth co-indexing entry note]: "Identification of Host Volatile Compounds for Monitoring Blueberry Maggot Fly." Liburd, Oscar E. Co-published simultaneously in *Small Fruits Review* (Food Products Press, an imprint of The Haworth Press, Inc.) Vol. 3, No. 3/4, 2004, pp. 307-312; and: *Proceedings of the Ninth North American Blueberry Research and Extension Workers Conference* (ed: Charles F. Forney, and Leonard J. Eaton) Food Products Press, an imprint of The Haworth Press, Inc., 2004, pp. 307-312. Single or multiple copies of this article are available for a fee from The Haworth Document Delivery Service [1-800-HAWORTH, 9:00 a.m. - 5:00 p.m. (EST). E-mail address: docdelivery@haworthpress.com].

Digital Object Identifer: 10.1300/J301v03n03_07

**KEYWORDS.** Blueberry maggot, *Vaccinium corymbosum*, blueberries, volatile

## INTRODUCTION

Blueberry maggot, *Rhagoletis mendax* Curran, is a key pest of blueberries, *Vaccinium* spp., from Nova Scotia southward into Georgia and northern Florida. Current monitoring practices for *R. mendax* rely on both visual and olfactory stimuli and include the use of ammonium-baited Pherocon AM yellow sticky panels (Prokopy and Coli, 1978) and green sticky spheres (Liburd et al., 1998). A principal problem with ammonium baits is that they attract non-target insects including beneficials (Liburd et al., 2000). Furthermore, ammonium acetate appears to lose its attractiveness to *R. mendax* as flies reach sexual maturity. Our objective was to increase selectivity and late-season captures of blueberry maggot fly (BMF) using a two-phase study involving both field and laboratory experiments. Phase I will determine if *R. mendax* is responsive to volatiles emitted from ripening blueberries and phase II would identify and screen synthetic host-volatile compounds for their attractiveness to *R. mendax* in the field as well as to evaluate release rates of potential volatile compounds for monitoring *R. mendax*.

## MATERIALS AND METHODS

*Phase I.* In order to determine if blueberry maggot flies were responsive to host volatiles emitted from ripening blueberries, two treatments consisting of (1) blueberries and (2) marbles (control) were placed in collapsible cages and hung ~ 2 m from stripped (blueberries removed) highbush blueberry bushes. Collapsible cages containing blueberries or marbles were fitted with four green sticky spheres for monitoring *R. mendax* flies. The experimental design was a completely randomized block with five replicates. Flies were counted and sexed twice per week and treatments within blocks were rotated twice per week.

*Phase II.* Volatile compounds were collected from ripening blueberries using Solid Phase Micro-extraction techniques (SPME) then identified using Gas Chromatography (HP-6980, Hewlett-Packard Co.) and Mass Spectrometry (Pegasus II, LECO Corp., St. Jos) techniques. Based on previous research (Liburd, 1997) and the compounds identified from our laboratory blueberry volatile profile, five synthetic compounds consisting of butyl butanoate, cis-3-hexen-1-ol, geraniol, alpha-terpiniol, and trans-2-hexenal were purchased from Aldrich Chemical Co., Milwaukee, WI. Volatile compounds were loaded (0.3 ml) into polyethylene slow-release BEEM® capsules (Ted Pella, Inc., Redding, CA). BEEM® capsules containing synthetic compounds were attached to green sticky sphere traps during field evaluations. Volatile compounds were compared with the standard ammonium acetate for their attractiveness to *R. mendax* throughout a typical blueberry growing season.

In experiments to compare the various load-rates, the two most active volatile compounds (based on attraction to *R. mendax*) that included; butyl butanoate and cis-3-hexen-1-ol were compared against the standard ammonium acetate (Figure 1). Each sticky green sphere had either one or five BEEM® capsules containing the respective volatile treatment. Fly attraction to baited spheres was recorded throughout the season.

## RESULTS AND DISCUSSION

Significantly more *R. mendax* flies were caught on green sticky spheres placed on collapsible cages and baited with blueberries compared with cages baited with marbles (surrogate blueberries) (Figure 2). Early in the season (6-17 July), ammonium acetate was significantly more attractive to *R. mendax* compared with other compounds evaluated (Figure 3). More than 150 *R. mendax* were captured on spheres baited with ammonium acetate (Figure 3). During the screening of volatile compounds, none of the five host volatile compounds evaluated were significantly more attractive to *R. mendax* early in the season compared with the control (Figure 3). Butyl butanoate was significantly more attractive than geraniol and alpha-terpiniol late in the season [2-16

FIGURE 1. Chemical structures of the three most attractive volatile compounds to *R. mendax* adults based on field studies conducted in Rhode Island and Michigan (1999-2001).

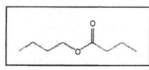
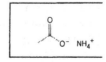

cis-3-hexen-1-ol          n-butyl butyrate          ammonium acetate

FIGURE 2. Response of *R. medax* files to collapsible cages baited with blueberries and marbles. Means followed by the same letter are not significantly different (*P* ≤ 0.05, LSD test).

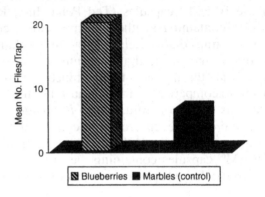

August] (Figure 3). Butyl butanoate and cis-3-hexen-1-ol were as attractive as ammonium acetate (1 BEEM® capsule) late in the production season when the majority of *R. mendax* were sexually mature (Figure 4). Increasing the load-rates of butyl butanoate and cis-3-hexen-1-ol did not significantly increase trap captures (Figure 4).

Five major host volatile compounds, including butyl butanoate, cis-3-hexen-1-ol, alpha-terpiniol, geraniol, and trans-2-hexenal were identified from ripening blueberries. Of these compounds, butyl butanoate and cis-3-hexen-1-ol were the most attractive compounds to *R. mendax* in the field. The attractiveness of these two compounds was more apparent later in the season when these two compounds were equally attractive to *R. mendax* compared with ammonium acetate.

## CONCLUSION

Ammonium acetate was the most attractive compound to *R. mendax* flies early in the season, although its attractiveness decreased as flies reached sexual maturity. Butyl butanoate, cis-3-hexen-1-ol, appear to be good candidates to be used in a mixed volatile blend or individually for monitoring mature *R. mendax* flies. However, appropriate devices for releasing volatile compounds and proper release rates must be identified before these compounds can be used effectively in any monitoring program. Future research should examine devices for releasing volatile compounds and optimum trapping distances.

FIGURE 3. Screening synthetic volatile compounds for attraction to *R. medax* adults. Mean number of flies captured on 9-cm diameter green spheres baited wtih volatile treatments. Means followed by the same letter are not significantly different (*P* ≤ 0.05, LSD test).

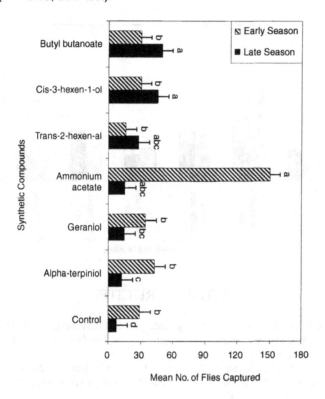

## GROWER BENEFITS

The identification of host volatile compounds will improve monitoring efficiency for growers. Currently, ammonium acetate is used for baiting monitoring traps for *R. mendax*. The problem with ammonium acetate is that it attracts beneficial insects, which can help to regulate populations of *R. mendax*. In addition, ammonium acetate does not work well late in the season because sexually mature flies are not highly attracted to this compound. Growers will be able to better time their insecticide applications, and subsequently use less insecticides if monitoring protocol for *R. mendax* is improved.

FIGURE 4. Attraction of *R. mendax* to synthetic volatile compounds at different load-rates. Mean number of flies captured on 9-cm diameter green spheres baited with volatile treatments. Means followed by the same letter are not significantly different ($P \leq 0.05$, LSD test).

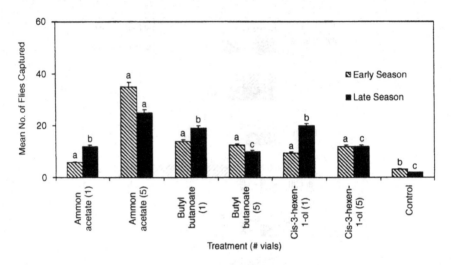

## LITERATURE CITED

Liburd, O. E., S. R. Alm, R. A. Casagrande and S. Polavarapu. 1998. Effect of trap color, bait, shape and orientation in attraction of blueberry maggot flies. J. Econ. Entomol. 91: 243-249.

Liburd, O. E., S. Polavarapu, S. A. Alm, and R. A. Casagrande. 2000. Effect of trap size, placement, and age on captures of blueberry maggot flies (Diptera: Tephritidae). J. Econ. Entomol. 93: 1452-1458.

Liburd, O. E. 1997. Behavioral management techniques for broccoli and blueberry pests. PhD Diss. Univ. Rhode Island, Kingston, R.I. Diss. Abstract No. 9831111.

Prokopy, R. J., and W. M. Coli. 1978. Selective traps for monitoring *Rhagoletis mendax* flies. Protection. Ecol. 1:45-53.

# Investigation
# of a Possible Sexual Function Specialization
# in the Lowbush Blueberry
# (*Vaccinium angustifolium* Aiton. Ericaceae)

Marina Myra
Kenna MacKenzie
S. P. Vander Kloet

**SUMMARY.** Commercial lowbush blueberry fields contain several clonal plants; of these *Vaccinium angustifolium* (Aiton) is frequently the dominant. The *V. angustifolium* component is typified by a mosaic of distinct genotypes, commonly called clones, which tend to vary dramatically in

---

Marina Myra is Masters Student, Department of Biology, Acadia University, Wolfville, Nova Scotia, Canada, B4P 2R6 (E-mail: marina.myra@acadiau.ca).

Kenna MacKenzie is Research Scientist, Atlantic Food and Horticulture Research Centre, Agriculture and Agri-Food Canada, 32 Main Street, Kentville, Nova Scotia, Canada, B4N 5G4.

S. P. Vander Kloet is Professor Emeritus, Department of Biology, Acadia University, Wolfville, Nova Scotia, Canada, B4P 2R6.

Thanks are given to Mike Hardman, Sherry Fillmore, and Ken MacRae for their statistical support; Melissa Reekie, Sarah Benjamin, Walter Wojtas, and Stephanie Chaisson for their technical assistance and Andrew Jamieson for his advice and blueberry plantings.

Research was supported by the Matching Investment Initiative of Agriculture and Agri-Food Canada and Bragg Lumber Company.

[Haworth co-indexing entry note]: "Investigation of a Possible Sexual Function Specialization in the Lowbush Blueberry (*Vaccinium angustifolium* Aiton. Ericaceae)." Myra, Marina, Kenna MacKenzie, and S. P. Vander Kloet. Co-published simultaneously in *Small Fruits Review* (Food Products Press, an imprint of The Haworth Press, Inc.) Vol. 3, No. 3/4, 2004, pp. 313-324; and: *Proceedings of the Ninth North American Blueberry Research and Extension Workers Conference* (ed: Charles F. Forney, and Leonard J. Eaton) Food Products Press, an imprint of The Haworth Press, Inc., 2004, pp. 313-324. Single or multiple copies of this article are available for a fee from The Haworth Document Delivery Service [1-800-HAWORTH, 9:00 a.m. - 5:00 p.m. (EST). E-mail address: docdelivery@haworthpress.com].

Digital Object Identifer: 10.1300/J301v03n03_08

fruit production from one to another. The objective of this study was to examine the seed and berry production of forty-three select clones of *V. angustifolium*, which exhibit a wide range of berry yield. For each clone, seed set and pollen viability were measured. A significant positive relationship was found between berry weight and the number of large plump seeds per berry. However, a significant negative relationship was found between the production of large plump seeds and viable pollen tetrads. The results of this study suggest that these select clones of *V. angustifolium* may be separating female floral function from male floral function. *[Article copies available for a fee from The Haworth Document Delivery Service: 1-800-HAWORTH. E-mail address: <docdelivery@haworthpress.com> Website: <http://www.HaworthPress.com> © 2004 by The Haworth Press, Inc. All rights reserved.]*

**KEYWORDS.** *Vaccinium angustifolium* Aiton, clone, polymorphism, tetraploid, reproductive characteristics, pollen tetrad viability, open pollination

## INTRODUCTION

The lowbush blueberry (*Vaccinium angustifolium* Aiton) is a very important crop in eastern North America, and especially in Nova Scotia (Hall, 1978), where it is the largest fruit crop in value and acreage (McIsaac, 1999). The lowbush blueberry is unique among cultivated crops, naturally invading barren fields by way of an existing seed bank or seed deposits in animal scat (Hall et al., 1979). Common management practices of tree removing and elimination of grass and shrubs by burn-pruning and/or herbicide application, create barren fields for *V. angustifolium* colonization. The expanding rhizome system of established *V. angustifolium* plants grow radially at different speeds, which results in colonies differing in stem densities, heights and areas.

Phenotypic variation of clones within fields has been noted since the beginning of blueberry cultivation and has been classified as having differing forms within fields (Hall et al., 1979). An important reason for the variability of plant characteristics in *V. angustifolium* resides in its tetraploid genome (2n = 4x = 48) (Hall and Aalders, 1961; Vander Kloet, 1976). Polyploidy is conducive to polymorphism, since the doubling of the genome and sorting of duplicate genes increases heterozygosity and genetic variation (Soltis and Soltis, 1999; 2000). Polymorphic species are often defined as having varying allelic states, thereby expressing phenotypic traits in many different ways.

Productivity of blueberry fields depends on the mix of plants contained therein. Fields composed largely of unproductive clones will have low berry yield, and conversely, fields with many productive clones will have high berry yield. A possible reason for low berry yield within fields may be due to kinship of neighbouring blueberry clones. A high degree of relatedness among the clones may lead to inbreeding depression, since *V. angustifolium* is not self-compatible (Aalders and Hall, 1961; Jackson et al., 1972; Vander Kloet, 1976). Along with low seed set and berry yield, fields having low genetic diversity may also have high mutation rates. For example, Hall and Aalders (1964) found that four of eight mutants resulted from selfing partially self-compatible clones, and that these mutations usually persisted in families.

Furthermore, Aalders and Hall (1961; 1963) found that male and female sterility vary within populations of *V. angustifolium*; some clones are completely male or female sterile, where others produce high amounts of viable pollen or ovules (Hall et al., 1966). This article sets out to investigate further how male and female fertility is distributed in selected clones of *V. angustifolium*.

## MATERIALS AND METHODS

This study was carried out on a 1970 planting of select *Vaccinium angustifolium* clones (Hall, 1979) located at the Agriculture and Agri-Food Canada (AAFC) substation in Sheffield Mills, Nova Scotia. Average productivity figures (Andrew Jamieson, unpublished data) were used to choose 43 blueberry clones, ranging from low (1.4-5.1 t ha$^{-1}$) to high yields (9.55-14.3 t ha$^{-1}$) for this study.

Rhizome cuttings with budding stems were taken from each clone and brought into the AAFC greenhouse and allowed to flower. Pollen from five randomly chosen flowers was shaken onto a slide using a 440 Hz tuning fork, and stained with Alexander's stain (Alexander, 1969). This stain was used to evaluate the viability of each pollen grain within the pollen tetrad, colouring living protoplasm red and pollen walls blue-green. A maximum of 400 and a minimum of 34 pollen tetrads were counted per slide, depending on the pollen productivity of that certain clone (Table 1). Pollen grain stainability was used to represent the number of viable grains per tetrad. Four tetrad fertility types were counted, varying in number from zero to four viable pollen grains within the tetrad (Table 1).

TABLE 1. Distribution of pollen tetrad fertilities for each of the forty-three lowbush blueberry clones sampled in this study. The blueberry clones are arranged in ascending order from lowest to highest pollen viability.

| Clone Name | # Tetrads | 0/4 | 1/4 | 2/4 | 3/4 | 4/4 | % Viable |
|---|---|---|---|---|---|---|---|
| 73-29 | 35 | 34 | 1 | 0 | 0 | 0 | 0.007 |
| 73-12 | 125 | 119 | 5 | 1 | 0 | 0 | 0.014 |
| 841 | 293 | 117 | 75 | 100 | 1 | 0 | 0.237 |
| 871 | 255 | 81 | 96 | 73 | 5 | 0 | 0.252 |
| 803 | 116 | 40 | 34 | 24 | 9 | 9 | 0.313 |
| 548 | 253 | 36 | 82 | 85 | 48 | 2 | 0.399 |
| 70-27 | 226 | 20 | 45 | 145 | 14 | 2 | 0.426 |
| 73-10 | 260 | 4 | 50 | 204 | 2 | 0 | 0.446 |
| 322 | 57 | 5 | 9 | 24 | 17 | 2 | 0.509 |
| 381 | 284 | 15 | 72 | 90 | 76 | 31 | 0.532 |
| Me3 | 272 | 48 | 28 | 76 | 63 | 57 | 0.549 |
| 527 | 260 | 6 | 48 | 96 | 90 | 20 | 0.567 |
| 390 | 328 | 4 | 39 | 144 | 94 | 47 | 0.607 |
| Fundy | 278 | 13 | 40 | 93 | 70 | 62 | 0.615 |
| 11 | 311 | 38 | 31 | 68 | 94 | 80 | 0.618 |
| 73-21 | 313 | 3 | 22 | 112 | 100 | 76 | 0.679 |
| 281 | 287 | 9 | 19 | 75 | 120 | 64 | 0.684 |
| Brunswick | 253 | 0 | 10 | 99 | 69 | 75 | 0.707 |
| 229 | 394 | 35 | 15 | 65 | 120 | 159 | 0.724 |
| 73-16 | 267 | 44 | 12 | 25 | 20 | 166 | 0.736 |
| 448 | 387 | 8 | 42 | 77 | 70 | 190 | 0.753 |
| 360 | 291 | 19 | 28 | 37 | 43 | 164 | 0.762 |
| 520 | 400 | 17 | 23 | 55 | 60 | 245 | 0.808 |
| 443 | 301 | 6 | 12 | 34 | 70 | 179 | 0.836 |
| 361 | 287 | 2 | 9 | 27 | 95 | 154 | 0.840 |
| 396 | 383 | 3 | 3 | 50 | 106 | 221 | 0.852 |
| 454 | 290 | 4 | 7 | 21 | 73 | 185 | 0.869 |
| 510 | 347 | 19 | 12 | 12 | 44 | 260 | 0.870 |
| 517 | 321 | 2 | 5 | 30 | 67 | 217 | 0.883 |
| 534 | 265 | 7 | 8 | 12 | 43 | 195 | 0.888 |
| 357 | 356 | 1 | 6 | 33 | 58 | 258 | 0.897 |
| 70-36 | 269 | 0 | 8 | 23 | 16 | 222 | 0.920 |
| Augusta | 265 | 13 | 3 | 4 | 7 | 238 | 0.928 |
| 516 | 164 | 0 | 0 | 6 | 20 | 138 | 0.951 |
| 744 | 227 | 3 | 1 | 6 | 17 | 200 | 0.952 |
| Chignecto | 252 | 6 | 1 | 1 | 9 | 235 | 0.962 |
| Me4161 | 328 | 0 | 0 | 11 | 28 | 289 | 0.962 |
| 73-7 | 282 | 0 | 1 | 8 | 23 | 250 | 0.963 |
| 554 | 251 | 0 | 1 | 3 | 27 | 220 | 0.964 |
| 12 | 310 | 0 | 0 | 4 | 22 | 284 | 0.976 |
| 519 | 232 | 0 | 0 | 0 | 22 | 210 | 0.976 |
| 73-15 | 212 | 0 | 0 | 1 | 6 | 205 | 0.991 |
| 73-19 | 264 | 0 | 0 | 1 | 5 | 258 | 0.993 |

The selected clones were open pollinated and their berries were harvested when ripe. Berry ripeness was characterised by dark blue skin, lacking any immature red or green hues. Berries were then weighed and seeds were counted. Three seed classes were recognised using the classification observed by Bell (1956): (1) *perfect seeds* (large seeds that were plump and well formed), (2) *lignified seeds* (small seeds which were not as well formed), and (3) *collapsed seeds* (aborted seeds which were very small and concaved in shape). Seed counts represent the minimum number of ovules pollinated in each berry, and were used to represent reproductive success of each clone.

Analysis of variance (ANOVA) was used to test the null hypothesis of no difference in berry weight, proportion of large seed, and proportion of complete pollen tetrads per clone (SAS Systems, 1982). Simple linear regression was used to investigate the relationships among seed set, pollen viability, and berry weight (Genstat 5 Committee, 1993).

## *RESULTS*

Clonal differences were statistically significant for all the independent variables, with berry weight, proportion of large seed, proportion of small seed, and proportion of aborted seed all being accounted for by the differences among clones (Table 2). Large plump seed was used as an indication of female reproductive success since regression analysis showed that it accounted for 44.1% of the difference between clones. This is compared to 40.7% for poor small seed and 30.8% for aborted seed (Genstat 5 Committee, 1993).

According to the linear regression, berry weight had a significantly positive relationship with the percent of large seeds per berry (slope =

TABLE 2. ANOVA Summary table for the independent variables of berry weight and proportion of different seed types grouped by clone for open pollinated lowbush blueberry flowers. $\alpha = 0.05$.

| Variable | Mean Square | df Model | df Error | F-Value |
|---|---|---|---|---|
| Berry Weight | 0.2757 | 42 | 366 | 4.92 |
| Proportion of Large Seed | 0.1075 | 42 | 366 | 4.79 |
| Proportion of Small Seed | 0.3215 | 42 | 366 | 4.40 |
| Proportion of Aborted Seed | 0.2126 | 42 | 366 | 4.70 |

FIGURE 1. Linear regression of berry weight (g) in relation to (A) percent of large seed/berry with a slope of +0.44 was significant to p = 0.0002, (B) percent of small seed/berry with a slope of −0.35 was significant to p = 0.003, and (C) percent aborted seed/berry with a slope of −0.16 was not significant (p = 0.35), for all forty-three lowbush blueberry clones used in this study.

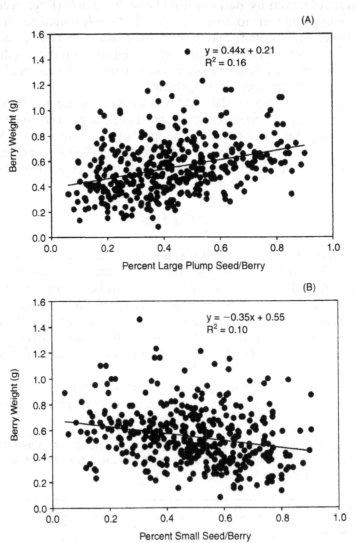

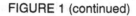

FIGURE 1 (continued)

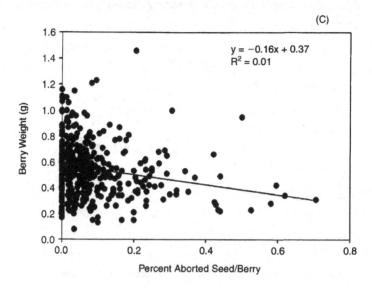

(C)

$y = -0.16x + 0.37$
$R^2 = 0.01$

Berry Weight (g)

Percent Aborted Seed/Berry

+0.44, p = 0.0002) (Figure 1A), and a significantly negative *relationship* with the percent of small seeds per berry (slope = $-0.35$, p = 0.003) (Figure 1B). Berry weight also had a somewhat negative relationship with the percent of aborted seed per berry (slope = $-0.15$, p = 0.35) (Figure 1C), however this was not significant.

A considerable amount of variability in pollen viability was observed between the forty-three blueberry clones; plants with low pollen viability also had low pollen output (Table 1). In comparing the relationship between pollen viability, berry weight, and good large seed, the proportion of completely formed tetrads was used because the data points were distributed normally. The remaining four pollen viability measures (three, two, one, and zero viable pollen grains) were greatly skewed and were not used in this analysis. No significant relationship was found between berry weight and percent of complete tetrads per clone (slope = 0.01, p = 0.82) (Figure 2). However, the percent large seed was negatively, but non-significantly, related to the proportion of tetrad pollen per flower (slope = $-0.24$, p = 0.16) (Figure 3). Overall, *V. angustifolium* plants having high seed set tended to have a low proportion of fully formed pollen tetrads.

FIGURE 2. Linear regression of berry weight (g) compared with the percent of fully viable pollen tetrads for all 43 blueberry clones gave a non-significant slope of 0.01.

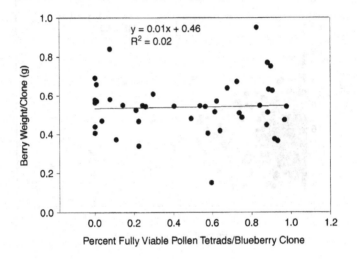

FIGURE 3. Linear regression of percent large seed/berry compared with the percent fully viable pollen tetrads for each of the 43 lowbush blueberry clones. The slope of $-0.24$ was non-significant ($p = 0.16$).

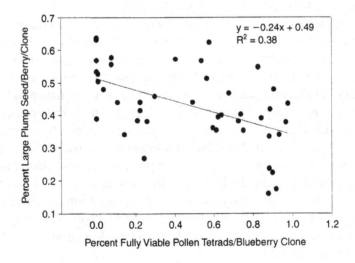

## *DISCUSSION*

As in previous studies of the lowbush blueberry (Aalders and Hall, 1961; Jackson et al., 1972), the relationship between berry weight was positively correlated with seed number. However, according to Bell (1956), large plump seeds contain embryos, whereas small and aborted seeds do not. Therefore, this study used proportion of good plump seed, instead of total seed, since it was deemed more important to the success of the berry, and also because good plump seed accounting for 44.1% of the difference between clones as compared to 40.7% for small seed and 30.8% for aborted seed.

The amount of variability in pollen viability among the sample population of blueberry clones used in this study was striking. The range extended from clones which produce miniscule amounts of incomplete pollen tetrads to those that almost exclusively release abundant fully viable tetrads. The severe variation in the production of male gametes seems to be counter-intuitive, since abundance of pollen would ensure optimum dispersal of genetic information. However, along with a range of pollen viability, a wide range of seed production was also found within this sample population of blueberry clones, with a trend towards a negative relationship between the two variables. The larger the proportion of fully viable pollen tetrads, the lower the proportion of large plump seeds per berry, and vice-versa. This suggests a trade-off between the two sexes within a plant, so that nutrient resources are allotted to one or the other, but not both, which may deplete resources. However, this negative trend is not always the case, since in a similar study with *Vaccinium corymbosum* L., the highbush blueberry, a positive relationship was observed, where plants with low seed production were found to have low pollen viability (Vander Kloet, 1983).

Interestingly, a few clones show a high incidence of diad production. Megalos and Ballington (1987) have suggested that diads represent unreduced pollen sporads or diploid gametes: a not uncommon event in *Vaccinium* (Hall and Galletta, 1971; Ortiz et al., 1992). In fact, it has been suggested that *V. angustifolium* itself has evolved from the joining of diploid gametes in autopolyploidy to form its now tetraploid form (Vander Kloet, 1977; Ortiz et al., 1992). This observed trend towards functional male and female flowers may be due to the speciation of *V. angustifolium* from its diploid ancestors (Vander Kloet, 1977). Indeed, it has been hypothesised by Hill and Vander Kloet (1983) that the *Vaccinium* section *Cyanococcus* is a young taxon, based upon the similarity of isozymes among its species.

Interestingly, however, as Figure 3 shows, there is a distinct gap in the distribution of individual clone data points. A suggestion for this apparent split may be due to a divergence in the population between plants that are good berry producers and poor pollen producers, and those that are poor berry producers and good pollen producers. Hence, molecular techniques such as DNA fingerprinting could be used in the future to identify and eliminate poor yielding plants at the seedling stage.

## CONCLUSION

A strong correlation among berry weight, seed type, and pollen viability in open pollinated lowbush blueberry has been demonstrated in this study. *V. angustifolium* clones that produce large berries were found to have many good large seeds, and alternately, clones that produce small berries, produced few good large seeds. Clones which had a large proportion of fully viable pollen tetrads, tended to produce fewer large viable seeds than those which produced proportionately less fully viable pollen tetrads. Therefore, these data suggest that among the selected clones of *V. angustifolium* at the AAFC research farm, several bear hermaphroditic flowers that are either functionally female or functionally male.

## GROWER BENEFITS

Good berry set in the lowbush blueberry, a xenogamic species, necessarily requires fields composed of many un-related clones. Furthermore, this study suggests that, through natural selection, the lowbush blueberry may be developing a strategy of functionally male and functionally female plants within populations. If this is indeed true, then not only is it expected that not all blueberry clones within a field will produce good berry yields, but that low yielding clones are needed as excellent pollen donors for the fertilization of high yielding clones. This accentuates the need for variable clonal makeup within fields, and advises that the maintenance of a diverse planting containing many clones within a field could provide for excellent berry production. However, since low yielding clones are not desired within fields, the potential is there to investigate DNA fingerprinting techniques to identify and possibly eliminate certain undesirable clones from fields before they are firmly established.

# LITERATURE CITED

Aalders, L.E. and I.V. Hall. 1961. Pollen incompatibility and fruit set in lowbush blueberries. Can. J. Genet. Cytol. 3:300-307.

Aalders, L.E. and I.V. Hall. 1963. The inheritance and morphological development of male-sterility in the common lowbush blueberry *Vaccinium angustifolium* Ait. Can. J. Genet. Cytol. 5:380-383.

Alexander, M.P. 1969. Differential staining of aborted and nonaborted pollen. Stain Technol. 44(3):117-122.

Bell, H.P. 1956. The development of the blueberry seed. Can. J. Bot. 35:139-153.

Genstat 5 Committee. 1993. Genstat 5 Release 3. Reference Manual. Clavendon Press. Oxford U.K.

Hall, I.V. 1978. *Vaccinium* species of horticultural importance in Canada. Commonwealth Bureau Hort. Plantation Crops, Hort. Abs. 48(6):441-445.

Hall, I.V. 1979. The cultivar situation in lowbush blueberry in Nova Scotia. Fruit Var. J. 33(2):54-46.

Hall, I.V. and L.E. Aalders. 1961. Note on male-sterility in the common lowbush blueberry, *Vaccinium angustifolium* Ait. Can. J. Plant. Sci. 41:865.

Hall, I.V. and L.E. Aalders. 1964. Note on some putative lowbush blueberry mutants. Can. J. Bot. 42:122-125.

Hall, I.V. and L.E. Aalders, and G.W. Wood. 1966. Female sterility in the common lowbush blueberry, *Vaccinium angustifolium* Ait. Can. J. Genet. Cytol. 8:269-299.

Hall, I.V., L.E. Aalders, N.L. Nickerson and S.P. Vander Kloet. 1979. The biological flora of Canada. 1. *Vaccinium angustifolium* Ait., sweet lowbush blueberry. Can. Field-Naturalist 93(4):415-430.

Hall, I.V. and G.J. Galletta. 1971. Comparative chromosome morphology of diploid *Vaccinium* species. J. Amer. Soc. Hort. Sci. 96:289-292.

Hill, N.M. and S.P. Vander Kloet. 1983. Zymotypes in *Vaccinium* section *Cyanococcus* and related groups. Proc. N.S. Inst. Sci. 33:115-121.

Jackson, L.P., L.E. Aalders, and I.V. Hall. 1972. Berry size and seed number in commercial lowbush blueberry fields of Nova Scotia. Le Naturaliste Can. 99:615-619.

Megalos, B.S. and J.R. Ballington. 1987. Pollen viability in five southern United States diploid species of *Vaccinium*. J. Amer. Soc. Hort. Sci. 112:1009-1012.

McIsaac, D. 1999. Wild blueberry production and marketing in Nova Scotia, a situation report. N.S. Dept. Agr. and Mktg., Nappan, N.S.

Ortiz, R., L.P. Bruederle, T. Laverty, and N. Vorsa. 1992. The origin of polyploids via 2n gametes in *Vaccinium* section *Cyanococcus*. Euphytica 61:241-246.

SAS Systems. 1982. Analysis of Variance, pp. 113-229, In A.A. Ray (ed.). SAS user's guide: Statistics. SAS Institute, Inc., Cary, NC.

Soltis, D.E. and P.S. Soltis. 1999. Polyploidy: Recurrent formation and genome evolution. Trends Ecol. Evol. 14:348-352.

Soltis, P.S. and D.E. Soltis. 2000. The role of genetic and genomic attributes in the success of polyploids. Proc. Nat. Acad. Sci. USA 97:7051-7057.

Vander Kloet, S.P. 1976. Nomenclature, taxonomy and biosystematics of *Vaccinium* section *Cyanococcus* (the blueberries) in North America. I. Natural barriers to gene exchange between *Vaccinium angustifolium* Ait. and *Vaccinium corymbosum* L. Rhodora 78:503-515.

Vander Kloet, S.P. 1977. The taxonomic status of *Vaccinium boreale*. Can. J. Bot. 55:281-288.

Vander Kloet, S.P. 1983. The relationship between seed number and pollen viability in *Vaccinium corymbosum* L. HortScience 18:225-226.

# Chilling Requirement Studies in Blueberries

James M. Spiers
Donna A. Marshall
John H. Braswell

**SUMMARY.** Two separate studies were initiated in January 1998 on 'Tifblue' rabbiteye (*Vaccinium ashei* Reade) and 'Magnolia' southern highbush (*V. corymbosum* L.) blueberry plants that had received > 500 chilling hours (7°C). In each study, the terminal 2 flower buds from both excised stems and paired stems on plants were forced in a greenhouse (17-23°C and natural daylength) and rated for floral bud development. In 'Magnolia', leaf removal from excised stems and intact stems resulted in an increased rate of floral bud development. After 7 days of forcing, excised stems with leaves removed showed no significant differences in floral bud development from intact plants with unabscised leaves. After 40 days of forcing, flower buds from all treatments had reached stage 3, an easily recognizable stage of development. 'Tifblue' stems with 5 mm basal sections removed weekly to retard vascular blockage did not differ in floral bud development from stems without basal pruning. Excised stems did not differ from matching intact stems for the first 3.5 weeks of forcing. At this time 'Tifblue' flower buds had reached 3.7 on the floral development scale. After ≥4.5 weeks of forcing in the greenhouse floral

James M. Spiers is Research Horticulturalist and Donna A. Marshall is Horticulturalist, USDA-ARS Small Fruits Research Station, Poplarville, MS 39470. John H. Braswell is Extension Specialist, Mississippi Cooperative Extension Service, Poplarville, MS 39270.

[Haworth co-indexing entry note]: "Chilling Requirement Studies in Blueberries." Spiers, James M., Donna A. Marshall, and John H. Braswell. Co-published simultaneously in *Small Fruits Review* (Food Products Press, an imprint of The Haworth Press, Inc.) Vol. 3, No. 3/4, 2004, pp. 325-330; and: *Proceedings of the Ninth North American Blueberry Research and Extension Workers Conference* (ed: Charles F. Forney, and Leonard J. Eaton) Food Products Press, an imprint of The Haworth Press, Inc., 2004, pp. 325-330. Single or multiple copies of this article are available for a fee from The Haworth Document Delivery Service [1-800-HAWORTH, 9:00 a.m. - 5:00 p.m. (EST). E-mail address: docdelivery@haworthpress.com].

Digital Object Identifer: 10.1300/J301v03n03_09

bud development ratings on excised stems were significantly lower than those on intact stems. These studies indicate that with sufficient chilling, floral bud development in excised stems without leaves approximate that of intact plants for a period of 3.5 weeks and reach at least stage 3, an easily identifiable floral stage. *[Article copies available for a fee from The Haworth Document Delivery Service: 1-800-HAWORTH. E-mail address: <docdelivery@haworthpress.com> Website: <http://www.HaworthPress.com> © 2004 by The Haworth Press, Inc. All rights reserved.]*

**KEYWORDS.** Rabbiteye blueberry, *Vaccinium ashei* Reade, southern highbush blueberry, *Vaccinium corymbosum* L., chill hour requirement

## INTRODUCTION

Rabbiteye blueberries (*Vaccinium ashei* Reade) are the major type (species) planted and have dominated the field in the Southeastern United States. Chilling requirements are an aspect that is necessary to insure the plants are grown in the proper region. Rabbiteye plants usually require 400 to 700 hours of chilling at 6-7°C for budbreak (Spiers and Draper, 1974; Spiers 1976). To determine chilling requirements, plants are typically moved into a greenhouse after receiving increasing amounts of chilling hours. A number of methods have been used to determine budbreak and thus fulfillment of chilling hour requirements. Austin and Bondari (1987) measured the diameter of the buds two weeks after cutting. Norvell and Moore (1982) considered chilling requirements met when four buds on a stem exhibited 1.0 cm of growth. Spiers (1976) defined budbreak as the period when a bud broke and elongated and the corolla opened on at least one flower.

When developing and evaluating new blueberry germplasm for potential cultivar releases, a method to quickly determine the chilling requirements using the minimum number of plants would be beneficial. It has been found that one-year-old rooted cuttings behave similarly as whole plants and can be used to determine chilling requirements of rabbiteye selections (Austin et al., 1982). This study was initiated to determine if excised stems could be used to mimic floral bud development in whole plants. Since juvenile 'Magnolia' plants grown in pots tend to retain their leaves through the winter more than mature plants grown in field conditions, the effects of leaf removal was also investigated.

## MATERIALS AND METHODS

*Study 1:* On January 21, 1998 after accumulation of 750 h of natural chilling (< 7°C), 20 two-year-old 'Magnolia' plants grown in 4-L containers were chosen based on uniformity of size and flower bud density. These plants were divided evenly into 2 groups. On each of 10 plants (group 1), 2 stems were excised and placed into floral aquapics filled with tap water. One excised stem was left with unabscised leaves intact, but on the other stem all unabscised leaves were removed. These excised stems were paired with 2 stems of a plant (group 2) having approximately the same number of buds per stem at the same developmental stage. The stems on the plants also had unabscised leaves left intact on one and removed on the other. All treatment stems were tagged and labeled. Plants and cuttings were placed in a greenhouse with a temperature range of 16.7-22.2°C and natural lighting where they were allowed to bloom. Floral bud development ratings (Figure 1, Spiers; 1978) were taken on the terminal 2 buds every 7-10 d. Data were analyzed as a split plot design using SAS (SAS Insitute, 1985).

*Study 2:* At the same time, 20 two-year-old 'Tifblue' plants grown in 4-L containers were chosen based on size and flower bud density. These plants were divided evenly into 2 groups. On each of 10 plants (group 1) 2 stems were excised and placed into floral aquapics filled with tap water. These excised stems were paired with 2 stems of a plant (group 2) having the same number of buds per stem at the same developmental stage. Essentially all leaves of 'Tifblue' had abscised at initiation of the study. Each week, approximately 2-5 mm of tissue were removed from the excision end on one of the 2 excised stems (in an attempt to clear blocked vascular tissue). All treatment stems were tagged and labeled. Plants and excised stems were placed in a greenhouse with a temperature range of 16.7-22.2°C and natural lighting. Floral bud development ratings were taken on the terminal 2 buds every 7-10 days. Data were analyzed as a split plot design using SAS (SAS Institute, 1985).

## RESULTS

In study 1, leaf removal increased the rate of floral bud development (Table 1). After 40 d of forcing, flower buds from all treatments had reached stage 3 (Spiers, 1978), an easily identifiable stage of development. After 23 d of forcing, the rate of floral bud development was significantly greater in intact plants with leaves removed than intact plants

FIGURE 1. Stages of floral-bud development in rabbiteye blueberry: 1 = no visible swelling, bud scales completely enclose inflorescence; 2 = visible swelling
of bud, scales separating, flowers still completely enclosed; 3 = bud scales
separated, apices of flowers visible; 4 = individual flowers distinguishable, bud
scales abscise; 5 = individual flowers distinctly separated, corollas unexpand-
ed and closed; 6 = corollas completely expanded and open; 7 = corollas
dropped. From Spiers, J.M. 1978. J. Amer. Hort. Sci. 103:452-455.

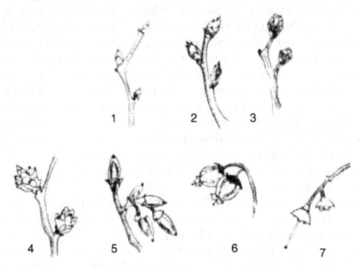

TABLE 1. Floral bud development of 'Magnolia' excised stems and intact stems,
1998.

| | Days of forcing | | | | | |
|---|---|---|---|---|---|---|
| | 7 | 16 | 23 | 30 | 40 | 44 |
| Plant | | | | | | |
| Intact leaves | 1.1 b$^z$ | 1.8 a | 2.3 b | 2.6 b | 3.3 b | 3.3 b |
| Leaves removed | 1.3 a | 1.9 a | 3.0 a | 3.9 a | 5.0 a | 5.4 a |
| Excised stems | | | | | | |
| Intact leaves | 1.0 b | 1.5 b | 1.6 c | 2.1 b | 3.3 b | 2.6 b |
| Leaves removed | 1.3 a | 1.9 a | 2.5 b | 2.6 b | 3.2 b | 2.9 b |

$^z$Means separation within columns. P < 0.05 level.

with no leaves removed. Excised stems with leaves removed exhibited a
faster rate of flower bud development than excised stems with no leaf
removal for the first 23 days of forcing, but no differences were found
for the remainder of the study. When unabscised leaves were left intact,
there were no differences between excised stems and intact plants in flo-

ral bud development for the first 7 days, but excised stems with leaves were significantly slower in development than intact plants after forcing for 16 and 23 d. For the rest of the study (30 to 44 d), there were no significant differences between excised stems and intact plants when leaves were left on.

When leaves were removed, there were no differences between excised stems and intact stems through 16 d of forcing, but throughout the rest of the study (23 to 44 d), flower bud development was significantly slower in excised stems compared to intact plants.

After 7 d of forcing, excised stems with leaves removed as compared to intact stems with unabscised leaves showed no significant differences in floral bud development. It appears that, 'Magnolia' excised stems, with leaves removed, exhibit the same rate of flower bud development as intact plants for a period of 4 to 6 weeks.

In study 2, floral bud development of excised stems remained equivalent to the stems on the intact plant for 23 d of forcing, and had reached or exceeded stage 3 of development (Table 2). After 30 d of forcing excised stems began to lag behind in development, and this continued throughout the rest of the study. Throughout the study, cutting the tips did not affect the rate of flower bud development.

## *CONCLUSION*

These studies indicate that with sufficient chilling, floral bud development in excised stems approximate that of intact plants for a period of 3.5 weeks if leaves are removed. Further studies are needed to ascertain that if flower buds reach stage 3 (an easily identifiable stage) within a

TABLE 2. Floral bud development of 'Tifblue' excised stems and intact stems, 1998.

| | Days of forcing | | | | | |
| | 7 | 16 | 23 | 30 | 40 | 44 |
|---|---|---|---|---|---|---|
| Plant | 1.1 a[z] | 2.2 a | 3.7 a | 4.7 a | 6.0 a | 6.1 a |
| Excised stems | | | | | | |
|   Cut tip | 1.2 a | 2.4 a | 3.6 a | 3.9 b | 4.2 b | 4.8 b |
|   Not cut | 1.5 a | 2.5 a | 3.5 a | 3.7 b | 4.3 b | 4.8 b |

[z]Means separation within columns. P < 0.05 level.

3.5 weeks period, then their chilling requirements have been fulfilled, and they will flower normally. If these criteria are met, we hope that this method will allow a quick determination of the chilling requirements using a minimum number of plants.

## LITERATURE CITED

Austin, M.E. and K. Bondari. 1987. Chilling hour requirement for flower bud expansion of two rabbiteye and one highbush blueberry shoots. HortScience 22:1247-1248.
Austin, M.E., B.G. Mullinix, and J.S. Mason. 1982. Influence of chilling on growth and flowering of rabbiteye blueberries. HortScience 17:768-769.
Norvell D.J. and J.N. Moore. 1982. An evaluation of chilling models for estimating rest requirements of highbush blueberries (*Vaccinium corymbosum* L.). J. Am. Soc. Hort. Sci. 107:54-56.
SAS Institute, Inc. 1985. SAS user's guide: Statistics. Version 5 Edition. Cary, NC.
Spiers, J.M. and A.D. Draper. 1974. Effect of chilling on bud break in rabbiteye blueberry. J. Amer. Soc. Hort. Sci. 99:398-399.
Spiers, J.M. 1976. Chilling regimes affect bud break in 'Tifblue' rabbiteye blueberry. J. Amer. Soc. Hort. Sci. 101:88-90.
Spiers, J.M. 1978. The effect of stage of bud development on cold injury in rabbiteye blueberry. J. Amer. Hort. Sci. 103:452-455.

# The Effect of Pruning
# on Blueberry Stem Gall Wasp

Kenna MacKenzie
David Hayman
Edward Reekie

**SUMMARY.** The influence of commercial pruning regimes on the survival of wasps inhabiting stem galls of lowbush blueberry (*Vaccinium angustifolium* Ait.) was studied in 1999 and 2000. Three commercial fields in Nova Scotia were used to examine the effect of mow pruning on wasp survival. Galls were removed from blueberry stems in fall 1999 and placed either above or within the leaf litter in a small blueberry plot or held at 2°C over the winter. Another group of galls was collected in spring 2000 from within and above leaf litter. Wasp emergence was not

Kenna MacKenzie is Research Scientist and David Hayman was Research Affiliate, Atlantic Food and Horticulture Research Centre, Agriculture and Agri-Food Canada, 32 Main Street, Kentville, Nova Scotia B4N 1J5 Canada. Edward Reekie is Professor, Department of Biology, Acadia University, Wolfville, Nova Scotia B0P 1X0 Canada.

The authors thank Dan Ryan for statistical support; Alana Hayman, Walter Wojtas, Stephanie Chaisson, and Jack Haggerty for excellent technical assistance; Andrew King for allowing them to borrow the plot burner from the Wild Blueberry Institute of Nova Scotia; and Keith Crowe, Bruce Mowatt, John Bragg, Clyde Blois, and Laurie Hanna for allowing the authors to work on their properties.

Supported by the Agri-Focus 2000 Technology Development Program of the Nova Scotia Department of Agriculture and Marketing.

affected by any treatment showing that mow pruning has no effect on wasp survival. In a second study, mowing was compared to mowing plus burning in either fall or spring. Burning did not affect gall number, but did affect wasp emergence. The poorest emergence from galls was seen in spring burning and those burnt in the fall had lower emergence than from galls that were only mowed. Thus, a spring burn is recommended if growers are concerned about stem gall populations in their fields. *[Article copies available for a fee from The Haworth Document Delivery Service: 1-800-HAWORTH. E-mail address: <docdelivery@haworthpress.com> Website: <http://www.HaworthPress.com> © 2004 by The Haworth Press, Inc. All rights reserved.]*

**KEYWORDS.** Blueberry stem gall, *Hemadas nubilipennis*, lowbush blueberry, *Vaccinium angustifolium*, pruning, cultural control

## *INTRODUCTION*

Galls, abnormal plant tissue growth, on the stems of wild blueberry (*Vaccinium angustifolium* Aiton) are caused by a small wasp, the blueberry stem gall wasp (Hymenoptera, Chalcidoidea, Pteromalidae, *Hemadas nubilipennis* Ashmead) (Shorthouse et al., 1986). Some six other species of chalcid wasps, which may be inquilines or parasitoids, also have been found inhabiting the galls (Brooks, 1993; Driggers, 1927; Hayman, 1998; McAlister and Anderson, 1932; Shorthouse et al., 1990).

Within wild blueberry fields in Nova Scotia, stem gall numbers appear to be increasing possibly due to changes in management strategies (MacKenzie, unpublished data). While growers are concerned that high levels of stem galls may reduce crop yields, the biggest problem with galls is that they have been found to contaminate both fresh and processed product. Even though concerns have been raised about gall populations, no control tactics have been developed for this insect (Crozier, 1997). Lowbush blueberry crop management involves a two-year cropping cycle. After harvest the fields are pruned by either mowing or burning to encourage vegetative growth the following year (McIsaac, 1997). While it has been suggested that pruning may reduce the numbers of stem galls by affecting blueberry stem gall wasp survival (Crozier, 1997), the issue has never been scientifically examined. This study was set up to determine if pruning reduces the survival of wasps inhabiting blueberry stem galls.

## MATERIALS AND METHODS

*Removal of galls from stems.* This study was done in three commercial fields in Nova Scotia. Thirty galls, collected from each field on 5 October 1999, were randomly divided into three groups. These were (1) held at 2°C, or put into mesh bags and placed (2) above or (3) within the leaf litter of a wild blueberry plot located in Kentville, Nova Scotia. On 4 May 2000, 10 galls from within and from above the leaf litter were collected from the three fields. The five treatment groups were then incubated at 21°C, 50%RH and constant fluorescent light with an average photosynthetic photon flux of 72.3 ± 3.6 µmoles $m^{-2}s^{-1}$. All emerged wasps were removed from cups daily. The wasps were identified to species. After emergence was completed the galls were dissected and all dead inhabitants counted.

For each field, the proportion of emerged wasps that were blueberry stem gall wasps was calculated. To examine the effects of treatment on wasp survival (includes all wasp species), survival [(emerged + 0.5)/total − emerged + 0.5)] was log transformed and subjected to a one-way ANOVA. Means were then back transformed.

*Pruning technique.* Five commercial fields in Nova Scotia were used for this study. After commercial mow pruning was completed in October 1999, the experiment was set up. Six parallel 100-meter transects each separated by 7 m were staked off, and two of these were randomly assigned to each of three treatments: mow only, mow and fall burn, and mow and spring burn. A small diesel-fueled plot burner was used to apply the burn treatments with fall burning done in November and December 1999, and spring burning in April 2000.

In May 2000, all galls above and within the leaf litter in a 0.5 $m^2$ quadrat were collected at five meter intervals along each transect. One randomly selected gall formed in 1999 from each quadrat where galls were present was incubated and evaluated as described above with the exception that emerged wasps were removed weekly rather than daily.

The effect of treatment on the total number of stem galls per quadrat were tested using a split plot ANOVA. For each field, the proportion of emerged wasps that were blueberry stem gall wasps was calculated. Data analyses was the same as the previous section except that a split plot ANOVA was performed.

## RESULTS AND DISCUSSION

*Removal of galls from stems.* It was common for more than one species to emerge from a single gall with a total of six species (*H.*

*nubilipennis, Eurytoma solenozopheriae* Ashmead, *Sycophila vaccinii-cola* Balduf, *Orymus vacciniicola* Ashmead, *Eupelmus vesicularis* Ritzius and an unidentified species of *Pteromalus*) present in this study. Blueberry stem gall wasps emerged from 46%, 26%, and 28% of the incubated galls from the three sites. In terms of community structure, blueberry stem gall wasps made up 27%, 15%, and 17% of all wasps in the three sites, respectively.

Wasp survival was significantly affected by treatment ($p < 0.05$) (Figure 1). The poorest survival was seen in galls that were held at 2°C and greatest survival was from galls wintered at Kentville within the leaf litter. Emergence was high in this study at 80% for galls wintered in the field plot at Kentville and 70% for those wintered in the fields. Differences in survival are likely due to climatic conditions as Kentville has milder winter temperatures than the commercial blueberry fields. This work shows that removal of galls by mow pruning is likely to have little affect on Blueberry stem gall wasp populations.

*Pruning technique.* The six species of chalcid wasps mentioned

FIGURE 1. The effect of gall removal from lowbush blueberry stems on survival of wasps inhabiting blueberry stem galls. Treatments are designated as follows: 2C–removed from stems in fall and held at 2°C over winter, ra–removed from stems in fall and placed in field above the leaf litter at Kentville, N.S., rwi–removed from stems in fall and placed in field within the leaf litter at Kentville, N.S., fa–wintered in field above leaf litter, and fwi–wintered in field below leaf litter.

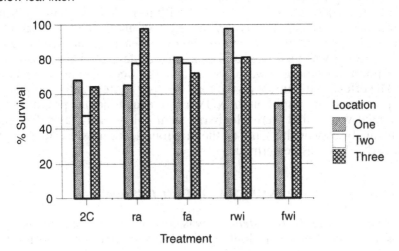

above were also collected in this study. In the five fields, the Blueberry stem gall wasp made up 24%, 27%, 28%, 51%, and 70% of the wasps that emerged from the incubated galls. Differences in co-inhabitant levels as large as this suggest that parasitoids may be involved in regulating Blueberry stem gall wasp populations and it would be interesting to further examine this issue.

There were no significant differences in the effect of treatments in the five fields, and thus, data was pooled for analysis. The total number of galls was similar for the three pruning treatments indicating that the burning treatment did not physically destroy galls. Wasp survival was greatest in the mowed only plots, with fall burning intermediate and spring burning showing the lowest survivor rates ($p < 0.05$) (Figure 2). Galls above the leaf litter tended to have poorer survival than those within it.

The wasp community associated with mature blueberry stem galls from this study is similar to that seen in other research. Similar species were reared from galls collected from *V. angustifolium*, *V. atrococcum* (Gray), and *V. corymbosum* L. (Brooks, 1993; Hayman, 1998; Judd, 1959; McAlister and Anderson; 1932, Shorthouse et al., 1990). Except for two fields where it dominated, *H. nublipennis* made up a small proportion of the wasps emerging from galls in this study. Similar high

FIGURE 2. The effect of pruning method on survival of wasps inhabiting blueberry stem galls. For each prune treatment and position, letters above the bars indicate statistically significant differences in means ($p < 0.05$).

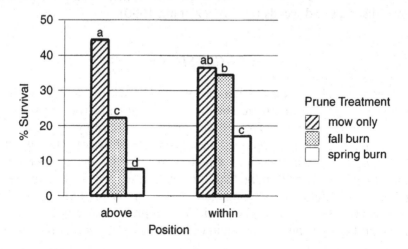

parasitism was seen by Shorthouse et al. (1990). They felt that the parasitoids delay their development until after blueberry stem gall wasp larvae have almost completed development and the gall is near maturity. Thus, although parasitoids may influence stem gall populations in the future by reducing gall wasp populations, the presence of galls that contaminate the berries could still remain a problem.

The removal of galls from stems in the fall had no effect on survival as compared to galls removed from the plant the following spring. Blueberry stem gall wasp larvae mature in mid-August and survive the winter as prepupae (West and Shorthouse, 1989). Thus, this insect has reached the prepupal wintering stage well before the time of fall pruning and is able to survive removal from the plants. Burning reduced survival as compared to mowing. This was an effect of high temperature as wasps in galls within the leaf litter had better survival than those on the top of the litter that were directly exposed to the burn. Other work has also found that insects living within or near the ground are not as affected by fire as those in the above ground vegetation (Seastedt and Reddy, 1991). While the switch to mow pruning has been on-going within the lowbush blueberry industry due to better land-leveling and the higher costs of burning (Kinsman, 1993), it now appears that other problems such as increasing numbers of blueberry stem galls could be exacerbated by this management change. Other benefits of the use of a burn prune every three or four production cycles could include reductions in other insect pests such as blueberry leaftier (Ponder and Seabrook, 1994) and blueberry spanworm (DeGomez et al., 1990), and some diseases and weeds (DeGomez et al., 1990).

## CONCLUSIONS

This research has shown that the survival of wasps inhabiting blueberry stem galls is not affected by the removal of galls from the stem by mow pruning in the fall. The pruning practice of burning, especially in the spring, significantly reduced wasp survival. However, the fact that the number of galls was not affected means that the temperatures reached above and within the leaf litter by burning was sufficient to kill the insects without destroying gall tissues. It is recommended that lowbush blueberry producers consider burn pruning as a pest management strategy for the management of blueberry stem gall populations.

## GROWER BENEFITS

Lowbush blueberry growers have concern about the number of blueberry stem galls within their fields. This research shows that it is possible to reduce gall wasp populations by using a spring burn for pruning. Growers should monitor populations within their fields, and if levels are of concern, they should plan their pruning treatments accordingly.

## LITERATURE CITED

Brooks, S. 1993. Geographic variation of the parasitoid complex associated with galls induced by *Hemadas nubilipennis* (Hymenoptera: Pteromalidae). Honours Thesis, Laurentian Univ., Sudbury, Ont.

Crozier, L. 1997. The blueberry stem gall. Lowbush blueberry fact sheet. N.S. Dept. Agric. Mktg., Truro, N.S.

DeGomez, T., D. H. Lambert, H. Y. Forsythe, Jr., and J. A. Collins. 1990. The influence of pruning methods on disease and insect control. Wild blueberry fact sheet No. 218. Univ. Maine Coop. Ext., Orono, Maine.

Driggers, B. F. 1927. Galls on stems of cultivated blueberry (*Vaccinium corymbosum*) caused by a chalcidoid, *Hemadas nubilipennis* Ashmead. J. N. Y. Entomol. Soc. 34:253-259.

Hayman, D. I. 1998. Species composition, development and dispersal of wasps in lowbush blueberry stem galls. Thesis for the Certificate of Honours Equivalency. Saint Mary's Univ., Halifax, N.S.

Judd, W. W. 1959. Studies of the Byron bog in southwestern Ontario. vii. Wasps reared from the blueberry stem gall on *Vaccinium atrococcum*. Trans. Amer. Microsc. Soc. 78:212-214.

Kinsman, G. 1993. The history of the lowbush blueberry industry in Nova Scotia: 1950-1990. Blueberry Producer's Assoc. N.S., N.S. Dept. Agric. Mktg., Truro, N.S.

McAlister, L. C. and W. H. Anderson. 1932. The blueberry stem gall in Maine. J. Econ. Entomol. 25:1164-1169.

McIsaac, D. 1997. Growing wild blueberries in Nova Scotia. Lowbush blueberry fact sheet, N.S. Dept. Agric. Mktg., Truro, N.S.

Ponder, B. M. and W. P. Seabrook. 1994. The effect of pruning of *Vaccinium angustifolium* on the *Croesia curvalana* larval population. J. Small Fruit Viticul. 2(2):57-64.

Seastedt, T. R. and M. V. Reddy. 1991. Fire, mowing and insecticide effects on soil Sternorrhyncha (Homoptera) densities in tall-grass prairie. J. Kansas Entomol. Soc. 64:238-242.

Shorthouse, J. D., I. F. Mackay, and T. J. Zmijowshiyj. 1990. Role of parasitoids associated with galls induced by *Hemadas nubipennis* (Hymneoptera: Pteromalidae) on lowbush blueberry. Environ. Entomol. 19:911-915.

Shorthouse, J. D., A. West, R. W. Landry, and P.D. Thibideau. 1986. Structural damage by female *Hemadas nubilipennis* (Hymenoptera: Pteromalidae) as a factor in gall induction on lowbush blueberry. Can. Entomol. 118:249-254.

West, A. and J. D. Shorthouse. 1989. Initiation and development of the stem gall induced by *Hemadas nubilipennis* (Hymenoptera: Pteromalidae) on lowbush blueberry, *Vaccinium angustifolium* (Ericaceae). Can. J. Bot. 67:2187-2198.

# The Effects of Chill Hour Accumulation on Hydrogen Cyanamide Efficacy in Rabbiteye and Southern Highbush Blueberry Cultivars

Stephen J. Stringer
Donna A. Marshall
Blair J. Sampson
James M. Spiers

**SUMMARY.** A controlled environment study was conducted to evaluate the effects of chill hour accumulation on the time of application and the resulting efficacy of the plant growth regulator, hydrogen cyanamide ($H_2CN_2$) in both rabbiteye (*Vaccinium asheii* Reade) and southern highbush (*V. corymbosum*) blueberry cultivars. Application of $H_2CN_2$ at the interval in which accruement of 75% of the individual chill-hour requirements of 'Bladen', 'Jubilee', 'Premier', and Tifblue' blueberry cultivars resulted in greater vegetative bud break than the 50% chill-hour application timing, or than their untreated checks. The 75% timing also resulted in a significant increase in the terminal growth of stems in

---

Stephen J. Stringer is Research Geneticist, Donna A. Marshall is Horticulturist, Blair J. Sampson is Research Entomologist, and James M. Spiers is Research Horticulturalist; all with the Department of Agriculture-Agricultural Research Service, Small Fruit Research Station, P.O. Box 287, Poplarville, MS 39470.

Address correspondence to: Stephen J. Stringer (E-mail address: sjstringer@ars.usda.gov).

[Haworth co-indexing entry note]: "The Effects of Chill Hour Accumulation on Hydrogen Cyanamide Efficacy in Rabbiteye and Southern Highbush Blueberry Cultivars." Stringer, Stephen J. et al. Co-published simultaneously in *Small Fruits Review* (Food Products Press, an imprint of The Haworth Press, Inc.) Vol. 3, No. 3/4, 2004, pp. 339-347; and: *Proceedings of the Ninth North American Blueberry Research and Extension Workers Conference* (ed: Charles F. Forney, and Leonard J. Eaton) Food Products Press, an imprint of The Haworth Press, Inc., 2004, pp. 339-347. Single or multiple copies of this article are available for a fee from The Haworth Document Delivery Service [1-800-HAWORTH, 9:00 a.m. - 5:00 p.m. (EST). E-mail address: docdelivery@haworthpress.com].

'Bladen' and 'Premier'. Consideration of a blueberry cultivar's exposure to chill-hours in application timing decisions should provide a greater degree of precision in optimizing vegetative bud break with $H_2CN_2$. *[Article copies available for a fee from The Haworth Document Delivery Service: 1-800-HAWORTH. E-mail address: <docdelivery@haworthpress.com> Website: <http://www.HaworthPress.com>]*

**KEYWORDS.** Chill hours, dormancy, Dormex®, hydrogen cyanamide

## INTRODUCTION

Deciduous fruit crops including blueberries annually undergo the physiological phase of development known as dormancy. The release of dormancy and subsequent development of floral and vegetative buds requires a period of exposure to winter chilling temperatures followed by a subsequent rise in temperatures in the spring (Coville, 1921; Fuchigami et al., 1982; NeSmith and Bridges, 1992; Suare, 1985). The amount of chilling (hours of exposure to temperatures below 7.2°C) required to break dormancy and induce floral and vegetative bud break varies among blueberry cultivars (Austin and Bondari, 1987; Darrow, 1942). Both rabbiteye (*Vaccinium ashei* Reade) and southern highbush (*Vaccinium corymbosum* L.) blueberries are grown in the Southeastern US and have chilling requirements ranging from 200 to 650 h. Mild winters occur frequently in this region and often result in abnormal floral and vegetative bud development in some blueberry cultivars (Lyrene and Williamson, 1997; NeSmith and Bridges, 1992). The adverse effects of delayed spring leafing in blueberries include reductions in fruit set and quality, and delays in maturity (Williamson and Lyrene, 1995). Stresses associated with insufficient leaf canopy may also result in early growth cessation, inhibiting the development of shoots, and floral and leaf buds necessary for the subsequent year's crop (Erez, 1987). Utilization of growth regulating compounds provides a mechanism for circumventing effects of insufficient chilling following mild winters by breaking leaf bud dormancy and promoting earlier spring leaf development. This relatively new cultural practice has importance as a management tool for Southern blueberry growers who previously without recourse faced risks to crop production in winters of insufficient chilling.

Dormex® (SKW, Trostberg, Germany; 49% hydrogen cyanamide, $H_2CN_2$), has been demonstrated to, in effect, substitute for inadequate chilling and promote vegetative bud break and advanced leaf development in numerous deciduous fruit crops (Dokoozlian and Williams, 1995; Erez, 1987; Shulman et al., 1986; Siller-Cepeda et al., 1992). When applied properly to blueberries, its utilization has been demonstrated to advance vegetative bud break, increase leaf : fruit ratios, and hasten fruit maturity (Williamson et al., 2002). However, $H_2CN_2$ is phytotoxic to flower buds in more advanced stages of development, and in some fruit crops its use is actually recommended as a crop thinning agent (Falhi et al., 1998). Thus, the proper timing of $H_2CN_2$ applications in blueberries is critical in optimizing its efficacy for promoting leaf development while concurrently minimizing injury to those developing floral buds desired for fruit production (NeSmith, 1998). Current application timing recommendations from Florida and Georgia suggest that the material should be applied during the dormant season after "significant" chilling has occurred and before a significant number of flower buds have reached stage 3 (Spiers, 1978) of development. Although $H_2CN_2$ usage following these recommendations successfully improves spring leafing, the effects of chill hour accumulation on $H_2CN_2$ efficacy are not completely understood. Thus, the objectives of this research were to evaluate the effects of utilizing values for chill hour accumulation on $H_2CN_2$ efficacy, and assess its effects on crop initiation of rabbiteye and southern highbush blueberry cultivars.

## MATERIALS AND METHODS

The experiment was conducted under greenhouse conditions in Poplarville, MS during 2001 and 2002. One-year-old potted 'Bladen' and 'Jubilee' southern highbush and 'Premier' and 'Tifblue' rabbiteye blueberries were used for this study. These blueberry cultivars are known to have chill hour accumulation (CHA) requirements of approximately 600, 500, 450, and 600 h below 7.2°C, respectively. Forty plants of each cultivar were placed in a cold room and sixteen of each were differentially removed after exposure to chilling of 50%, 75%, and 100% of their respective chill hour requirements. At each removal period, a solution of either 0 or 1.0% $H_2CN_2$ and 0.25% Surfaid, a non-ionic surfactant, was applied to point of drip to the entire canopy of eight plants per cultivar utilizing a knapsack sprayer. Plants allowed to accumulate 100% of their chill hour values were not treated with $H_2CN_2$ but instead were used only as controls for comparative purposes. Treatments were arranged in a split-plot design with cultivar as the main effect and chill hour accumulation at time of application as sub-effects. Thus, each of eight plants represented a replication for each cultivar/timing/treatment combination. The developmental stages of vegetative and flower buds at the time of application were rated 1.0 (no visible bud swelling) and 2.0 (visible swelling of buds, scales starting to separate, flowers still completely in scales), respectively. To further promote bud break, following $H_2CN_2$ applications, plants were moved to, and grown under greenhouse conditions in which temperatures ranged from 20 to 25°C.

Ten stems per plant were randomly selected and tagged for evaluation. All vegetative buds on each stem were rated for stage of physiological development (NeSmith et al., 1998) every 7 to 10 d until 56 d after treatment. Flower buds were also assessed for evidence of mortality. At 56 d after treatment, the terminal growth of 10 stems per plant was measured. No fruit or yield data were included in the evaluations since natural pollination did not occur in the greenhouse environment. Data were subjected to mean and standard error calculations or ANOVA procedures (SAS Institute Inc., Cary, NC).

## RESULTS AND DISCUSSION

No injury to flower buds was detected in any cultivar at either $H_2CN_2$ application timing (data not shown). Results of observations on vegetative bud development for the respective blueberry cultivars are presented in Figures 1-4. At each application timing, the 1.0% $H_2CN_2$ treatment resulted in increased rate of vegetative bud development in each of the four blueberry cultivars when compared to 0% $H_2CN_2$. Mean bud stage ratings observed on the last evaluation dates indicated that when $H_2CN_2$ was applied at the point at which cultivars had accumulated 50% of their chill hour requirement the vegetative bud ratings of 'Bladen', 'Jubilee', 'Premier', and 'Tifblue' advanced by 0.2, 0.4, 0.5, and 0.7 stages, respectively. $H_2CN_2$ applications made when these cultivars had accrued 75% of their chill hour requirements resulted in physiological development rating increases of 0.9, 1.0, 1.0, and 1.2, respectively. Comparing the 50% CHA, 1.0% $H_2CN_2$ application to the control (100% CHA, 0% $H_2CN_2$) at these evaluation dates, bud development ratings of the respective cultivars differed by −0.1, 0.4, 0.3, and

FIGURE 1. Effect of chill hour accumulation on efficacy of $H_2CN_2$ in vegetative budbreak and development in 'Bladen' southern highbush blueberry. Within the individual 50%, 75%, and 100% CHA (chill hours accumulated) figures, control = untreated.

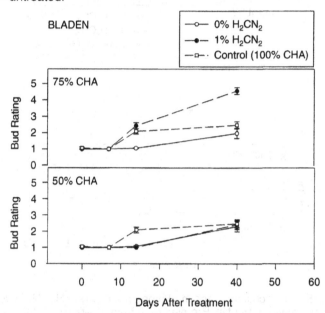

0.3 stages. At 75% CHA + $H_2CN_2$, advances in bud development were substantially greater with rating increases of 1.9, 2.2, 1.5, and 1.0, respectively, over that of the control. Results of this study demonstrated that 1.0% $H_2CN_2$ applied to blueberry cultivars having artificially accrued 75% of their chill hour requirements, advanced vegetative bud development to a greater extent than applications at 50% CHA. This same treatment generated even greater vegetative bud development than that observed on cultivars having accrued 100% of their chill requirement, but lacking a supplemental $H_2CN_2$ application.

Results of evaluations on the effects of CHA at time of $H_2CN_2$ application on terminal growth of blueberry stems are presented in Table 1. At the 50% CHA application timing the control resulted in significantly greater terminal stem growth than either 0% or 1.0% $H_2CN_2$ in 'Bladen', 'Jubilee', and 'Premier'. Terminal growth resulting from 1.0% $H_2CN_2$ application at 50% CHA did not differ significantly from that from that of 0% $H_2CN_2$ in any cultivar. At 75% CHA, terminal stem growth of

FIGURE 2. Effect of chill hour accumulation on efficacy of $H_2CN_2$ in vegetative budbreak and development in 'Jubilee' southern highbush blueberry. Within the individual 50%, 75%, and 100% CHA (chill hours accumulated) figures, control = untreated.

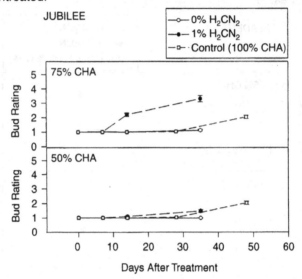

FIGURE 3. Effect of chill hour accumulation on efficacy of $H_2CN_2$ in vegetative budbreak and development in 'Premier' rabbiteye blueberry. Within the individual 50%, 75%, and 100% CHA (chill hours accumulated) figures, control = untreated.

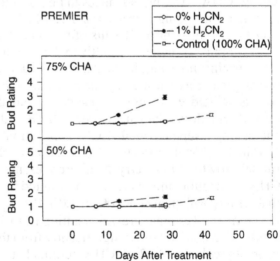

FIGURE 4. Effect of chill hour accumulation on efficacy of $H_2CN_2$ in vegetative budbreak and development in 'Tifblue' rabbiteye blueberry. Within the individual 50%, 75%, and 100% CHA (chill hours accumulated) figures, control = untreated.

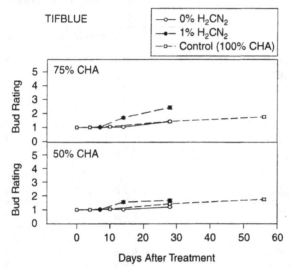

both 'Bladen' and 'Premier' resulting from 1.0% $H_2CN_2$ was significantly greater than that resulting from 0% $H_2CN_2$, and that of 'Bladen' was also significantly greater than the control. Terminal growth on stems of 'Tifblue' was not affected by chill hour accumulation at the time of $H_2CN_2$ application.

## CONCLUSIONS AND GROWER BENEFITS

The effects of chill hour accumulation on the efficacy of $H_2CN_2$ in blueberry cultivars have not been completely elucidated. This study demonstrates that the consideration of the degree of chilling a blueberry cultivar has accrued in application timing decisions should optimize the efficacy of $H_2CN_2$ and increase vegetative bud-break and subsequent bud development. Since the quality of artificial chilling differs from actual field conditions, field studies will be conducted to assess the effects of natural chill hour accumulation on the efficacy of $H_2CN_2$ in the induction of bud break in blueberry cultivars.

TABLE 1. The effect of CHA and $H_2CN_2$ on terminal growth of stems of blueberry cultivars.

| | Chill Hour Accumulation (%) | |
| | 50 | 75 |
|---|---|---|
| | ------------Terminal growth (mm)------------ | |
| Bladen | | |
| 1.0% $H_2CN_2$ | 4.8 b[z] | 15.5 a |
| 0.0% $H_2CN_2$ | 4.4 b | 4.9 c |
| Control[y] | 9.3 a | 9.3 b |
| Jubilee | | |
| 1.0% $H_2CN_2$ | 1.5 b | 14.4 a |
| 0.0% $H_2CN_2$ | 0.7 b | 7.1 a |
| Control | 7.9 a | 7.9 a |
| Premier | | |
| 1.0% $H_2CN_2$ | 4.0 b | 22.7 a |
| 0.0% $H_2CN_2$ | 0.3 b | 6.8 b |
| Control | 12.7 a | 12.7 ab |
| Tifblue | | |
| 1.0% $H_2CN_2$ | 8.0 a | 8.0 a |
| 0.0% $H_2CN_2$ | 7.5 a | 7.8 a |
| Control | 8.6 a | 8.7 a |

[y]Means separation within cultivar and within columns, $P < .05$ level, N = 10 stems per cultivar.
[z]100% chilling used as control.

# LITERATURE CITED

Austin, M.E. and K. Bondari. 1987. The effect of chilling temperature on flower bud expansion of rabbiteye blueberry. Sci. Hortic. 31:71-79.

Coville, F.V. 1921. The influence of cold stimulating the growth of plants. J. Agr. Res. 20:151-160.

Darrow, G.M. 1942. Rest requirements for blueberries. Proc. Amer. Soc. Hort. Sci. 41:189-194.

Dokoozlian, N.K. and L.E. Williams. 1995. Chilling exposure and hydrogen cyanamide interact in breaking dormancy in grape buds. HortScience 30:1244-1247.

Eck, P. 1988. Plant development. In: Blueberry science. Paul Eck (ed.). Rutgers Univ. Press, New Brunswick, NJ.

Erez, A. 1987. Chemical control of bud break. HortScience 22:1244-1247.

Fallahi, E., R.R. Lee, and G.A. Lee. 1998. Commercial-scale use of hydrogen cyanaide for apple and peach blossom thinning. HortTechnology 8:556-560.

Fuchigami, L.H., C.J. Weiser, K. Kobayashi, R. Timmis, and L.V. Gusta. 1982. A degree growth stage (°GS) model and cold acclimation in temperate woody plants, pp. 93-116. In: P.H. Li, and A. Sakai (eds.). Plant cold hardiness and freezing stress, vol. 2, Academic Press, NY.

Lyrene, P.M. and J.G. Williamson. 1997. Highbush blueberry varieties for Florida. Proc. Fla. State Hort. Soc. 110:171-174.

NeSmith, D.S. 1998. The effect of timing of Dormex applications on blueberry leafing and flower mortality. HortScience 33:601.

NeSmith, D.S., G. Krewer, and J.G. Williamson. 1998. A leaf bud development scale for rabbiteye blueberry (*Vaccinium ashei*, Reade). HortScience 33:757.

NeSmith, D.S. and D.C. Bridges. 1992. Modeling chilling effects on cumulative flowering: A case study using 'Tifblue' rabbiteye blueberry. J. Amer. Soc. Hort. Sci. 117:698-702.

SAS Institute, 1989. SAS/STAT user's guide. 4th ed. Ver. 6. SAS Inst., Cary, NC.

Shulman, Y., G. Nair, and S. Lavee. 1986. Oxidative processes in doremancy and the use of hydrogen cyanamide in breaking bud dormancy. Acta Hort. 179:141-148.

Siller-Cepeda, J.H, L.H. Fuchigami, and T.H.H. Chen. 1992. Hydrogen cyanamide-induced bud break and phytotoxicity in 'Redhaven' peach buds. HortScience 27: 874-876.

Spiers, J.M. 1978. Effect of stage of bud development on cold injury in rabbiteye blueberry. J. Amer. Soc. Hort. Sci. 103:452-455.

Suare, M.C. 1985. Dormancy release in deciduous fruit trees. Hort. Rev. 7:239-300.

Williamson, J.G., G. Krewer, B.E. Maust, and P.E. Miller. 2002. Hydrogen cyanamide accelerates vegetative bud break and shortens fruit development period. HortScience 37:539-542.

Williamson, J.G., B.E. Maust, and D.S. NeSmith. 1998. Timing and concentration of hydrogen cyanamide affect blueberry bud development and flower mortality. Hort-Science 36:922-924.

Williamson, J.G. and P.M. Lyrene. 1995. Commercial blueberry production in Florida. Univ. Florida Coop. Ext. Serv. Bul. SP-0129, Gainsville, FL.

Williamson, J.G., G. Krewer, B.E. Maust, and P.E. Miller. 2002. Hydrogen cyanamide accelerates vegetative bud break and shortens fruit development period. HortScience 37:539-542.

Williamson, J.G., G. Krewer, B.E. Maust, and P.E. Miller. 2002. Hydrogen cyanamide accelerates vegetative bud break and shortens fruit development period. HortScience 37:539-542.

# Evaluation of New Approaches for Management of Japanese Beetles in Highbush Blueberries

Rufus Isaacs

Zsofia Szendrei

John C. Wise

**SUMMARY.** The Japanese beetle, *Popillia japonica*, can be a pest of highbush blueberries because of direct feeding on berries and leaves, and the risk of contaminating harvested fruit. To determine where beetles are most abundant and whether cultural controls have potential for use against *P. japonica* in blueberry, soil was sampled for grubs during 2001 and 2002 in and around fifteen blueberry fields. Densities of

Rufus Isaacs is Assistant Professor and Small Fruits Extension Specialist, Zsofia Szendrei is Graduate Student, and John C. Wise is Research Specialist and Director of the Trevor Nichols Research Complex, Department of Entomology, Michigan State University, East Lansing, MI 48824 USA.

Address correspondence to: Dr. Rufus Isaacs, Department of Entomology, Michigan State University, East Lansing, MI 48824 (E-mail: isaacsr@msu.edu).

This research was supported by the USDA-CSREES Crops at Risk program, Project GREEEN, and the Michigan Blueberry Growers Association. The authors thank Keith Mason, Nikhil Mallampalli, and Ann Hanley for assistance with this research. They also acknowledge Dave Trinka of Michigan Blueberry Growers Association and the Michigan blueberry growers who made this work possible by providing access to their fields. Mention of these products does not constitute an endorsement over other similar products.

[Haworth co-indexing entry note]: "Evaluation of New Approaches for Management of Japanese Beetles in Highbush Blueberries." Isaacs, Rufus, Zsofia Szendrei, and John C. Wise. Co-published simultaneously in *Small Fruits Review* (Food Products Press, an imprint of The Haworth Press, Inc.) Vol. 3, No. 3/4, 2004, pp. 349-360; and: *Proceedings of the Ninth North American Blueberry Research and Extension Workers Conference* (ed: Charles F. Forney, and Leonard J. Eaton) Food Products Press, an imprint of The Haworth Press, Inc., 2004, pp. 349-360. Single or multiple copies of this article are available for a fee from The Haworth Document Delivery Service [1-800-HAWORTH, 9:00 a.m. - 5:00 p.m. (EST). E-mail address: docdelivery@haworthpress.com].

overwintering *P. japonica* were greater under permanent sod outside fields than in the soil between the rows. When grub density inside the fields was compared between clean-cultivated rows and those with sodded row middles, cultivated rows had significantly lower grub densities than those with sod. Bioassays with pyrethrum insecticides against adult beetles indicated their potential for removal of beetles from bushes just prior to harvest. An integrated strategy including elimination of suitable habitat by cultivation and use of chemical controls to remove beetles before harvest is under development for reducing populations and minimizing contamination of blueberry by this pest. *[Article copies available for a fee from The Haworth Document Delivery Service: 1-800-HAWORTH. E-mail address: <docdelivery@haworthpress.com> Website: <http://www. HaworthPress.com> © 2004 by The Haworth Press, Inc. All rights reserved.]*

**KEYWORDS.** Japanese beetle, *Popillia japonica*, blueberry, cultural control, integrated pest management

## INTRODUCTION

The Japanese beetle, *Popillia japonica* Newman (Coleptera: Scarabaeidae) was introduced accidentally from Japan to the eastern coast of North America in 1916 (Fleming, 1972). Since its arrival in southern New Jersey, this mobile insect has gradually increased in geographic distribution. *P. japonica* has become established as far west as Minnesota and Kansas (Anonymous, 2000) and has been sporadically detected and eradicated in some western US states (Fleming, 1972; Potter and Held, 2002). Adult beetles can cause extensive feeding damage to a broad range of ornamental and fruit plants by feeding on their leaves and fruit (Fleming, 1972; Potter and Held, 2002), and this species has become a major pest of managed turf due to the preference of larval stages (grubs) for feeding on grass roots (Vittum et al., 1999). Beetles can cause feeding damage on leaves and ripe berries of highbush blueberries, *Vaccinium corymbosum* L. if other plants are not present. However, it is the presence of adult beetles on plants at harvest that is of greatest concern, due to the risk of contamination of harvested berries. Japanese beetles were first detected in Michigan in the early 1930s near Detroit and even though early eradication programs included lead arsenate applications at 1000 pounds per acre, the beetle has increased in range and is now found throughout the lower tier of Michigan (Cappaert and Smitley, 2002), including some regions with highbush blueberry production. Insecticide applications continue to be the foundation of management strategies for Japanese beetle in blueberries and many other fruit crops, as growers strive to meet the market's demand for contamination-free fruit. Broad-spectrum insecticides are effective against beetles that are treated directly (Wise and Isaacs, unpublished data), but most of these products have long pre-harvest intervals creating a potential for immigrating beetles to re-infest fields as residue activity declines. Over 70% of Michigan's 7,285 ha (18,000 acres) of blueberry is harvested mechanically, and because beetles can be present on bushes at the time of harvest, strategies that eliminate the risk of adult beetles contaminating the fruit are needed.

Conditions in and around crop fields can favor development of this insect if primary requirements for Japanese beetle population development are met (Vittum et al., 1999). Ground covers of seeded grass or the naturally-invading mix of grass and broadleaved weeds are commonly used in blueberry fields to maintain soil structure, provide conditions for agricultural machinery to drive across during wet conditions, and reduce soil erosion. Japanese beetle grubs feed on the roots of grasses and some broadleaved weeds (Crutchfield and Potter, 1995) and blueberry fields often have areas with these ground covers within and around the bushes. In addition, sufficient soil moisture for development of first instar grubs is present in many commercial blueberry fields in July and August because of natural rainfall and irrigation, and soil temperatures above $-9.4°C$ in winter allow grub survival. These conditions are not unique to Michigan, and Japanese beetle has become established in most of the eastern US states (Allsopp, 1996). Although biological control agents could provide population suppression, a recent survey indicates that natural enemies are at extremely low levels in the Michigan population of Japanese beetle (Cappaert and Smitley, 2002), compared to those found in areas near to the original site of introduction. This is presumably because native and introduced biological control agents have yet to become established in this leading edge of the beetle's expanding geographic distribution.

Use of a short-lived insecticide with rapid 'knock-down' activity, such as a pyrethrum may be an effective approach for removing live beetles from bushes immediately before harvest. Extracted from chrysanthemum flowers, pyrethrums create immediate paralysis in treated insects, causing them to drop from plants (Casida and Quistad, 1995). With their short residual activity combined with minimal pre-harvest and re-entry restrictions, pyrethrums may provide growers with a tool for meeting the target of insect-free fruit, even in situations where immigrating Japanese beetles are challenging the performance of broad-spectrum insecticides.

Recent experience indicates that one-dimensional strategies based solely on conventional insecticides may not be an effective long-term approach for achieving the required pest management goals for Japanese beetle in blueberry. The combination of market demands and pest abundance makes it imperative that management strategies include approaches for reduction of overall beetle populations. An integrated Japanese beetle management program should include (1) consideration of the surrounding habitat as part of the production system, (2) reducing the attraction of blueberry crop habitats to adult beetles, (3) establishing a suppressive habitat for grub development, (4) use of effective insecticides at the appropriate time in the production cycle, and (5) optimizing post-harvest beetle removal. Biological control should also be included as part of a long-term strategy, and future introductions may be one approach to aid in the long-term suppression of populations of *P. japonica* within the blueberry production landscape.

As part of our research toward developing the integrated program outlined above, this study aimed to determine the relative abundance of Japanese beetles in and around blueberry fields. Within fields, the effect of tillage for control of Japanese beetle was determined by comparing grub abundance in row middles that were rotovated with those covered by permanent sod. In addition, we report on bioassays to compare the activity of biological insecticides that can be applied immediately before harvest to remove beetles from bushes before harvest.

## MATERIALS AND METHODS

*Grub sampling.* During April 2001 and May 2002, soil samples were taken at 15 commercial blueberry farms in southwest Michigan (in Allegan, Berrien, Muskegon, Ottawa, and Van Buren counties) to evaluate the density of Japanese beetle larvae. Soil was sampled to 15 cm depth from the perimeter and from row middles at each of the farms using a golf cup cutter (10 cm$^2$ area). Eighty perimeter samples were taken from between the edge of the blueberry field and the woods. To sample grubs within each field, a row was selected at random from each of ten equal sections across the field width. Within each selected row, grubs were sampled from the mid-point of the row middle, at six positions spaced along the length of the field. Soil cores were examined in the field for white grubs, and all grubs were placed in plastic bags with a small amount of soil and transported back to the laboratory for species identification using the diagnostic rastral pattern in which Japanese beetle has 14 or fewer hairs in a V-shaped pattern near the anus (Vittum et al., 1999). The number of Japanese beetle grubs was recorded and the number per unit area was calculated for each sample. Data were analyzed to compare abundance in the interior and the perimeter of the fields. Row middle samples were analyzed separately to compare larval densities in rotovated and sodded row middles. Due to the non-normality of the data, statistical comparisons were made using a Mann-Whitney U test.

*Bioassays.* To test the effect of pyrethrin insecticides on Japanese beetle, blueberry shoots were cut from untreated *V. corymbosum* cv. Rubel bushes at the Trevor Nichols Research Complex, Fennville, MI (TNRC), and brought to the laboratory. Shoots were adjusted to a standard size of 10 leaves and 5 berries, and these were treated by spraying to runoff with aqueous solutions of either Pyganic EC 1.4 (1.4% pyrethrum) or Evergreen EC 60-6 (6% pyrethrum and 60% piperonyl butoxide). Solutions were prepared to be equivalent to a 0.473 L/ha rate of Pyganic (16 oz/acre) or 0.118 L/ha of Evergreen (4 oz/acre), to provide nearly equivalent rates of active ingredient in the two solutions (7% difference). Solutions of these insecticides were prepared at the appropriate rate, placed into spray bottles, and mixed thoroughly before application. After shoots were treated to runoff, they were shaken to remove excess solution, then placed into bioassay chambers (1.8 L plastic cups with ventilation holes in the wall) containing 5 cm depth of floral oasis on the bottom that was covered with paraffin wax. Adult beetles were collected during the two hours prior to bioassays from an unsprayed grassy field at TNRC using baited Japanese beetle traps (Trecé Inc., Salinas, CA). Beetles were then held on blueberry foliage in a ventilated container until needed. Ten beetles were placed in the bottom of each chamber immediately after foliage was in place, and the cups were sealed with a lid punctured with holes for ventilation. The beetles' position and behavior within the chamber was observed after 1, 24, and 120 hours of exposure. Beetles were scored for mobility, and whether they exhibited sub-lethal toxicity symptoms. After the 120 h observations, chambers were opened, and visual assessments were made of feeding damage by the beetles to foliage and fruit, using 5% increments. Bioassay data were analyzed by analysis of variance on arcsine transformed percentages, followed by Fisher's protected least square difference test to compare between treatments.

## RESULTS AND DISCUSSION

A majority of the blueberry fields at farms in this study had a mix of grass and broad leaf weeds in the row middles, showing that this is a common ground cover in Michigan blueberry fields where Japanese beetle is reported. All sampled blueberry fields were surrounded by grass or a grass-weed mix in the area where spraying and harvesting machinery are driven. These fields varied in whether rotovated soil or permanent sod was between the rows of blueberry bushes. Soil sampling for Japanese beetle grubs within and around blueberry fields showed that grub abundance was significantly greater in the ground cover surrounding fields than in the row middles inside fields (Table 1). A similar pattern of greater grub abundance outside managed fields was documented in ornamental crops by Smitley (1996) and in soybean and cornfields by Hammond and Stinner (1987). Grass is the optimal habitat for survival of Japanese beetle grubs, and female beetles select grassy areas near their feeding sites for egglaying (Fleming, 1972). In the absence of grass, grubs can survive on a diet of weeds, though they perform less well (Crutchfield and Potter, 1995). Because of this, the presence of a grass and weed mix ground cover around all fields is a significant risk factor for Japanese beetle infestation. In a recent study of movement by Japanese beetles, recapture of marked beetles released from the surrounding headlands was greatest on nearby blueberry border rows, indicating that some beetles do not move far into fields once an acceptable plant host has been found (Z. Szendrei, unpublished data). This further indicates the importance of removing habitats suitable for grub development near to blueberry fields to minimize the risk of beetle immigration into fields.

When grub abundance was compared between samples taken from rotovated row middles and those covered permanently with plant cover,

TABLE 1. Abundance of Japanese beetle grubs (average grubs per square meter ± S.E.) in soil samples taken from the perimeter and from the row middles of blueberry fields (n = 15) during spring 2001 and 2002. Values in a column are not significantly different if followed by the same letter ($P \leq 0.05$).

| Sampling position | Spring 2001 | Spring 2002 |
|---|---|---|
| Field perimeter | 44.45 ± 8.07 a | 28.95 ± 5.06 a |
| Row middle | 11.03 ± 3.44 b | 10.33 ± 4.52 b |

grubs were significantly more abundant in the latter habitat (Table 2). At rotovated sites, row middles were kept without any ground cover by using mechanical soil management techniques. Row middles that were rotovated as a cultural control had over 90% fewer Japanese beetle grubs in the soil during 2001 when the study started. During 2001-2, some of the study sites that had previously had permanent sod were changed to clean cultivation by the growers, although some weeds remained in these transitional sites. Because of this, the difference was not as great between the row middle types as previously found, though the difference remained significant (Table 2). Using a similar control strategy, Chittenden (1916) showed that cultural disruption of the rosechafer, *Macrodactylus subspinosus*, during the pupation period led to significant control of this pest in vineyards. The relative impact of direct mortality from rotovation and from removal of host plants remains to be determined. Previous studies have reported that cultivation causes significant mortality of eggs (Smith, 1924) and grubs (Cory and Langford, 1955; Smith, 1924) of Japanese beetle, though the impact of host plant removal on Japanese beetle oviposition remains unknown. Timing of rotovation for maximum impact on this pest has been reported to be in the early fall and late spring when grubs are near the surface (Fleming, 1976), and curative management may require tillage at both timings.

Permanent removal of host plants is a potentially effective strategy for reducing beetle infestation, but negative horticultural impacts may reduce the likelihood of adoption. Bare ground can be wet in spring, preventing tractor access to the field, and bare ground can be dusty at the time of harvest, potentially reducing fruit quality. In addition, erosion and pesticide leaching are also likely to be greater when soil is cultivated (Elliot et al., 2000). Rather than constant clean cultivation, acid-tolerant cover crops are a potential method of maintaining soil structure after tillage. Although the impact of planted cover crops on

TABLE 2. Abundance of Japanese beetle grubs (average grubs per square meter ± S.E.) in row middles that had permanent plant cover or were rotovated to create clean cultivation, during spring 2000 and 2001. Values in a column are not significantly different if followed by the same letter ($P < 0.05$).

| Row middle type | Spring 2001 | Spring 2002 |
| --- | --- | --- |
| Permanent sod | 16.36 ± 4.63 a | 11.84 ± 0.65 a |
| Clean cultivation | 1.08 ± 0.43 b | 8.93 ± 0.11 b |

Japanese beetle is not well understood (Potter and Held, 2002), variation in grub abundance under different plant covers has been reported (Hawley, 1944), indicating that this could be exploited to reduce beetle abundance within and around blueberry fields. Planting cover crops in blueberry row middles is currently under evaluation to provide a component of a sustainable approach to Japanese beetle control in blueberries by targeting the most vulnerable stages of this insect's life cycle.

Application of soil insecticides to infested grassy areas within the blueberry field is an alternative non-cultural control approach for grub control, and various insecticides have been shown to have activity against Japanese beetle in nursery crops (Mannion et al., 2001) and turf (Potter, 1998). New selective insecticides such as imidacloprid from the neonicotinoid class show promise because of their high activity against young grubs of Japanese beetle and their relatively low environmental impact (Potter and Held, 2002), though registration is still pending for use in blueberry.

Both pyrethrum insecticide formulations tested caused over 90% of beetles to drop from the blueberry foliage within one hour after application (Table 3). However, after 24 h, beetles had recovered from exposure to the Pyganic treatment, whereas the effect of the Evergreen treatment remained. Five days after exposure to the residues, the activity of both treatments had declined to the point where only 20% of beetles were immobile in the containers containing Pyganic-treated foliage, compared to over 50% in the Evergreen treated foliage. This greater activity of Evergreen compared to Pyganic, and its extended period of activity, was likely caused by the presence of piperonyl butoxide rather than the difference in amount of active pyrethrum because there was only a 7% difference in the rates of pyrethrum between the two treatments. Although there was no significant difference between the insecticide treatments on the number of beetles found alive on the container, the numeric values increased over time, becoming twice as large after 120 h on the untreated and Pyganic-treated containers than in containers with Evergreen-treated shoots, suggesting that there is some difference in the rate of recovery by adult beetles from the two treatments. The number of live beetles on blueberry shoots were significantly reduced by the initial application of both pyrethrum compounds, and no beetles were found on the shoots within 1 and 24 h after application. After 5 d of exposure to the residues, a small number of beetles had begun to move back onto the foliage, though there was still no significant difference between the treatments tested.

TABLE 3. Response of adult Japanese beetle to residues of two pyrethrum insecticides on blueberry foliage, at different times after initial exposure. Data are average percent beetles observed in the different categories of mortality and sublethal response at 1, 24, and 120 hours after initial exposure.

| Treatment | Rate (L/Ha) | Immobile on container | | | Live on the container | | | Live on the plant | | |
|---|---|---|---|---|---|---|---|---|---|---|
| | | 1 h | 24 h | 120 h | 1 h | 24 h | 120 h | 1 h | 24 h | 120 h |
| Untreated | 0 | 0.0 ± 0.0 b | 12.5 ± 12.5 b | 0.0 ± 0.0 a | 5.0 ± 2.9 a | 2.5 ± 2.5 a | 65.0 ± 17.6 a | 95.0 ± 2.9a | 72.5 ± 24.3 a | 35.0 ± 17.6 a |
| Pyganic | 0.473 | 92.5 ± 4.8 a | 27.5 ± 13.8 b | 25.0 ± 15.0 a | 0.0 ± 0.0 a | 2.5 ± 2.5 a | 65.0 ± 21.8 a | 0.0 ± 0.0b | 0.0 ± 0.0 b | 2.5 ± 2.5 a |
| Evergreen | 0.118 | 97.5 ± 2.5 a | 82.5 ± 7.5 a | 52.5 ± 13.8 b | 0.0 ± 0.0 a | 0.0 ± 0.0 a | 27.5 ± 4.8 a | 0.0 ± 0.0b | 0.0 ± 0.0 b | 7.5 ± 4.8 a |

In addition to the direct effects on the beetles, fruit treated with both pyrethrum formulations received significantly less damage after 5 d of exposure to beetles than did the untreated fruit (Figure 1). Although the reason for this result is not known, only Evergreen was able to protect the foliage over this same period, while the level of beetle feeding on leaves was not significantly affected by residues of Pyganic. During the 5 d of exposure, more beetles were observed to recover from their initial sub-lethal knockdown symptoms after exposure to the Pyganic-treated, compared to the Evergreen-treated foliage (Table 3). These end-point results show that addition of the agonist piperonyl butoxide is an important component of effectively using pyrethrums for control of *P. japonica* and protection of the crop from feeding damage. From data presented

FIGURE 1. Percent feeding (by area) on foliage and fruit of blueberry treated either with Pyganic, Evergreen, or a water only control after five days of exposure to ten Japanese beetles. Damage to fruit or foliage was significantly different between treatments (P ≤ 0.05) if bars are topped by different letters.

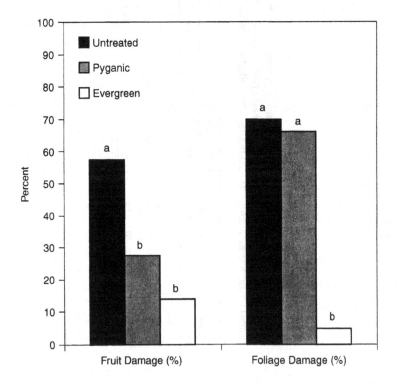

above, formulations that contain piperonyl butoxide are expected to provide the required short-term protection against beetle presence if applied the day before harvest, and trials are underway at commercial farms to verify this expectation under field conditions.

## CONCLUSION AND GROWER BENEFITS

Effective in-field management of Japanese beetle in highbush blueberry requires that growers implement an integrated strategy across their farms. This should include attention to areas around and within fields that have habitat for Japanese beetle grub development. It is expected that this will have the double benefit of reducing the pest pressure and enabling foliar insecticides to perform more effectively due to the lower levels of beetle immigration. Use of foliar treatments for adult control should remain a component of this integrated strategy, with selection of pyrethrum products or other insecticides with short pre-harvest intervals if beetles must be removed from the bushes prior to harvest. Although not discussed above, post-harvest strategies for removing contaminants are an additional step that should be considered to ensure that growers can meet the market's demand for contaminant-free berries.

## LITERATURE CITED

Allsopp, P.G. 1996. Japanese beetle, *Popillia japonica* Newman (Coleoptera: Scarabaeidae): Rate of movement and potential distribution of an immigrant species. The Coleopt. Bull. 50: 81-95.

Anonymous 2000. National Agricultural Pest Information System: Japanese beetle (http://ceris.purdue.edu/napis/pests/jb/index.html).

Cappaert, D.L. and D.R. Smitley. 2002. Parasitoids and pathogens of Japanese beetle (Coleoptera: Scarabaeidae) in Southern Michigan. Environ. Entomol. 31: 573-580.

Casida, J.E. and G.B. Quistad. 1995. Pyrethrum flowers: Production, chemistry, toxicology and uses. Oxford Univ. Press, New York.

Chittenden, F.H. 1916. The rose chafer: A destructive garden and vineyard pest. USDA Farmer's Bull. 712: 1-8.

Cory, E.N. and G.S. Langford.1955. The Japanese beetle retardation program in Maryland. Maryland Univ. Ext. Bull. 156.

Crutchfield, B.A. and D.A. Potter. 1995. Feeding by Japanese beetle and southern masked chafer grubs on lawn weeds. Crop Sci. 35:1681-1684.

Elliott, J.A., A.J. Cessna, W. Nicholaichuk, and L.C. Tollefson. 2000. Leaching rates and preferential flow of selected herbicides through tilled and untilled soil. J. Environ. Quality 29:1650-1656.

Fleming, W.E. 1972. Biology of the Japanese beetle. USDA Tech. Bull. 1449. U.S. Gov. Print. Office, Washington, DC.

Fleming, W.E. 1976. Integrating control of the Japanese beetle–a historical review. USDA Tech. Bull. 1545. U.S. Gov. Print. Office, Washington, DC.

Hammond, R.B. and B.R. Stinner. 1987. Soybean foliage insects in conservation tillage systems: Effects of tillage, previous cropping history, and soil insecticide application. Environ. Entomol. 16:524-531.

Hawley, I.M. 1944. Notes on the biology of the Japanese beetle. U.S. Bur. of Entomol. Plant Quarratine E-615.

Mannion, C.M., W. McLane, M.G. Klein, J. Moyseenko, J.B. Oliver, and D. Cowan. 2001. Management of early instar Japanese beetle (Coleoptera:Scarabaeidae) in field-grown nursery crops. J. Econ. Entomol. 94:1151-1161.

Potter, D.A. 1998. Destructive turfgrass insects. Biology, diagnosis, and control. Ann Arbor Press, Chelsea, MI.

Potter, D.A. and D.W. Held. 2002. Biology and management of the Japanese beetle. Ann. Rev. Entomol. 47:175-205.

Smith, L.B. 1924. The Japanese beetle. New Jersey Dept. Agr. Bul. 41:55-63.

Smitley, D. 1996. Incidence of *Popillia japonica* (Coleoptera: Scarabidae) and other scarab larvae in nursery fields. Hort. Entomol. 89:1262-1266.

Vittum, P.J., M.G. Villani, and H. Tashiro. 1999. Turfgrass insects of the United States and Canada. Cornell University Press, Ithaca, NY.

# Effects of Kaolin Clay Application on Flower Bud Development, Fruit Quality and Yield, and Flower Thrips [*Frankliniella* spp. (Thysanoptera: Thripidae)] Populations of Blueberry Plants

James D. Spiers
Frank B. Matta
Donna A. Marshall
Blair J. Sampson

**SUMMARY.** Three separate studies were conducted to report the effects of kaolin applications (Surround WP) on southern highbush blueberries (*Vaccinium corymbosum* L.) and rabbiteye (*V. ashei* Reade) blueberries. When applied to mature blueberry plants, kaolin clay emulsion dried to form a white reflective film and affected bud development, fruit set and development, plant growth, and fruit yield, but had no effect on fruit quality parameters. When kaolin was applied before fruit set,

James D. Spiers is Graduate Student in Horticulture and Frank B. Matta is Professor of Horticulture, Mississippi State University, MS 39762. Donna A. Marshall is Horticulturalist, and Blair J. Sampson is Research Entomologist, USDA-ARS Small Fruit Research Station, Poplarville, MS 39470.

[Haworth co-indexing entry note]: "Effects of Kaolin Clay Application on Flower Bud Development, Fruit Quality and Yield, and Flower Thrips [*Frankliniella* spp. (Thysanoptera: Thripidae)] Populations of Blueberry Plants." Spiers, James D. et al. Co-published simultaneously in *Small Fruits Review* (Food Products Press, an imprint of The Haworth Press, Inc.) Vol. 3, No. 3/4, 2004, pp. 361-373; and: *Proceedings of the Ninth North American Blueberry Research and Extension Workers Conference* (ed: Charles F. Forney, and Leonard J. Eaton) Food Products Press, an imprint of The Haworth Press, Inc., 2004, pp. 361-373. Single or multiple copies of this article are available for a fee from The Haworth Document Delivery Service [1-800-HAWORTH, 9:00 a.m. - 5:00 p.m. (EST). E-mail address: docdelivery@haworthpress.com].

yield was increased with no significant residue left on the fruit. Surround WP consistently reduced the number of flower thrips (*Frankliniella* spp.) within the canopy of rabbiteye blueberry plants by approximately 50%. Kaolin applications were not phytotoxic to blueberry buds, flowers, leaves, or fruit and were harmless to foraging bees. *[Article copies available for a fee from The Haworth Document Delivery Service: 1-800-HAWORTH. E-mail address: <docdelivery@haworthpress.com> Website: <http://www. HaworthPress.com> © 2004 by The Haworth Press, Inc. All rights reserved.]*

**KEYWORDS.** Kaolin, southern highbush blueberry, thrips, fruit yield, fruit quality, crop protectant

## *INTRODUCTION*

Hydrophobic particle film technology was introduced in 1999. The hydrophobic particle film based on the inert clay mineral called kaolin has been formulated as a crop protectant and has a sublethal impact on fungal pathogens and insects (Glenn et al., 1999). Simple dust applications of particle films are not sufficiently durable for long service. For this reason, the hydrophobic kaolin clay particles are pre-mixed with methanol to increase their miscibility in water and surfactants are added (Glenn et al., 2001). This hydrophilic particle film is now marketed as Surround WP. Surround WP, based on a nontoxic kaolin particle and a spreader-sticker made of natural materials, has been officially recognized as "organic" by the Organic Materials Review Institute (ATTRA-a). Kaolin particle film has proven to be very effective in protecting fruit trees from various insect pests, but many horticultural benefits have been reported as well.

Flower thrips including *Frankliniella bispinosa* (Morgan), *F. occidentalis* (Pergande), *F. tritici* (Fitch), *Scirtothrips ruthveni* Shull are emerging as pests of cultivated blueberries in New Jersey and the Southern States. Severe feeding and oviposition injury to flowers and fruit is thought to seriously curtail blueberry yield in certain years (Parker et al., 1995). Both adult and nymphal flower thrips typically take refuge between the crevices that exist between various floral organs and are difficult to kill with conventional contact insecticides.

Kaolin-based emulsions have multiple modes of action that can offer broader protection to blueberry crops from flower thrips and other blueberry pests (Glenn et al., 1999; Knight et al., 2000; Lapointe, 2000;

Puterka et al., 2000; Unruh et al., 2000). Kaolin clay particles adhere to plant surfaces with the aid of wetting agents. These particles form a physical barrier over the flowers, leaves, stems, and fruit. If these particles attach to the waxy cuticles of pest insects they interfere with feeding, as well as cause irritation and desiccation (Glenn et al., 1999; Swamiappan et al., 1976). Also, kaolin has no nutritional value to herbivorous insects and can potentially reduce the digestibility of plant tissue (Howe and Westley, 1988). Induced plant resistance to pest insects is also possible, as plants sprayed with kaolin can become unrecognizable as suitable hosts (Bar-Joseph and Frenkel, 1983; Glenn et al., 1999). Host suitability to flower thrips can diminish with intensification of reflected UV-light (350 to 390 nm) visible to flower thrips (Glenn et al., 1999; 2002; Lewis, 1997; Parker et al., 1995).

Particle film sprays such as kaolin have been recommended to lower the temperature of apple fruit; thereby reducing sunburn and improving red fruit color in situations where temperatures are higher than optimal (Heacox, 1999; Stanley, 1998; Warner, 2000a; Werblow, 1999). Yield was increased in some cases, yet was unaffected in other studies. The increase in yield was reportedly due to a decrease in pre-harvest fruit drop. This also was evident when kaolin particle film was applied to Seckel pear trees (Puterka et al. 2000).

Solar injury was suppressed with spray applications of a reflective, processed-kaolin particle film material on apple [*Malus sylvestris* (L.) Mill var *domestica* (Borkh.) Mansf.]. Glenn et al. (2002) reported that fruit surface temperature was reduced by the application of reflective particles and the amount of temperature reduction was proportional to the amount of particle residue on the fruit surface. It was also noted that the processed-kaolin film material was highly reflective to the ultraviolet wavelengths and this characteristic may be important in reducing solar injury to both fruit and leaves (Glenn et al. 2002).

Much research has been done on the effects of kaolin clay particle films on apple (*Malus* spp.), pear (*Pyrus* spp.), etc., but no research has been reported for blueberry plants. The objectives of this research were: (1) to determine the effects of kaolin clay on blueberry yield, quality, size, development, fruit set, and plant growth; (2) to determine at what stage of flower or fruit development would the application of kaolin clay be the most beneficial; and (3) to determine the impact of kaolin clay application on the populations of flower thrips (genus *Frankliniella*).

## MATERIALS AND METHODS

*Fruit quality and yield.* Mature southern highbush blueberry (*Vaccinium corymbosum* L.) plants grown on Ruston fine sandy loam soil (fine-loamy, siliceous, thermic Typic Paleudult) at the USDA Small Fruits Research Station site in Stone County, Mississippi were used in the following 2 studies. Prior to and throughout the studies, all plants received uniform fertilizer and irrigation rates (Braswell et al., 1991). All treated plants were sprayed with a mixture of 1077 g of kaolin clay (= 1133.95 g Surround WP, Englehard Corporation, Iselin, NJ) and 59 ml of the non-ionic surfactant Silken (Riverside/Terra Corp., Sioux City, Iowa) per 19 L of solution. Treated plants were sprayed using a 12 L pressurized backpack sprayer (SOLO™ Kleinmotoren GmbH, Sindelfingen, Germany). The plants were sprayed until drip and were visually inspected and re-sprayed to ensure complete coverage.

*Effects of kaolin clay on bloom development.* 'Cooper' blueberry plants established in 1995 were arranged in a randomized complete block design consisting of 9 replications with each replication having 2 treatments (sprayed and unsprayed) and border plants between treatments. Treated plants were sprayed with kaolin on March 8, 2001. Ten buds from each plant were rated according to the blueberry flower bud rating scale (Spiers, 1978) prior to treatment, with the goal of getting 2 of each rating 3-7 on each plant. Tags were attached next to the buds and labeled with the rating and a letter (3a, 3b, etc.). Flower buds were rated 10 days following application, and the number of flowers per bud was recorded. Fruit diameter was measured, and the number of fruit per bud was counted on 3 occasions (April 17, May 23, and May 30). The number of flowers/bud, berries present, and fruit picked was analyzed to assess fruit set. The rate of development and size of the fruit was also recorded. All data was analyzed by ANOVA using SAS (SAS Institute, 1985).

*Effects of application time of kaolin clay.* 'Magnolia' blueberry plants established in 1994 were arranged in a randomized complete block design consisting of 4 replications with 15 plants (5 treatments and 3 plants per treatment) per replication. The 5 treatments consisted of an unsprayed control and a single spray of kaolin applied at 4 different stages of fruit development; pre-fruit (March 8, 2001; approximately 50% flowering), early fruit set (March 29), mid-maturity (April 17), and pre-harvest (May 3). The volume (length × width × height) of the plant size was obtained prior to treatments. Fruit was harvested from the middle plant in each block when ripe and the yield was measured. The harvesting dates were May 14, May 23, May 31, and June 7, 2001.

On each harvesting date, randomly selected berries were measured for chemical analysis, compression, and turbidity (residue). Thirty berries from each plant were swirled in 100 ml of distilled water for 1 min. to remove the residue from the berries. The water with residue was tested for turbidity using a Perkin-Elmer Lambda 3B UV/VIS spectrophotometer. Standards of 0, 10, and 100 ppm of kaolin clay (Surround WP) were measured with the spectrophotometer prior to measuring the water with unknown amounts of residues. The standards were used to set up a linear regression and calculate the amount of kaolin clay residue left on the berries in ppm. A 30 g sample of berries was homogenized 1 min with a Waring blender. Soluble solids concentration (SS) of the homogenate was measured by filtering homogenate through 2 layers of cheesecloth and measuring the juice with a (Bausch and Laumb) hand held refractometer. Total solids (TS) were obtained by drying a 10 g aliquot of homogenate in a forced air drying oven at 72°F for 24 h. A 10 g aliquot diluted to 100 ml with distilled water was used for pH measurement.

Compression force was measured with a QTS 25 (Stevens Mechtric) texture press using a spherical probe. For each treatment and replication 10 berries were compressed. Compression force was applied parallel to the stem-calyx axis until 150 g of pressure was applied. Modulus (g/s) and deformation at peak load (mm) were recorded.

After the final harvest, the volume of the plants was calculated from measuring the height and a cross-sectional width (length × width × height). Growth was calculated against preliminary measurements and was analyzed with ANOVA using SAS (SAS Institute, 1985). Yield and chemistry data were analyzed with GLM using SAS.

*Flower thrips populations.* Blueberry host plants were established at the USDA-ARS Small Fruits Research Station located in Poplarville, Mississippi. Three rabbiteye blueberry (*V. ashei* Reade) cultivars were used: 'Delite', 'Tifblue', and 'Woodard'. Plants were established in 1988 in a Ruston fine sandy loam soil, and were planted according to a 3 × 3 Latin square. There were 6 replications, 2 treatments (kaolin and unsprayed), and 9 plants per experimental plot (total N = 108 plants). The application was a mixture of 1077 g of kaolin (= 1134 g Surround WP, Engelhard Corporation, Iselin, NJ) and 59 ml of a non-ionic surfactant (Silkin, Riverside/Terra Corp., Sioux City, IA) per 5 gallons of water. Therefore, each blueberry bush received about 20 g kaolin and control plants went unsprayed (0 g kaolin). Three applications were repeatedly made to the same plants during bloom (March 7, 2001), post-bloom (April 20), and pre-harvest (May 17). Liquid applications were applied with a pressurized 12 L backpack sprayer (SOLO™ Kleinmotoren GmbH, Sindelfingen, Germany). The plants were sprayed until drip and were visually inspected and re-sprayed to ensure complete coverage.

Yellow Pherocon AM sticky traps (23 cm × 28 cm: Tréce Inc. Salinas, CA) were placed on the center plant in each experimental block in the middle of the canopy, on the same day of application after the kaolin had completely dried. The 36 sticky traps were collected exactly one week after making the kaolin applications. Thrips abundance was estimated by counting the total number of thrips in five randomly chosen quadrants on a 3 × 5 counting grid. Multiplying this tally by 3 and adding the number of thrips that fell outside the boundaries of the grid gave an estimate of the total number of thrips per trap. Thrips abundance data were analyzed using a 3 × 3 factorial ANOVA and sequential sums of squares for calculating F-values (SAS Institute, 1985).

## RESULTS AND DISCUSSION

*Bud development.* The number of flowers/bud, berries present, and berries picked were significantly greater for the plants treated with H kaolin (Table 1). Therefore, fruit set was increased with the application of Surround. This is in accordance with previous studies on apples and pears (Glenn et al., 1999; Glenn et al., 2001; Puterka et al., 2000). The rate of bud development was slower for the treated plants, possibly due to lowered heat stress and lowered direct photosynthesis. Increased fruit set with the application of kaolin could be due to insect control or horticultural benefits such as reduced heat stress and increased net photosynthesis (ATTRA-b). The berry size was smaller for the treated plants. Previous studies on apples and pears found fruit size to be equal or increased with the application of kaolin particle films (Glenn et al., 1999; Glenn et al., 2001; Puterka et al., 2000). In this study, the decrease in size correlates with the increase in fruit set. When there is an increase in fruit set there are fewer nutrients available per berry. Also, reductions in fruit size could possibly be due to an increase in the amount of reflected light (Glenn et al., 1999; Schupp et al., 2002). When fruit trees were not exhibiting heat stress, Glenn et al. (2001) showed that kaolin sprays reduced $CO_2$ assimilation and resulted in no size increase. Surround was found to reduce fruit size under environmental conditions where light

TABLE 1. Impact of kaolin on fruit set, development, and size of 'Cooper' southern highbush blueberry fruit.

| Treatment | Number of flowers/bud | Floral bud development[z] rating | Berries/bud[y] | | Berries/bud[x] | |
|---|---|---|---|---|---|---|
| | | | Number | Size (mm) | Number | Size (mm) |
| Spray | 5.72 a[w] | 6.30 b | 5.36 a | 1.03 b | 3.45 a | 1.61 b |
| No spray | 4.86 b | 6.45 a | 4.44 b | 1.07 a | 1.90 b | 1.70 a |

[z] Floral bud development scales from Spiers (1978) 10 days after kaolin treatment.
[y] Unharvested berries on bush, 17 April 2001.
[x] Harvested berries, 30 May 2001.
[w] Means separation within columns at $P \leq 0.05$ level.

was more likely to be limiting to $CO_2$ assimilation than was high temperature (Schupp et al., 2002).

*Effects of application time.* The soluble solids, total solids, pH, and compression measurements were not significantly different for the treated and untreated plants (Table 2). These results support previous studies where kaolin sprays did not affect fruit quality (Glenn et al., 1999; Glenn et al., 2001; Puterka et al., 2000; Schupp et al., 2002). Turbidity (residue on fruit) measurements increased linearly from the control to the last date kaolin applications were made. However, plants treated during the pre-fruit (~50% bloom) stage were not significantly different from control plants in the amount of residue on the fruit (Table 2). The yield for the plants sprayed during the pre-fruit stage was significantly greater than the unsprayed and other treatments (Figure 1). Similar yield results were found when kaolin particle films were applied to apples, peaches, and pears (Glenn et al., 1999; Glenn et al., 2001; Puterka et al., 2000).

Plant growth increased with the application of kaolin (Figure 2). The earlier applications achieved more growth than the later applications. The control plants had the least amount of growth. Previous studies have found no reduction in shoot growth when applied to fruit trees, but have not observed increased growth (Glenn et al., 1999; Puterka et al., 2000). Increased plant growth could be due to kaolin particles protecting the plants from heat stress and insect pests. Reducing the heat load possibly increased the net photosynthesis and resulted in more plant growth. No phytotoxicity was detected from the application of kaolin particle film possibly because of the decreased particle size ($\leq 2$ μm) that allowed for gas exchange. Generic kaolin (1.6-4.8 μm) acts as an antitranspirant and can result in phytotoxicity (Eveling and Eisa, 1976).

TABLE 2. Impact of kaolin on fruit quality characteristics of 'Magnolia' southern highbush blueberry.

| Application stage | Soluble solids (Brix) | Total Solids (%) | pH | Turbidity | Modulus (g/s) | Deformation (mm) |
|---|---|---|---|---|---|---|
| Control | 13.34 a[z] | 14.40 a | 3.28 a | 12.86 d | 20.73 a | 1.96 a |
| Pre-fruit | 13.33 a | 14.58 a | 3.24 a | 13.27 d | 20.37 a | 1.91 a |
| Early fruit set | 13.61 a | 14.60 a | 3.27 a | 22.64 c | 21.52 a | 1.80 a |
| Mid-maturity | 13.43 a | 14.51 a | 3.31 a | 35.25 b | 21.34 a | 1.87 a |
| Pre-harvest | 13.29 a | 14.43 a | 3.23 a | 61.38 a | 21.41 a | 1.81 a |

[z] Means separation within columns at $P \leq 0.05$ level.

FIGURE 1. Impact of kaolin particle film on the average yield of 'Magnolia' southern highbush blueberry plants. Means by Duncan's Multiple Range Test ($P \leq 0.05$).

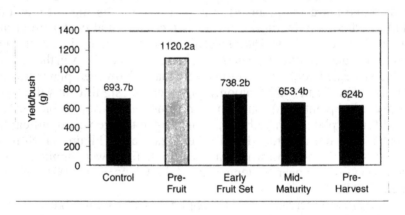

*Flower thrips population.* Thrips populations in the experimental rabbiteye blueberry field increased rapidly from mid-bloom (7 March 2001) to the onset of harvest (17 May 2001). Aerial applications containing kaolin reduced the abundance of flower thrips associated with *V. ashei* bushes by about 50% (Table 3, Figure 3). No interaction was observed between cultivar and treatment, showing that the efficacy of kaolin to reduce thrips abundance was consistent and not seriously impeded by minor cultivar differences (Table 3). The effectiveness of kaolin as a deterrent to adult thrips increased on those application dates

FIGURE 2. Impact of kaolin clay particle film on plant growth of 'Magnolia' southern highbush blueberry plants.

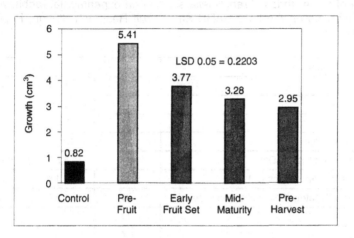

when thrips counts were the highest, suggesting that the mode of action for Surround WP is to some extent density-dependent.

## CONCLUSIONS AND GSROWER BENEFITS

*Fruit quality and yield.* These results indicate that an application of kaolin clay particle film at pre-fruit (50% bloom) can provide horticultural benefits to blueberry plants. Yield enhancement can be obtained without any significant residue on the berries when applied before fruit set. Later single applications were not more beneficial and a kaolin residue, although organic, was evident on berries. Kaolin clay can be used to increase fruit set without affecting fruit quality. The size of the fruit was decreased, probably because of the increased number of fruit. The application of kaolin can promote growth of blueberry plants without affecting pollination.

*Flower thrips populations.* Kaolin applications invariably reduced the number of adult flower thrips associated with three cultivars of rabbiteye blueberry by about 50%. The efficacy of kaolin increased as adult thrips became more numerous. This demonstrated that kaolin might sufficiently alter host suitability, forcing larger numbers of thrips to move to adjacent, unsprayed bushes. Kaolin treatments broadcast

TABLE 3. Summary of ANOVA results for three experimental factors (Cultivar, Treatment and Application Date) as well as interactions affecting the abundance of flower thrips (*Frankliniella* spp.) at an experimental rabbiteye blueberry planting on the USDA-ARS Small Fruit Research Station, Poplarville, Mississippi.

| Source of Variation | df | F | P |
|---|---|---|---|
| Cultivar | 2 | 11.13 | < 0.0001 |
| Treatment | 1 | 11.56 | 0.0010 |
| Cultivar × Treatment | 2 | 3.09 | 0.0503 |
| Application Date | 2 | 10.40 | < 0.0001 |
| Cultivar × Application Date | 4 | 4.57 | 0.0021 |
| Treatment × Application Date | 2 | 5.11 | 0.0079 |
| Cultivar × Treatment × Application Date | 4 | 1.82 | 0.1315 |
| Error | 90 | | |
| Total | 107 | | |

over a much wider area could effectively control flower thrips in rabbiteye blueberry fields, especially around peak bloom and early fruiting when damage is more severe (Polavarapu, 2001).

These preliminary results clearly demonstrate that kaolin particle barriers can alter the foraging patterns of adult thrips within patches of flowering and fruiting blueberries. Additional studies are needed to verify the more direct effects of kaolin on thrips oviposition, feeding behavior and survival on blueberry host plants. Kaolin formulated as Surround WP mixed with a non-ionic surfactant was not phytotoxic to blueberry plants and had no adverse impact on pollinator activity (Sampson et al., unpublished data) and yield.

The Food Quality Protection Act (FQPA) and other regulations are encouraging the discovery, development, and implementation of reduced-risk alternatives to the toxic pesticides currently in use on fruit crops. A crop protectant like kaolin-based emulsions can become an increasingly important element of blueberry integrated pest management (IPM). Our results show kaolin clay particle films to be an effective method for reducing thrips populations associated with rabbiteye blueberry plants. Kaolin that is not amended with prohibited substances can also be a viable method of crop protection for certified organic farmers.

FIGURE 3. The impact of kaolin applications on the seasonal abundance of adult flower thrips associated with three varieties of rabbiteye blueberry. Bars represent mean ± 1 SD.

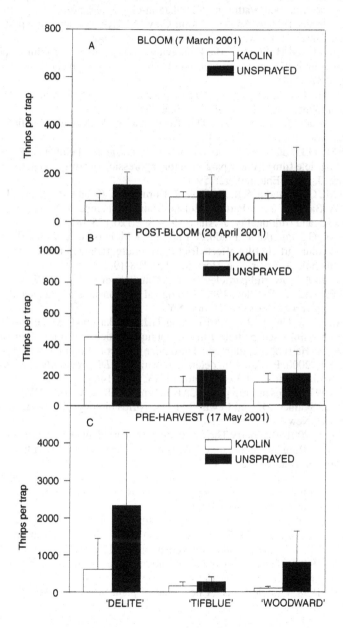

# LITERATURE CITED

ATTRA-a. Considerations in organic apple production. ATTRA On-line publications. Internet: www.attra.org/attra-pub/PDF/kaolin-clay-apples.pdf.

ATTRA-b. Insect IPM in Apples: Kaolin Clay. ATTRA Reduced-risk pest control factsheet. Internet: www.attra.org/attra-pub/omapple.pdf.

Bar-Joseph, M. and H. Frenkel. 1983. Spraying citrus plants with kaolin suspensions reduces colonization by the spiraea aphid (*Aphis citricola* van der Goot). Crop Protection 2:371-374.

Braswell, J.H., J.M. Spiers, and C.P. Hegwood, Jr. 1991. Establishment and maintenance of blueberries. Miss. Agr. For. Expt. Sta. Bul. 1758.

Eveling, D.W. and Eisa, M.Z. 1976. The effects of a cuticle-damaging kaolin on herbicidal phytotoxicity. Weed Res. 16:15-18.

Glenn, D.M., G.J. Puterka, T. Vanderzwet, R.E. Byers, and C. Feldhake. 1999. Hydrophobic particle films: A new paradigm for suppression of arthropod pests and plant diseases. J. Econ. Entomol. 92:759-771.

Glenn, D.M., G.J. Puterka, S.R. Drake, T.R. Unruh, A.L. Knight, P. Baherle, E. Prado, and T.A. Baugher. 2001. Particle film application influences apple leaf physiology, fruit yield, and fruit quality. J. Amer. Hort. Sci. 126:175-181.

Glenn, D.M., G.J. Puterka, E. Prado, A. Erez, and J. McFerson. 2002. A reflective processed kaolin particle film affects fruit temperature, radiation reflection, and solar injury in apple. J. Amer. Soc. Hort. Sci. 127:188-193.

Heacox, L. 1999. Powerful particles. Amer. Fruit Grower (Feb.): 16-17.

Howe, H.F. and L.C. Westley. 1988. Ecological relationships of plants and animals. Oxford University Press, New York, NY.

Knight, A.L., T.R. Unruh, B.A. Christianson, G.J. Puterka, and D.M. Glenn. 2000. Effects of kaolin-based particle films on obliquebanded leafroller, *Choristoneura rosaceana* (Harris), (Lepidoptera: Tortricidae). J. Econ. Entomol. 93: 744-749.

Lapointe, S.L. 2000. Particle film deters oviposition by *Diaprepes abbreviatus* (Coleoptera: Curculionidae). J. Econ. Entomol. 93: 1459-1463.

Lewis, T. 1997. Thrips as crop pests. CAB International, New York, NY.

Parker, B.L., Skinner, M., and Lewis, T. 1995. Thrips biology and management. Plenum Press. New York.

Polavarapu, S. 2001. Blueberry Thrips: An emerging problem in New Jersey highbush blueberries. Rutgers Cooperative Extension and New Jersey Agricultural Experiment Station. Blueberry Bull. 17:2.

Puterka, G.J., D.M. Glenn, D.G. Sekutowski, T.R. Unruh, and S.K. Jones. 2000. Progress toward liquid formulations of particle films for insect and disease control in pear. Environ. Entomol. 29:329-339.

SAS Institute, 1985. SAS/STAT user's guide. 4th ed. Ver. 6. SAS Inst., Cary, NC.

Schupp, J.R., E. Fallahi, and I. Chun. 2002. Effect of particle film on fruit sunburn, maturity and quality of 'Fuji' and Honeycrisp' apples. HortTechnology 12:87-90.

Spiers, J.M. 1978. Effect of stage of bud development on cold injury in rabbiteye blueberry. J. Amer. Hort. Sci. 103:452-455.

Stanley, D. 1998. Particle films: A new kind of plant protectant. Agr. Res. 46(11): 16-19.

Swamiappan, M., C. S. Deivavel, and S. Jayaraj. 1976. Mode of action activated kaolin on the pulse beetle, *Callosobruchus chinensis*. Madras Agr. J. 63: 576-578.

Unruh, T.R., A.L. Knight, J. Upton, D.M. Glenn, and G.J. Puterka. 2000. Particle films for suppression of the codling moth [*Cydia pomonella* (L.)] in apple and pear orchards. J. Econ. Entomol. 93: 737-743.

Warner, G. 2000. Look for more Surround around. Good Fruit Grower 51(5): 16-17.

Werblow, S. 1999. Favorable film. Ore. Farmer-Stockman. (April): 8-10.

# Flowering and Leafing
# of Low-Chill Blueberries in Florida

## Paul M. Lyrene

**SUMMARY.** Obtaining high yields of high-quality blueberries early in the season requires cultivars and growing systems that result in plants that have a full canopy of healthy leaves for at least the last 70% of the flowering-to-ripening interval. Florida's two native evergreen blueberry species, *V. darrowi* and *V. myrsinites*, accomplish this by maintaining healthy leaves from the previous growing season through the time of berry maturity the following year and by producing strong new growth flushes in late winter at or shortly after the time of flowering. Southern highbush cultivars and selections, which are complex hybrids between evergreen *V. darrowi* and deciduous *V. corymbosum*, range from highly deciduous to mostly evergreen when grown in commercial fields in north Florida. Many selections that maintain most of their leaves through the winter will defoliate before fruit ripening the following spring, leaving the plant without enough foliage to mature a high-quality crop. Profuse vegetative growth in early spring can be promoted by cultivar selection, winter pruning to reduce the number of flower buds, application of hydrogen cyanamide or other defoliants during the dormant season, and control of blueberry gall midge, which kills the apical meristems of new growth flushes. *[Article copies available for a fee from The Haworth Document Delivery Service: 1-800-HAWORTH. E-mail address: <docdelivery@haworthpress.com> Website: <http://www.HaworthPress.com> © 2004 by The Haworth Press, Inc. All rights reserved.]*

Paul M. Lyrene is Professor, Horticultural Sciences Department, University of Florida, P.O. Box 110690, Gainesville, FL 32611.

[Haworth co-indexing entry note]: "Flowering and Leafing of Low-Chill Blueberries in Florida." Lyrene, Paul M. Co-published simultaneously in *Small Fruits Review* (Food Products Press, an imprint of The Haworth Press, Inc.) Vol. 3, No. 3/4, 2004, pp. 375-379; and: *Proceedings of the Ninth North American Blueberry Research and Extension Workers Conference* (ed: Charles F. Forney, and Leonard J. Eaton) Food Products Press, an imprint of The Haworth Press, Inc., 2004, pp. 375-379. Single or multiple copies of this article are available for a fee from The Haworth Document Delivery Service [1-800-HAWORTH, 9:00 a.m. - 5:00 p.m. (EST). E-mail address: docdelivery@haworthpress.com].

Digital Object Identifer: 10.1300/J301v03n03_14

**KEYWORDS.** *Vaccinium corymbosum, V. darrowi,* chilling requirement, southern highbush blueberry

Most Florida blueberry growers want their plants to flower in early
February and ripen fruit during April and early May. This requires
low-chill varieties with a short bloom-to-ripe interval. Most of the sugar
that goes into the fruit is manufactured by the leaves during the period
of berry development. Thus, production of high yields of high-quality
berries early in the season requires that the plants have an abundance of
healthy leaves during most of the fruit development period. Theoreti-
cally, this can be accomplished either by carrying the previous year's
leaves through the winter and spring to ripen the spring crop or by obtaining a strong flush of new growth at the time of flowering in the
spring. In eastern Australia, near sea level at latitude 30°S, Gary Wright
and Ridley Bell (personal communication) have selected southern highbush clones that mature a medium to heavy crop of high-quality fruit in
early spring on plants that produce few or no new leaves until after harvest is complete. After harvest, the plants are pruned to initiate new
shoot growth, which remains healthy through the fall, winter, and following spring.

In north Florida, we have not been able to find clones that give satisfactory yields and quality if they do not produce new spring leaves before
harvest. Leaf diseases in the fall and winter, combined with occasional
hard winter freezes (to −8°C) weaken or damage the leaves on many
clones that would otherwise be quite evergreen. The clones that are
most evergreen often have very low chilling requirements, and may begin heavy flowering as early as late December or early January in north
Florida. Evergreen commercial cultivars have not proved practical in
north Florida for several reasons. Growers do not want to begin freeze
protection (by overhead irrigation) before the first of February; March
berries tend to be less valuable than April berries; and low-chill evergreen plants tend to have a long, drawn-out harvest season.

Some low-chill rabbiteye and southern highbush selections have the
ability to flower and produce abundant new vegetative growth in early
spring, even while maintaining most of their old leaves from the previous year. This would seem to offer the best situation with respect to developing a large, early crop. Unfortunately, leaf diseases (particularly
blueberry leaf rust) tend to attack the new leaves, having overwintered
on the old. Another problem is that plants that leaf strongly before they
lose their old leaves may be more susceptible to damage in severe

late-winter freezes. Clones that flower and produce new leaves before the old leaves fall may be more practical in central and south Florida than in north Florida.

The man-made cultigen, southern highbush blueberry, is not unique in the leafing behavior it shows when grown in subtropical climates. Thomas (2000) defined a class of temperate and tropical trees, which he called "leaf exchanging," in which "the new leaves appear around the time the majority of the old leaves fall, either just before or just after (this can vary with the species), so that some specimens are almost leafless for a brief period." Species classified by Thomas as leaf exchanging include *Magnolia grandiflora, Cinnamomum camphora, Persea americana*, and *Quercus suber.*

One highly undesirable phenological pattern followed by some southern highbush selections in north Florida is to maintain the old leaves through the flowering period but then drop the old leaves during the fruit development period, meanwhile having made no new leaves. This leaves the plants with a crop of half-developed berries and few or no leaves to support it. The result is slowed berry development, reduced berry quality, and sometimes, dieback of the fruit-bearing branches.

In some clones, maintenance of old leaves through the winter appears to inhibit the sprouting of new leaf buds in the spring. In such clones, failure of the old leaves to drop during December is highly undesirable. Examination of several hundred southern highbush test selections in replicated field plots during December in Gainesville, Florida normally reveals a significant number of clones that appear to be highly evergreen, despite the fact that fertilization and irrigation have been managed during the fall to promote dormancy. A high percentage of these clones will overcrop and have trouble producing new leaves in early spring.

In some southern highbush clones, the cultivars 'Misty' and 'Marimba' being good examples, flower buds appear to inhibit leafing in the spring. Both cultivars produce a heavy crop of flower buds which open over an extended period during February and early March in north Florida. In many years, few or no vegetative buds sprout until all the flower buds have opened. By that time, the plants may be under heavy stress from rising temperatures, a heavy crop load, and depleted carbohydrates, and the stems may start dying from the tips downward. *Botryosphaeria dothidea*, the stem blight pathogen, and other fungi may contribute to this process. Even when the stems and plants do not die, the berries are delayed in ripening, and fruit quality is reduced by the excess crop and lack of leaves. The same plants in the same fields readily produce new leaves early in the spring if the flower buds are removed during the winter. Growers have abandoned 'Marimba' because of its tendency to over-fruit and under-leaf. Growers can sometimes manage 'Misty' successfully in Florida, using a combination of winter pruning to reduce the flower bud load and application of hydrogen cyanamide. In areas that receive high amounts of chilling (600 to 800 hours), 'Misty' opens all its flower buds in a short time and will then leaf without major problems.

Southern highbush plants that are growing vigorously in the fall are often hard to defoliate in Florida, even if sprayed with recommended concentrations of hydrogen cyanamide. On 2 December 2001, four-year-old plants of 8 southern highbush selections in a commercial field near Gainesville, Florida were sprayed to runoff with a mixture of 10% ammonium sulfate, 2% $ZnSO_4$, 2% oil, and water, in an effort to defoliate the plants 3 weeks before they received a 1% hydrogen cyanamide application. Four 5-plant plots were sprayed for each clone and four unsprayed 5-plant plots served as checks. Although natural defoliation was occurring in the field due to leaf rust and other factors, the salt-oil spray visibly increased the rate of defoliation in most clones. The weather in Gainesville was unusually warm during most of November and the first three weeks of December 2001. In some clones, by 20 December, both flower buds and vegetative buds showed more swelling in the sprayed plots than in the unsprayed plots. The hydrogen cyanamide that was sprayed on the plants in late December, and a hard freeze (ca $-7°C$) in early January, caused considerable flower bud damage and yield reduction in the plots that had been sprayed with the salt-oil mixture in early December. The two most evergreen clones in the test showed little or no defoliation before 1 March despite the salt sprays, the hydrogen cyanamide, and the freezes.

Since poor leafing in the spring is seldom a problem with blueberries in Michigan, New Jersey, or in the Pacific Northwest, it is tempting to consider the problem in Florida to be a manifestation of lack of chilling during the winter. We have found that dormant southern highbush blueberry plants (chilling requirements 100 to 500 hours below 7°C) leaf more quickly and prolifically after being placed in a warm greenhouse if they were previously chilled for 1200 hours at 2°C instead of at 6°C. According to Sherman and Beckman (2002), peach and blueberry varieties that leaf before they flower in the southeastern US are usually those with the lowest chill requirement. With blueberries in Florida, however, some counterexamples can be cited. For example, 'Bladen' and 'Legacy' (selected in North Carolina and New, Jersey, respectively) are clearly higher in chilling requirement than the Florida varieties 'Misty' and 'Marimba', since the former, when planted in Gainesville, flower and leaf 3 to 4 weeks later in the spring than low-chill Florida cultivars. Nonetheless, in most years in Gainesville, 'Bladen' and 'Legacy' leaf strongly at about the time they flower, whereas 'Misty' and 'Marimba' flower early but leaf poorly unless the flower buds have been thinned by freezes or winter pruning. All four cultivars are heavy flower bud producers, but only in 'Misty' and 'Marimba' do the flower buds seem to strongly inhibit leafing.

In some years, southern highbush cultivars growing in commercial fields in central Florida south of Orlando leaf better than the same varieties that appear to be under similar management 200 km farther north. This is contrary to what would be expected if lack of chilling were the primary reason for poor leafing. One explanation could be that leaf buds on the plants in the warmer area may not enter so deep a dormancy as those on plants farther north. Another possibility is that blueberries tend to produce fewer flower buds in central and south Florida due to higher temperatures in the fall, and the leaf buds therefore have less competition from flower buds as they open in the spring.

Some growers believe that variations in leafing on the same variety from field to field may relate to differences in the nutrient status of the plants during the winter. Some growers spray their plants during the winter and spring with dilute solutions of N, P, K, and minor elements with the intention of improving spring leafing, but the effectiveness of this practice has not been studied scientifically. A problem with grapes in California called RSG (reduced spring growth), whose symptoms include erratic bud break and irregular shoot growth in the spring, is attributed to insufficient carbohydrate accumulation in the fall (Brase, 2001). Maintaining a healthy vine canopy in the fall is said to minimize the potential for RSG. Whether some autumn management technique could improve spring leafing on blueberries in Florida is unknown.

## LITERATURE CITED

Brase, R. 2001. Fall irrigations are important. Amer. Fruit Grower. Nov. 2001. p. 25.

Sherman, W.B. and T.G. Beckman. 2002. Climatic adaptation in fruit crops. Acta Hort., in press.

Thomas, P. 2000. Trees: their natural history. Cambridge Univ. Press, Cambridge, UK.

Williamson, J.G., B.E. Maust, and D.S. NeSmith. 2001. Timing and concentration of hydrogen cyanamide affect blueberry bud development and flower mortality. HortScience 36:922-924.

# Screenhouse Evaluations of a Mason Bee *Osmia ribifloris* (Hymenoptera: Megachilidae) as a Pollinator for Blueberries in the Southeastern United States

Blair J. Sampson
Stephen J. Stringer
James H. Cane
James M. Spiers

**SUMMARY.** The behavioral traits and the pollination efficiency of *Osmia ribifloris* Cockerell (Hymenoptera: Megachilidae) were studied to assess the potential value of this bee for pollinating southern blue-

Blair J. Sampson is Research Entomologist, United States Department of Agriculture-Agricultural Research Service, Small Fruit Research Station, P.O. Box 287, Poplarville, MS 39470.

Stephen J. Stringer is Research Geneticist, United States Department of Agriculture-Agricultural Research Service, Small Fruit Research Station, P.O. Box 287, Poplarville, MS 39470.

James H. Cane is Research Entomologist, United States Department of Agriculture-Agricultural Research Service, Bee Biology and Systematics Laboratory, 5310 Old Main Hill, Logan, UT 84322-5310.

James M. Spiers is Research Horticulturalist, United States Department of Agriculture-Agricultural Research Service, Small Fruit Research Station, P.O. Box 287, Poplarville, MS 39470.

Address correspondence to: Blair J. Sampson (E-mail: bsampson@ars.usda.gov).

[Haworth co-indexing entry note]: "Screenhouse Evaluations of a Mason Bee *Osmia ribifloris* (Hymenoptera: Megachilidae) as a Pollinator for Blueberries in the Southeastern United States." Sampson, Blair J. et al. Co-published simultaneously in *Small Fruits Review* (Food Products Press, an imprint of The Haworth Press, Inc.) Vol. 3, No. 3/4, 2004, pp. 381-392; and: *Proceedings of the Ninth North American Blueberry Research and Extension Workers Conference* (ed: Charles F. Forney, and Leonard J. Eaton) Food Products Press, an imprint of The Haworth Press, Inc., 2004, pp. 381-392. Single or multiple copies of this article are available for a fee from The Haworth Document Delivery Service [1-800-HAWORTH, 9:00 a.m. - 5:00 p.m. (EST). E-mail address: docdelivery@haworthpress.com].

berry crops (*Vaccinium* sect. *Cyanococcus*). Emergence for this typically univoltine bee can be made to coincide with blueberry flowering, but there were some unusual patterns. For example, some bees delayed prepupal development and emerged one year late (i.e., parsivoltinism). Adults began foraging at blueberry flowers at air temperatures as cool as 9°C. Thirty-three to 67 percent of open-pollinated rabbiteye blueberry (*Vaccinium ashei* Reade) flowers visited solely by female *O. ribifloris* set seeded fruits. Bagged flowers that excluded *O. ribifloris* resulted in 0-5% seedless fruit. Calculations based on flower visits and fruit set for female bees suggest that the monetary return a farmer receives from each female bee that nests in the orchard is about $12.00 to $24.00. *[Article copies available for a fee from The Haworth Document Delivery Service: 1-800-HAWORTH. E-mail address: <docdelivery@haworthpress.com> Website: <http://www.HaworthPress.com>]*

**KEYWORDS.** Apoidea, non-*Apis* pollinator, *Vaccinium*, fruit set, pollination, parsivoltinism

## INTRODUCTION

Mason bees (*Osmia* spp.) are among the most promising pollinators of North American fruit crops (e.g., blueberries, apple, pears, and sweet cherries, Bosch and Kemp, 2002; Drummond and Stubbs, 1997), but are rare at flowering rabbiteye blueberries (*Vaccinium ashei* Reade: Ericaceae) in the southeastern United States (Cane and Payne, 1988, 1990). *Osmia ribifloris* Cockerell is a mason bee that efficiently pollinates rabbiteye blueberries (Sampson and Cane, 2000; Sampson et al. 1995), but does not naturally co-occur with this cultivated blueberry. *O. ribifloris* populations can be easily relocated to rabbiteye blueberry orchards by first encouraging females to nest in man-made domiciles; cylindrical cavities drilled into wooden blocks or straws arranged inside cardboard boxes. After the brood reach maturity inside their cocoons, they can easily be recovered from the trap-nests and then transported to where they are needed (Stubbs et al., 1994; Torchio, 1990). Female *O. ribifloris* are not aggressive, so blueberry growers can comfortably handle bees without the need for protective bee suits. We report here on aspects of *O. ribifloris* biology that will help with developing the bee as a pollinator for rabbiteye and perhaps southern highbush blueberries. We have studied the regional and environmental adaptation, nest provisioning behavior, pollination efficiency, and monetary value for *O. ribifloris*.

## MATERIALS AND METHODS

*Study area, pollinator management, and regional adaptability.* A population of *Osmia ribifloris* was maintained for four generations in a secure 15 m × 9 m × 5 m (L × W × H) screenhouse located at the USDA-ARS Small Fruit Research Station, Poplarville, MS. Cocoons stored for 90-120 days at 4-5°C were warmed at 21-29°C to break adult dormancy just before each release. Incubation conditions and seasonal emergence patterns for cocooned bees (i.e., emergence date, sex, parasites, and release date) were recorded daily for *O. ribifloris*. As part of the X-ray screening protocols for natural enemies, the parasitic wasps, *Monodontomerus* sp. and *Sapyga pumila* Cresson were removed, and healthy adult *Osmia ribifloris* transferred to the screenhouse for testing. Cocoons containing healthy prepupae (i.e., post defecating larvae) that had not emerged the first year were held a second year to test their viability. Methods for acquiring and managing wild bees were those described by Sampson and Cane (2000).

Environmental adaptation. Rates of flower visitation by female bees were recorded under favorable weather conditions between 8-15 March 2000. The number of *V. ashei* flowers visited by females was counted during 30-s foraging bouts. Each day, at least 15 bouts were tracked in the morning (0915-1030 HR) and in the afternoon (1304-1400 HR). One-way analysis of variance was used to test the hypothesis that female *O. ribifloris* foraged faster in the afternoons.

First counting male *O. ribifloris* cruising above the bushes gauged the daily onset of adult flight activity of bees released inside the screenhouse. Females were then counted as they visited *V. ashei* flowers. Bees were counted from 0700-1900 HR for 3 d, while corresponding weather readings were compiled by a portable weather station (GroWeather 7450/7455, Davis Instruments, Hayward, CA) in the screenhouse. Multivariate regression (MANOVA in PROC GLM, SAS Institute, 1985) identified which of four weather factors–air temperature (°C), percent relative humidity, solar radiation (W/m$^2$), and barometric pressure (kPa)–correlated with bee activity.

*Nest provisioning behavior.* Nesting and foraging behaviors, which are difficult to measure in the field, were more easily studied inside the screenhouse. The average flowers visited per foraging trip were calculated from five uninterrupted foraging sequences between the time bees departed nests and returned. Sequences were monitored between 10-16 March 2000. The number of trips needed to provision a nest cell was tallied for six females over a 1 or 2-d period from 20-23 March 2000. Six whole provision masses were frozen, later thawed and placed into pre-cleaned glass vials. Each vial contained 50 mL filtered ethanol and the resulting pollen suspension was dispersed in an ultrasonic bath. Half of the agitated subsample (25 mL) of the pollen suspension was counted using a particle counter (HIAC-ROYCO, Pacific Scientific Inc., Silver Spring, MD) (Cane et al., 1996). Doubling the pollen count gave an estimate of the total blueberry tetrads contained in a single *O. ribifloris* provision.

Of interest to those wanting to maintain healthy bee populations, propensities for nest reuse were evaluated for *O. ribifloris*. To test this, each nesting block (N = 15 blocks) contained six straws, two of which were clean and unused (controls), two were used before by *O. ribifloris*, and the remaining two were earlier used by *O. lignaria*. The contents of the used straws were pulverized and shaken in each straw. Large obstructions such as partitions and dead immature bees were removed from the used straws. The clean straws were sliced open and bound with clear, adhesive tape so as to duplicate as closely as possible the distortions caused by repairing the used straws before inserting them into the nesting blocks. The six straws were randomly assigned to the six holes in each block. If the used nest straws were overwhelmingly rejected by *O. ribifloris*, then the two additional nest blocks containing six clean straws were provided to alleviate possible nest ursurpation (total nest straws = 102). A chi-square test tested if the selection of the three straw types (new, old *O. ribifloris*, and old *O. lignaria* straws) by female bees was random.

Pollination efficiency and monetary value. Bees were provided with five species of blueberry (N = 26 clones) native to the Gulf Coastal Plain (USDA hardiness zone 8B). They were *Vaccinium ashei* Reade, diploid and tetraploid *V. corymbosum* L., *V. darrowi* Camp, *V. elliottii* Chapm., and *V. tenellum* Aiton. Primary food sources of the caged bees were 300 potted rabbiteye blueberry plants. A minimum of 25 terminal racemes per blueberry species, most of which had between 6-26 flowers (N = 335 racemes), were chosen on these plants of all vigorous clones and their flowers counted. Pollination efficiency and the level of autogamy (= parthenocarpy) for each blueberry species were based on percent green fruit set, percent ripe fruit set, and percent fruit drop measurements using the same protocols developed by Sampson and Spiers (2002), except that green fruit set was evaluated 35 d after flowers were pollinated. The percent fruit drop was the percentage of green berries that were present on d 35, but did not mature. Ripe berries were harvested daily from tagged clusters from 5 May 2001 to 11 July 2001. Statistical analyses of fruit set observations (see Table 1) were the same as those reported by Lyrene (1989), Sampson and Cane (2000), and Sampson and Spiers (2002).

## RESULTS AND DISCUSSION

*Regional adaptation.* Adult *O. ribifloris* emerged after chilling from 22 February to 31 March during 1999-2002; this coincided with rabbiteye blueberry bloom in Mississippi. For some cocoons, unplanned incubation stimulated earlier adult emergence (28 January-2 February). Under some circumstances, early emergence might be beneficial to this bee's pollination of southern highbush blueberries–hybrid blueberries that can bloom a month before rabbiteye blueberries (Sampson and Spiers, 2002). Although *O. ribifloris* is univoltine, delays in emergence (i.e., parsivoltinism) also occurred. In an isolated case, 1 in 5 bees passed the first winter as prepupae, and the second winter as adults. Both sexes had similar rates of parsivoltinism ($P^2 = 0.29$; df = 1, $P = 0.5912$). Parsivoltinism in *Osmia* appears to be a rare, but facultative condition perhaps induced by unusual climactic or rearing conditions (Bosch and Kemp 2000; Torchio and Tepedino, 1982; Wcislo and Cane, 1996).

Although male *O. ribifloris* are the first bees to emerge in the spring in larger numbers, female bees are the more valuable blueberry pollinators. Female bees can control the sex of their offspring, and the pro-

TABLE 1. Summary of ANCOVA statistics. Effect of *Vaccinium* species (SP), level of pollinator (*O. ribifloris*) visitation (TRT: no visitation/virgin), and their interaction on the reproductive characteristics of blueberries pollinated by *Osmia ribifloris*.

| | Green fruit set (%) | | | Ripe fruit set (%) | | | Fruit drop (%) | | |
|---|---|---|---|---|---|---|---|---|---|
| Source | df | F | P | df | F | P | df | F | P |
| SP | 5 | 38.37 | ** | 5 | 19.58 | ** | 4 | 1.69 | NS |
| TRT | 1 | 21.41 | ** | 1 | 40.05 | ** | 1 | 2.12 | NS |
| SP*TRT | 5 | 2.71 | 0 | 5 | 14.6 | ** | 1 | 1.5 | NS |
| Error df | 312 | | | 311 | | | 113 | | |

**P ≤ 0001, *0.0001 < P ≤ 0.05, NS – P > 0.05.

duction of male-biased progeny distinguishes solitary bees like *Osmia* from many of the social bees that exist in larger female groups (Sugiura and Maeta, 1989; Torchio, 1985, 1990). Male-biased sex ratios for *O. ribifloris* were initially high for bees imported from the west, but quickly fell to a more stable sex ratio for succeeding generations reared primarily on *V. ashei* pollen (Figure 1) (Torchio, 1990). There are many intraspecific factors that affect sex ratios in *Osmia*: offspring size, seasonality, locality (Bosch and Kemp, 2002), dimensions of nest cavities (Torchio, 1990), nest cell size, and quantities of stored food (Krombien, 1967). Lowering or stabilizing male:female ratios is desirable for commercial blueberry pollination, as *O. ribifloris* females visit and pollinate substantially more blueberry flowers than males (Sampson, personal observation). Further studies are planned to identify the sources of these variable sex ratios, and develop methods to maximize female production.

*Environmental adaptation. O. ribifloris* flew during most of the daylight hours (0700-1800 HR), on clear, sunny days. Temperature and illumination influenced ($P < 0.003$) foraging activity. Females flew at temperatures as cool as 9°C (Figure 2). This is a threshold several degrees lower than that reported for a related fruit pollinator, *O. lignaria* (Torchio, 1985). Foraging peaked at 1300 HR as air temperatures also peaked near 21-26°C (Figures 2 and 3). Except for rain, the bees' foraging activities were not correlated with fluctuations in relative humidity and barometric pressure (Figures 2 and 3, $P > 0.5$).

*O. ribifloris* visited *V. ashei* blooms at a uniform rate throughout the day ($4 \pm 1$ flowers every 30 s; F = 0.73; df = 1, 136, P = 0.3944). This rate equaled the rate of females foraging from northern highbush blueberries (*V. corymbosum*) in California (3-4 flowers/30s calculated from data published by Torchio, 1990). Females returned to their nests every 10.2 minutes (SD = 2.7, N = 17) with a scopal load of pollen gathered from an average of 55 *V. ashei* flowers (range: 43-63, N = 5). An egg being laid on the provision followed the completion of 18 pollen collecting trips (SD = 2, N = 6), and so, multiplying this number by the bees' average foraging tempo (55 flowers/trip), we can expect about 990 flowers were visited for each nest cell provisioned. Bees often needed 2 d to complete a nest cell. Partial nest provisions were often finished and, after receiving an egg, were sealed the next morning. Each provision contained a mixture of nectar and approximately $1.39 \times 10^6 \pm 1.60 \times 10^5$ blueberry tetrads (N = 6). Female *O. ribifloris* were not known to reuse natal nests for nest cell provisioning (Torchio, 1990) because the risk of fungal infections is greater. However, our findings indicate fe-

FIGURE 1. Emergence sex ratios (#males/#females) for *O. ribifloris* maintained in the screenhouse for four seasons (years 1999-2002). $P_1$ is the parental generation and $F_1 - F_3$ are the subsequent filial generations. The sample sizes (total adults emerged) for each generation and chi-square analysis are shown. Declining sample sizes were due to cocooned bees being repatriated for field trials in Oregon blueberry fields.

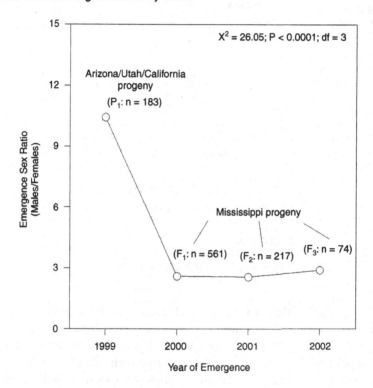

males cannot readily tell pristine nest straws from used straws ($X^2 = 1.143$; df = 2; P > 0.05). It is likely that older natal straws become more acceptable to females when their interiors were cleared of debris. Clean nesting materials should always be provided annually to *O. ribifloris* to maintain proper hygiene within nesting populations.

*Pollination efficiency and pollination value.* The potential value of *O. ribifloris* to blueberry production may be calculated from existing information. *O. ribifloris* pollinated cultivars of rabbiteye blueberries with an efficiency that equaled other established pollinator species in the Southeast (Cane and Payne, 1990; Payne et al., 1989, 1991; Sampson

FIGURE 2. The diel foraging activity of *O. ribifloris* as a function of ambient air temperature inside the screenhouse. Female (○) and male (+) responses to air temperature are separated. ANCOVA results and equations are shown for the two regressions.

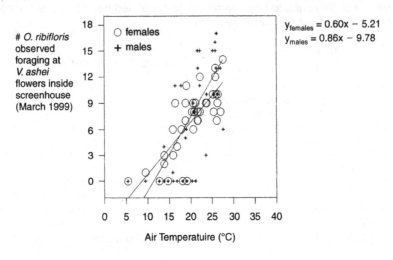

# *O. ribifloris* observed foraging at *V. ashei* flowers inside screenhouse (March 1999)

$y_{females} = 0.60x - 5.21$

$y_{males} = 0.86x - 9.78$

Air Temperatuire (°C)

ANCOVA results:
Sex: $F_{1,75} = 0.65$; $P = 0.4224$
Temperature: $F_{1,75} = 1086.30$; $P < 0.0001$

and Cane, 2000). Mean overall green fruit set of rabbiteye blueberries in these trials was 43.3% (range: 33%-67%); ripe fruit set was 33.6% (range: 24%-56%, Table 1 and Figure 4). If 990 flowers are visited by a female per nest cell and 34% of them set ripe fruit, then for each nest cell produced, the foraging efforts of a female *O. ribifloris* will yield 336 blueberries. These blueberries would weigh approximately 0.60 kg, and be worth between $1.00-$2.00 at recent wholesale prices (Williamson and Lyrene, 1997). A female *O. ribifloris* can provision about 12 nest cells in her lifetime (Torchio, 1990). So, upon her death she would have contributed to the production of $12.00-$24.00 worth of fresh blueberries. This value is similar to that calculated for the southeastern blueberry bee, *H. laboriosa* (Cane, 1997). The above calculations, although simplistic, highlight the potential value of *O. ribifloris* as a pollinator of rabbiteye and perhaps southern highbush blueberries. Sustaining larger populations of *O. ribifloris* on the farm might also prove to be lucrative for southeastern blueberry producers. Currently, supplemental blue-

FIGURE 3. The diel flight activity of male and female *O. ribifloris*. Observed (○) and predicted (+) values are given. Model equations containing four terms [i.e., the four weather factors: temperature (temp), solar radiation (sun), relative humidity (RH), and barometric pressure (BP)] from which predicted values and regression lines were derived, are provided.

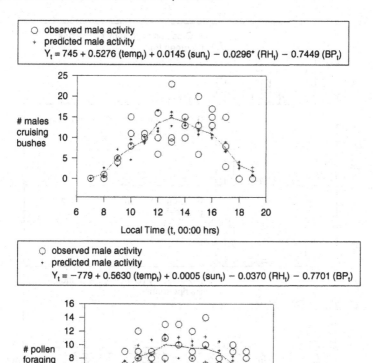

FIGURE 4. Percent of green and ripe fruit that set for tagged racemes resulting from two levels of visitation: open-pollination by *O. ribifloris* (4A, 4B), *O. ribifloris* excluded (4C, 4D). Fruit set values are shown for each species and the clones and cultivars comprising each species. Gaps indicate 0% fruit set for a given clone(s) and treatment(s). Bars represent ± 1 standard error.

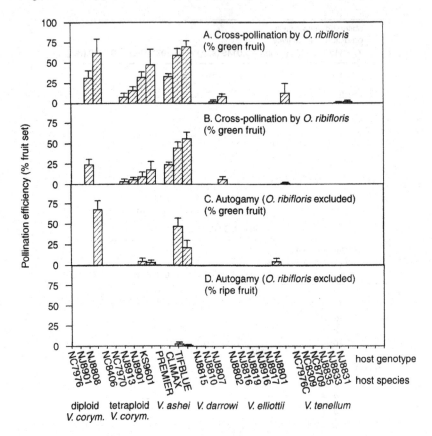

berry pollinators (e.g., honey bees and bumble bees) are obtained from off-farm sources, often with greater inconvenience or expense to the grower. For a bee like *Osmia ribifloris*, the grower may own and annually propagate the bee on the farm (Batra, 1982; Sampson et al., 1995) and if diapausal mortality is low, and adult dispersal is minimized, the grower can recover and sell the surplus *O. ribifloris* cocoons.

Flowers of the two most popular *V. ashei* cultivars 'Climax' and 'Tifblue' set more fruit after being pollinated by *O. ribifloris* (Figure 4) than those remaining unvisited. Unrestricted visitation yielded fruit sets that were no greater than the single-visit fruit set levels measured the year before (Sampson and Cane, 2000). Thus, a single visit by this bee is sufficient to set rabbiteye blueberry fruit. Northern and southern highbush blueberry clones (*V. corymbosum*) also received satisfactory pollination by *O. ribifloris*. Fruit sets might have been higher for *V. corymbosum* if an increased number of compatible plants were placed inside the screenhouse. Many clones of the wild blueberries, *V. elliottii, V. darrowi,* and *V. tenellum,* were largely ignored by *O. ribifloris,* were poorly pollinated (Figure 4), and, therefore, may not compete with cultivated blueberries for *O. ribifloris* pollinators.

## CONCLUSIONS AND GROWER BENEFITS

We are moving closer to our goal of eventually releasing *O. ribifloris* as a blueberry pollinator in the Southeast. Field releases and pathological studies are now underway, and the final step is to expand our propagation efforts to produce enough bees for a limited field release. Experimental releases have occurred in Oregon with the objective of testing the nesting success of *O. ribifloris* in an open orchard environment. These data are forthcoming.

## LITERATURE CITED

Batra, S. W. T. 1982. The hornfaced bee for efficient pollination of small farm orchards. In: H. W. Kerr and L. Knutson. (Eds.) Proceedings of a special symposium: Research for farms. USDA-ARS Misc. Publ. 1422: 116-120.

Bosch, J. and W. P. Kemp. 2000. Development and emergence of the orchard pollinator, *Osmia lignaria* (Hymenoptera: Megachilidae). Environ. Entomol. 29: 8-13.

Bosch, J. and W. P. Kemp. 2002. Developing and establishing bee species as crop pollinators: The example of *Osmia* spp. (Hymenoptera: Megachilidae) and fruit trees. Bull. Entomol. Res. 92: 3-16.

Cane, J. H. 1997. Lifetime monetary value of individual pollinators: the bee *Habropoda laboriosa* at rabbiteye blueberry (*Vaccinium ashei* Reade). Acta Hort. 446: 67-70.

Cane, J. H. and J. A. Payne. 1988. Foraging ecology of the bee *Habropoda laboriosa* (Hymenoptera: Anthophoridae), an oligolege of blueberries (Ericaceae: *Vaccinium*) in the Southeastern United States. Ann. Entomol. Soc. Amer. 81: 419-427.

Cane, J. H. and J. A. Payne. 1990. Native bee pollinates rabbiteye blueberry. Ala. Agri. Exp. Sta., Auburn Univ. Highlights Agric. Res. 37: 1,4.

Cane, J. H., D. Schiffhauer, and L. J. Kervin. 1996. Pollination, foraging, and nesting ecology of the leafcutting bee *Megachile* (*Delomegachile*) *addenda* (Hymenoptera: Megachilidae) on cranberry beds. Ann. Entomol. Soc. Amer. 89: 361-367.

Drummond, F. A. and C. S. Stubbs. 1997. Potential for management of the blueberry bee, *Osmia atriventris* Cresson. Acta Hort. 446:77-85.

Krombien, K. L. 1967. Trap-nesting wasps and bees: Life histories, nests, and associates. Smithsonian Press, Washington, DC.

Lyrene, P. M. 1989. Pollen sources influences fruiting of 'Sharpblue' blueberry. J. Amer. Soc. Hort. Sci. 114: 995-999.

Payne, J. A., J. H. Cane, A. A. Amis, and P. M. Lyrene. 1989. Fruit set, seed size, seed viability and pollination of rabbiteye blueberries (*Vaccinium ashei* Reade). Acta Hort. 241: 38-43.

Payne, J. A., D. L. Horton, J. H. Cane, and A. A. Amis. 1991. Insect pollination of rabbiteye blueberries, pp. 42-49. In: 1991 Missouri Small Fruit Conference, 19-20 Feb. 1991. Springfield, MO.

SAS Institute. 1985. SAS User's Guide: Statistics Version 5 edition. SAS Institute Inc, Cary, NC.

Sampson, B. J. and J. H. Cane. 2000. Pollination efficiencies of three bee (Hymenoptera: Apoidea) species visiting rabbiteye blueberry. J. Econ. Entomol. 93: 1726-1731.

Sampson, B., J. Cane, and J. Neff. 1995. Blue bees for blueberries. Alabama Agricultural Exp. Sta., Auburn Univ. Highlights Agric. Res. 42: 12,13,15.

Sampson, B. J. and J. M. Spiers. 2002. Evaluating bumblebees as pollinators of 'Misty' southern highbush blueberry growing inside plastic tunnels. Acta Hort. 574: 53-61.

Stubbs, C. S., F. A. Drummond, and E. A. Osgood. 1994. *Osmia ribifloris biedermannii* and *Megachile rotundata* (Hymenoptera: Megachilidae) introduced into lowbush blueberry agroecosystems in Maine. J. Kansas Entomol. Soc. 67: 173-185.

Sugiura, N. and Y. Maeta. 1989. Parental investment and offspring sex ratio in a solitary mason bee, *Osmia cornifrons* (Radoszkowski) (Hymenoptera: Megachilidae). Jpn. J. Entomo. 57: 861-875.

Torchio, P. F. 1985. Field experiments with the pollinator species, *Osmia lignaria propinqua* Cresson, in apple orchards: (1979-1980), methods of introducing bees, nesting success, seed counts, fruit yields (Hymenoptera: Megachilidae). J. Kans. Entomol. Soc. 58: 448-464.

Torchio, P. F. 1990. *Osmia ribifloris*, a native bee species developed as a commercially managed pollinator of highbush blueberry (Hymenoptera: Megachilidae). J. Kan. Ent. Soc. 63: 427-436.

Torchio, P. F. and V. J. Tepedino. 1982. Parsivoltinism in three species of *Osmia* bees. Psyche. 89: 221-238.

Wcislo, W. T. and J. H. Cane. 1996. Floral resource utilization by solitary bees (Hymenoptera: Apoidea) and exploitation of their stored foods by natural enemies. Annu. Rev. Entomol. 41: 257-286.

Williamson, J. and P. Lyrene. 1997. Florida's commercial blueberry industry. Bulletin HS 742. Hort. Sci. Dept., Fla. Coop Ext. Service, Inst. Food Agric. Sci., Univ. Florida, Gainesville, FL.

# *In Vitro* Culture of Lowbush Blueberry (*Vaccinium angustifolium* Ait.)

Samir C. Debnath

**SUMMARY.** Cultures of three lowbush blueberry (*Vaccinium angustifolium* Ait.) clones collected from natural stands in Newfoundland were established *in vitro* on a modified cranberry (*V. macrocarpon* Ait.) tissue culture medium containing zeatin (5 μM) or $N^6$-[2-isopentenyl]adenine (2iP) (10 μM). Shoot proliferation with respect to shoot number per explant differed among clones at various concentrations of zeatin over two culture periods. Best total shoot proliferation was obtained when basal nodal segments were cultured in the medium supplemented with 2-4 μM zeatin. In another experiment, nodal explants were more productive than shoot tips. Shoots growing for more than 12 weeks on media that contained more than 4 μM zeatin occasionally produced adventitious shoot masses, which appeared to arise from dense calli growing at the base of the shoots in the medium. The lower concentration of sucrose and lower irradiance improved shoot proliferation with respect to vigor

---

Samir C. Debnath is Research Scientist, Atlantic Cool Climate Crop Research Centre, Agriculture and Agri-Food Canada, P.O. Box 39088, 308 Brookfield Road, St. John's, Newfoundland, Canada A1E 5Y7.

Address correspondence to: Samir C. Debnath at the above address (E-mail: debnaths@agr.gc.ca).

The author gratefully acknowledges the cooperation of Boyd G. Penney. Thanks are also due to Sarah Devine, Glen R. Chubbs, and Sandra M. Cooke, for their excellent technical help. The manuscript is contribution no. 150 of the Atlantic Cool Climate Crop Research Centre.

[Haworth co-indexing entry note]: "*In Vitro* Culture of Lowbush Blueberry (*Vaccinium angustifolium* Ait.)." Debnath, Samir C. Co-published simultaneously in *Small Fruits Review* (Food Products Press, an imprint of The Haworth Press, Inc.) Vol. 3, No. 3/4, 2004, pp. 393-408; and: *Proceedings of the Ninth North American Blueberry Research and Extension Workers Conference* (ed: Charles F. Forney, and Leonard J. Eaton) Food Products Press, an imprint of The Haworth Press, Inc., 2004, pp. 393-408. Single or multiple copies of this article are available for a fee from The Haworth Document Delivery Service [1-800-HAWORTH, 9:00 a.m. - 5:00 p.m. (EST). E-mail address: docdelivery@haworthpress.com].

Digital Object Identifer: 10.1300/J301v03n03_16

compared to the control treatments (30 g $L^{-1}$ and 30 $\mu$mol $m^{-2}$ $s^{-1}$, respectively). In all experiments with subculture, there was an increase in shoot multiplication rate for all clones. A 50-100-fold multiplication rate was obtained every 3 months with the clones. *[Article copies available for a fee from The Haworth Document Delivery Service: 1-800-HAWORTH. E-mail address: <docdelivery@haworthpress.com> Website: <http://www.HaworthPress. com> © 2004 by The Haworth Press, Inc. All rights reserved.]*

**KEYWORDS.** *Vaccinium angustifolium*, micropropagation, shoot proliferation, rooting, tissue culture

## *INTRODUCTION*

The lowbush blueberry (*Vaccinium angustifolium* Ait.), a perennial, rhizomatous cross-pollinated shrub, is native to Newfoundland and Labrador with a diverse distribution ranging from northeastern to mid-North America (Vander Kloet, 1988). It is a commercially important crop in Maine, Quebec, and the Canadian Atlantic Provinces (Hoefs and Shay, 1981), where wild stands, made up of numerous heterogenous genetically diverse clones, are commercially managed and harvested. Their genetic variation for traits affecting yield (berry size, number of berries per cluster, stem density) results in more and less fruitful areas in a field. Furthermore, incomplete coverage results from inadvertent kills of plants from applied herbicides, erosion that had been prevented by weeds, and 'scalping' by machinery is very common in commercial lowbush blueberry fields (Metzger and Ismail, 1977; Morrison et al., 2000). Using high yielding clones to fill in bare spots in existing fields will make the current management practices more efficient and result in higher yields at a lower cost per pound.

Plants for establishing new blueberry fields can be produced from softwood or rhizome cuttings of selected clones. While stem or rhizome cuttings are relatively easy to root, their extreme precocity of flowering results in very slow establishment of plantings. This problem can be largely avoided by using plants produced through micropropagation (Smagula and Lyrene, 1984). *In vitro* cloning is a more demanding and potentially more effective method for improving lowbush blueberry fields, tissue culture-propagated plants exhibit the spreading growth habit of seedlings along with the uniform productivity characteristics of rooted cuttings (Frett and Smagula, 1983).

Although there are a number of reports of *in vitro* propagation of various *Vaccinium* species, few are available with *V. angustifolium* being propagated by tissue culture. *In vitro* culture in lowbush blueberry is still less developed and less efficient as compared with that of many other small fruit species. Shoot formation from hypocotyl or excised cotyledons of lowbush blueberry was reported by Nickerson (1978). Frett and Smagula (1983) used single-bud explants of mature tissue of lowbush blueberry to obtain multiple shoots, but the percentage of rooting was only 44%. In another study, new shoot growth was initiated on only two of 253 uncontaminated explants of *V. angustifolium* using very high concentration (49.2-73.8 µM) of N$^6$[2-isopentenyl]adenine (2iP) along with an auxin (11.4-22.8 µM indole-3-acetic acid, IAA) in the culture medium (Brissette et al., 1990). Excessive concentrations of auxins and/or cytokinins in media, however, can result in regenerated plants deviating morphologically from the normal (George, 1996). The appearance of somaclonal variation (Larkin and Scowcroft, 1981) in the *in vitro*-regenerated plants can be greatly reduced by using lower levels of cytokinin (Marcotrigiano and McGlew, 1991). While most of the early reports used the cytokinin, 2iP for initiating new growth from explants, zeatin, another cytokinin, has not been tested for shoot proliferation and maintenance of lowbush blueberry *in vitro* cultures. Zeatin was found to be more effective for shoot initiation in *Vaccinium* species (Reed and Abdelnour-Esquivel, 1991) and for shoot proliferation of highbush blueberry (*V. corymbosum* L.) (Chandler and Draper, 1986; Eccher and Noe, 1989) and lingonberry (*V. vitis-idaea* L.) (Debnath and McRae, 2001b).

The present study sought to develop an effective protocol for micropropagation of native lowbush blueberry clones using low levels of cytokinin without auxin in the culture media; three native Newfoundland clones were tested.

## MATERIALS AND METHODS

*Plant material, culture establishment, and general methods.* Young, actively growing lowbush blueberry clones designated as 'NFB1', NFB2', and 'NFB3' were harvested from plants being maintained in a greenhouse. Stem segments, 5-6 cm long, were rinsed in running water and washed in a solution of 2% (v/v) mild liquid detergent for 2 min. The shoots were surface sterilized by dipping in 70% ethanol for 30 s followed by treating in a solution of 0.79% sodium hypochlorite (15% commercial bleach) and 0.1% Tween 20 (polyoxyethylene sorbitan monolaurate) for 20 min, and then rinsed three times in sterilized deionized water. In all experiments, cultures grew five explant per 175 mL Sigma baby food glass jars with polypropylene clear caps (Sigma Chemical Co., St. Louis), which contained 35 mL of basal medium. The disinfected stem segments were aseptically cut into single-nodal sections and cultured, for shoot initiation, on the modified cranberry medium, a modification of a basal medium formulated in our laboratory for cranberry micropropagation (Debnath and McRae, 2001a) wherein $CaCl_2$ was replaced by 1.3 g $L^{-1}$ calcium gluconate and casein hydrolysate was increased to 100 mg $L^{-1}$ [referred to as basal medium hereafter] supplemented with 30 g $L^{-1}$ sucrose and either 5 µM zeatin or 10 µM 2iP. After the ingredients were combined, the media were adjusted to pH 5.0 and 3.5 g $L^{-1}$ Sigma A 1296 agar (Sigma Chemical Co., St. Louis) and 1.25 g $L^{-1}$ Gelrite (Sigma Chemical Co., St. Louis) were added. Culture jars were capped with clear permeable polypropylene caps and autoclaved at 121°C for 20 min. After explant transfer, culture jars were sealed with Parafilm, then placed upright and maintained at 20 ± 2°C under a 16-h photoperiod with, unless noted otherwise, a photosynthetic photon flux (PPF) density of 30 µmol $m^{-2} s^{-1}$ at the culture level provided by cool-white fluorescent lamps.

The vegetative shoots obtained *in vitro* from the axillary buds were excised and subcultured by transferring the nodal explants (with three nodes, leaves intact) of randomly selected shoots in each vial from the initial culture to fresh medium. This provided a new stock collection of shoots for micropropagation studies every 8 weeks. For each experiment, explants were grown for another 8 weeks on the basal medium that contained no plant growth regulators (PGR) to minimize the effect of PGR.

*Experiment 1. Effect of zeatin.* The purpose of this experiment was to determine the effect of zeatin concentration on shoot proliferation and callus growth (basal callusing) in three lowbush blueberry clones. The basal medium was supplemented with four zeatin concentrations. A 4 × 3 × 2 factorial experiment was used to study all possible combinations of 0, 1, 2, and 4 µM zeatin on the clones 'NFB1', 'NFB2', and 'NFB3' over two subculture periods. Eight-week-old *in vitro*-grown shoots were excised from their bases and cut into three-node basal sections with leaves intact. The tips and upper halves of the shoot were discarded. The sections were placed vertically on the medium supplemented with 30 g $L^{-1}$ sucrose. Subculturing was done with the nodal explants (with three nodes, leaves intact) of randomly selected shoots in each jar from the initial subculture. There were four jars for each clone-treatment combination, which were chosen in random order, and the experiment was repeated three times.

*Experiment 2. Effect of sucrose concentration and explant type.* A 4 × 3 factorial experiment (completely randomized) compared all combinations of four sucrose concentrations (0, 10, 20, and 30 g $L^{-1}$) and three explant types (basal node, middle node, and tip) in clone 'NFB1'. The aim was to determine if the cultured shoot tips were more responsive than the nodal stem sections when cultured on basal medium supplemented with different sucrose concentrations. Three-node shoot tips and nodal stem sections (basal section and the section taken from the middle of the shoot) with leaves intact were cultured on the basal medium supplemented with 2 µM zeatin. There were three jars for each treatment and the experiment was repeated four times.

*Experiment 3. Influence of irradiance.* To evaluate the effect of light on *in vitro* shoot proliferation and callus growth, clone 'NFB1' shoots were exposed to two irradiances (15 and 30 µmol $m^{-2}$ $s^{-2}$) over two subculture periods in a completely randomized experiment. Three-node basal stem sections (leaves intact), taken from actively growing *in vitro*-grown shoots, were cultured on the basal medium supplemented with 2 µM zeatin and 30 g $L^{-1}$ sucrose. There were three jars for each treatment. The experiment was repeated four times.

*Rooting of shoots.* Individual shoots with 2-3 expanded leaves were detached from the shoot clumps and dipped in 4.9 mM indole-3-butyric acid (IBA), and planted in 45-cell plug trays containing 2 peat:1 perlite (v/v) for *ex vitro* rooting. Trays were placed in a humidity chamber with a vaporizer (temperature 20 ± 2°C, humidity 95%, PPF = 55 µmol $m^{-2}$ $s^{-1}$, 16 h photoperiod) for rooting. Shoots were also allowed to grow for more than 10 weeks on the basal medium containing 1 and 2 µM zeatin to examine *in vitro* rooting. Rooted plantlets were acclimatized by gradually lowering the humidity over 2 to 3 weeks. Hardened-off plants were maintained in the greenhouse (temperature 20 ± 2°C, humidity 85%, maximum PPF = 90 µmol $m^{-2}$ $s^{-1}$, 16 h photoperiod). Percent survival was recorded at 10 weeks of culture in rooting medium.

*Data collection and statistical analysis.* After 8 weeks of culture in each experiment, growth characteristics of surviving explants were measured for each treatment. The following variates were determined: number of shoots (> 1.0 cm long) per responding explant, mean shoot length (for responsive explants), mean number of leaves per shoot, mean shoot vigor, and mean callus growth at shoot base. Vigor was determined by visual assessment, on a scale of 1 (strongly vitrified, necrotic and/or malformed shoots) to 8 (fully normal and healthy shoots). Callus growth was rated, on a scale of 1 (< 2mm diameter) to 8 (> 14 mm diameter). Mean data were subjected to analysis of variance with the SAS statistical software package (Release 8.2, SAS Institute, Inc., Cary, NC, USA). The square root transformation was used to stabilize the variance of the number of shoots in the ANOVA. When significant differences were found, means were separated by Duncan's multiple range test at P = 0.05. Sigma Plot 2001 (SPSS Inc., USA) was used to plot graphs and trend lines.

## RESULTS AND DISCUSSION

*Establishment of shoot culture.* The frequency of contamination varied from 5.5 to 9.5% whether the single-node stems were grown on the basal medium supplemented with zeatin or with 2iP. Nodal explants that were slightly tender and having greenish axillary buds responded efficiently for bud sprouting compared to hard nodal explants with brownish buds which showed no sign of growth. The explants produced elongated shoots on both zeatin and 2iP. However, better initiation rates (percentage of explants showing new shoot growth) were on zeatin-containing medium, zeatin allowed a greater number of shoots per explant to sprout than those on 2iP. Zeatin was also more effective than 2iP in the shoot initiation of *Vaccinium* species (Reed and Abdelnour-Esquivel, 1991), and for shoot proliferation of highbush blueberry (Chandler and Draper, 1986; Eccher and Noe, 1989) and lingonberry (Debnath and McRae, 2001b). As in other studies (Zimmerman and Broome, 1980; Eccher and Noe, 1989; Reed and Abdelnour-Esquivel, 1991), many differences existed between genotypes (data not shown); shoot initiation was best in 'NFB1' followed by 'NFB3' and 'NFB2'. Although callus formed at the base of the explants grown both on 2iP and zeatin supplemented media, shoots proliferated directly from the node via the axillary branching of buds from the original explants.

*Effect of zeatin.* Zeatin concentration significantly (P ± 0.05) affected shoot proliferation and callus growth (Table 1); shoots per explant and callus diameter increased in all three clones (Figure 1 a,e) while, in 'NFB1', shoot height, leaves per shoot, and shoot vigor decreased with concentration (Figure 1 b-d). In clone 'NFB1', though 4 μM produced the greatest number of shoots per explant, the rest of the characters were

TABLE 1. Mean values of the main effects across all treatments for combined effect of clone, zeatin concentration (ZN), and subculture on shoot proliferation and callus growth of *in vitro*-grown lowbush blueberry.[z]

| Treatment | Shoots (n/explant) | Shoot height (cm) | Leaves (n/shoot) | Vigor[y] (scale 1-8) | Callus growth[x] (scale 1-8) |
|---|---|---|---|---|---|
| Clone (C) | | | | | |
| 'NFB1' | 5.2c[w] | 3.1c | 9.0c | 4.7c | 4.3a |
| 'NFB2' | 2.1a | 2.2a | 5.2a | 3.3a | 5.8b |
| 'NFB3' | 2.6b | 2.6b | 6.8b | 3.6b | 5.7b |
| ZN (µM) | | | | | |
| 0 | 1.3a | 1.9a | 5.4a | 2.0a | 3.2a |
| 1 | 2.3b | 2.8b | 7.1b | 4.2b | 4.8b |
| 2 | 3.9c | 2.9b | 7.7c | 4.7c | 6.1c |
| 4 | 5.8d | 3.1c | 8.0c | 4.7c | 6.9d |
| Subculture (S) | | | | | |
| 1 | 3.1a | 2.6a | 6.9a | 3.7a | 5.2 |
| 2 | 3.5b | 2.7b | 7.2b | 4.0b | 5.3 |
| Significant effects[v] | C, ZN, S, C×ZN | C, ZN, S, C×ZN, C×S, ZN×S, C×ZN×S | C, ZN, S, C×ZN | C, ZN, S, C×ZN, ZN×S | C, ZN, C×ZN |

[z]Data were collected after 8 weeks in culture.
[y]Shoot vigor was scored on a scale from 1 to 8, with the poorest plant being 1 and 8 the best.
[x]Callus growth was scored on a scale from 1 to 8, with the smallest in diameter (> 2 mm) being 1 and 8 the largest (< 14 mm).
[w]Mean separation within rows by Duncan's multiple range test, P = 0.05. Rows with no letters indicate no significant differences.
[v]Significant effects (P ≤ 0.05).

the poorest of all concentrations (Figure 1). This treatment produced small leaves, the shortest shoots, and more callus tissue than other concentrations. Hu and Wang (1983) report that high concentrations of cytokinin can inhibit elongation of micropropagated shoots. However, higher levels of zeatin promoted shoot growth and development in 'NFS2' and 'NSB3' (Figure 1). While interactions between clones and zeatin concentrations were significant for all traits studied, shoot height also showed significant two- and three-way interactions among zeatin concentrations, clones, and subculture periods (Table 1). Nodal explants cultured on zeatin-free medium produced one or two unbranched axillary shoots each, and the shoot growth was not vigorous. This might be due

FIGURE 1. The influence of varying concentrations of zeatin and two subculture periods (S1, subculture 1 and S2, subculture 2) on shoot number per explant (a), shoot height (cm) (b), leaf number per shoot (c), shoot vigor (scored on a scale from 1 to 8, with the poorest shoot being 1 and 8 the best) (d), and callus growth (rated on a scale from 1 to 8, with the callus diameter > 2 mm being 1 and 8, < 14 mm) (e) of lowbush blueberry clones, 'NFB1', 'NFB2', and 'NFB3' grown *in vitro* for 8 weeks on the modified cranberry medium supplemented with 30 g $L^{-1}$ sucrose.

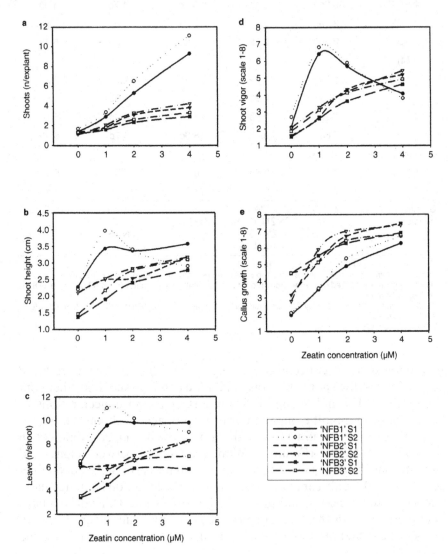

to persistence of a strong apical dominance, a major constraint in the development of efficient *in vitro* clonal propagation of some plant species (George and Sherrington, 1984). Indeed, axillary branching in nodal explants occurred only when cytokinin was applied exogenously in the present study. Cranberry and lingonberry respond similarly (Debnath and McRae, 2001a,b).

Significant (P ≤ 0.05) clonal differences were observed for shoot and callus development, where 'NFB1' was found to be the best followed by 'NFB3'; both were better than 'NFB2' (Table 1). This might be due to genotypic differences among the clones. Genotypic differences for number of *in vitro*-proliferated shoots were also noticed in cranberry (Marcotrigiano and McGlew, 1991; Smagula and Harker, 1997; Debnath and McRae, 2001a) and in lingonberry cultivars (Serres et al., 1994; Debnath and McRae, 2001b).

Shoot proliferation and development improved significantly (P < 0.01) with subculturing (Table 1); shoots per explant of rootable size (> 1.0 cm long) in the first subculture were fewer than in the second subculture (Table 1). Similar results were also reported by Marcotrigiano and McGlew (1991), Smagula and Harker (1997) and Debnath and McRae (2001a) in cranberry. Shoot height, leaf number, and shoot vigor were also better in second subculture, although callus growth was similar in both culture periods (Table 1).

Shoots growing for more than 12 weeks on media that contained more than 4 μM zeatin occasionally produced adventitious shoot masses, which appeared to arise from dense calli growing at the base of the shoots in the medium (data not shown). Adventitious bud formation in lingonberries was reported by Serres et al. (1994) and Debnath and McRae (2001b) for shoots grown on medium with high cytokinin concentrations. Since the formation of adventitious shoots are more likely to produce somaclonal variation (Larkin and Scowcroft, 1981), zeatin concentrations of more than 4 μM should be avoided if lowbush blueberries are to be cultured for more than 12 weeks on the same medium.

*Effect of sucrose concentration and explant type.* The sucrose concentration significantly (P ≤ 0.05) affected shoot proliferation and callus growth (Table 2). Significant (P ≤ 0.05) sucrose × explant type interactions occurred for shoot number per explant, leaves per shoot, shoot vigor, and callus growth (Table 2).

With respect to shoots proliferation and development, shoot tips and nodal segments responded positively to increasing levels of up to 20 g $L^{-1}$ sucrose (Figure 2 a-d), although callus growth increased with concentration up to 30 g $L^{-1}$ (Figure 2 e). Across all treatments, although

TABLE 2. Mean values of the main effects across all treatments for combined effect of sucrose concentration and explant type on shoot proliferation and callus growth of *in vitro*-grown lowbush blueberry clone 'NFB1'.[z]

| Treatment | Shoots (n/explant) | Shoot height (cm) | Leaves (n/shoot) | Vigor[y] (scale 1-8) | Callus growth[x] (scale 1-8) |
|---|---|---|---|---|---|
| Sucrose (Su) (g L$^{-1}$) | | | | | |
| 0 | 1.2a[w] | 1.7a | 6.1a | 3.3a | 1.0a |
| 10 | 2.7b | 2.8b | 7.4b | 5.2c | 1.9b |
| 20 | 7.0d | 3.1b | 9.2c | 6.1d | 3.6c |
| 30 | 4.7c | 3.0b | 8.6c | 4.9b | 6.0d |
| Explant (E) | | | | | |
| Shoot tip | 2.5a | 2.8 | 8.0b | 5.3c | 3.4c |
| Middle | 4.4b | 2.6 | 8.2b | 4.8b | 3.2b |
| Basal | 4.9c | 2.5 | 7.3a | 4.4a | 3.0a |
| Significant effects[v] | Su, E, Su×E | Su | Su, E, Su×E | Su, E, Su×E | Su, E, Su×E |

[z]Data were collected after 8 weeks in culture.
[y]Shoot vigor was scored on a scale from 1 to 8, with the poorest plant being 1 and 8 the best.
[x]Callus growth was scored on a scale from 1 to 8, with the smallest in diameter (> 2 mm) being 1 and 8 the largest (< 14 mm).
[w]Mean separation within rows by Duncan's multiple range test, P = 0.05. Rows with no letters indicate no significant differences.
[v]Significant effects (P ≤ 0.05).

shoot number and vigor were best at 20 g L$^{-1}$ sucrose in the culture medium, shoot height and number of leaves per shoot were statistically identical at 20 and 30 g L$^{-1}$ (Table 2). Nodal explants cultured on sucrose-free medium produced one or two unbranched axillary necrotic shoots each, and the shoot growth was very poor.

Separating the actively growing shoots according to segment position before plating significantly (P ≤ 0.05) influenced shoot proliferation in terms of shoot number, leaves per shoot, shoot vigor and callus growth when 2 µM zeatin was added to the culture medium (Table 2). Cuttings excised from basal nodes were the most productive explants and produced significantly (P ≤ 0.5) more shoots than those obtained from either shoot tips or middle nodes (Table 2). Nodal segments consistently produced more shoots than shoot tip explants. Apical dominance may explain the differences between the explants. The multiplication efficiency of shoot tips may be improved by removing the apex, and by

FIGURE 2. The influence of varying concentrations of sucrose on shoot number per explant (a), shoot height (cm) (b), leaf number per shoot (c), shoot vigor (scored on a scale from 1 to 8, with the poorest shoot being 1 and 8 the best) (d), and callus growth (rated on a scale from 1 to 8, with the callus diameter > 2 mm being 1 and 8, < 14 mm) (e) of lowbush blueberry clone, 'NFB1', grown *in vitro* for 8 weeks on the modified cranberry medium supplemented with 2 µM zeatin.

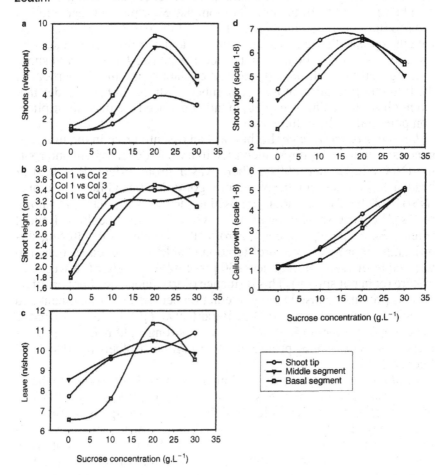

having more axillary buds in the nodal segments at a later stage of development. Differential morphogenic response between meristem-tips and single nodes has been reported by Gebhardt and Friedrich (1986) in lingonberries. When basal nodal segments were subcultured without removing basal callus, they were more vigorous in shoot proliferation than those without calli, as shown by greater shoot numbers per explant (data not shown). It may be suggested that the large surface of the callus tissue may partly compensate for the lack of root system. In addition, the positive effect of the attached callus may also be related to the lack of basal wounding stress. When faster multiplication is desired, the whole shoot may be divided into explants for subculture. When material is to be removed from the proliferation phase for plantlet regeneration and rooting, it is recommended that 1 cm of the basal shoots be excised with the basal callus attached and returned to the proliferation phase; the rest of the shoot can be used in the rooting phase. The observed shoot forming ability of all *in vitro*-derived explant types and, in particular, the frequent formation of a large number of shoots in basal nodes that did not have the callus removed, greatly enhanced the mass multiplication potential of the culture system.

*Influence of irradiance.* Light plays an essential role in the morphogenesis of *in vitro* cultures, such as axillary bud development and rooting (Noè and Eccher, 1994). There was a strong effect (P ≤ 0.5) of irradiance supplied *in vitro* on shoot vigor of lowbush blueberry clone, 'NSB1' (Table 3); the shoots exposed to the lower irradiance (15 $\mu$mol $m^{-2} s^{-1}$) clearly showed better shoot vigor both in subculture 1 and 2 (Figure 3). As in experiment 1, across all treatments, subculturing significantly improved shoot proliferation (Table 3). The leaves began to turn red at the control treatment (30 $\mu$mol $m^{-2} s^{-1}$) after 6-8 weeks of culture (data not shown). This influence of irradiance on shoot proliferation is probably caused by the interaction between light treatment and cytokinin, as was also reported by Baraldi et al. (1988). As compared to the control treatment (55 $\mu$mol $m^{-2} s^{-1}$), strong light had negative effects on *in vitro* shoot proliferation in highbush blueberry (Noè and Eccher, 1994).

*Rooting and acclimatization.* Elongated microshoots rooted in the *ex vitro* culture within 5-6 weeks of culturing with a frequency of 85% to 95% corroborating previous lowbush blueberry investigations (Brissette et al., 1990). Rooting was also obtained *in vitro* when shoots were allowed to grow for more than 10 weeks on the basal medium containing 1 and 2 $\mu$M zeatin (data not shown). Although *in vitro* rooting offers several advantages, including reduced exposure to disease and environmental stress during rooting process and the production of rooted plantlets for *in vitro* experiments, plantlets can be produced more rapidly with less cost by eliminating *in vitro* rooting (Zimmerman, 1987). Maene and Debergh (1983) pointed out the advantage of directly rooting microcuttings under *ex vitro* conditions. However, the process of *ex*

TABLE 3. Effect of light on shoot proliferation and callus growth of *in vitro*-grown lowbush blueberry clone 'NFB1' over two subculture periods.[z]

| Treatment | Shoots (n/explant) | Shoot height (cm) | Leaves (n/shoot) | Vigor[y] (scale 1-8) | Callus growth[x] (scale 1-8) |
|---|---|---|---|---|---|
| Irradiance (I) ($\mu$mol m$^{-2}$ s$^{-1}$) | | | | | |
| 15 | 5.2 | 3.2 | 10 | 5.2a[w] | 5.3 |
| 30 | 5.1 | 3.2 | 10 | 6.3b | 5.1 |
| Subculture (S) | | | | | |
| 1 | 5.0a | 3.1a | 9.8a | 5.6a | 5.1a |
| 2 | 5.3b | 3.3b | 10.2b | 5.9b | 5.3b |
| Significant effects[v] | S | S | S | I, S | S |

[z]Data were collected after 8 weeks in culture.
[y]Shoot vigor was scored on a scale from 1 to 8, with the poorest plant being 1 and 8 the best.
[x]Callus growth was scored on a scale from 1 to 8, with the smallest in diameter (> 2 mm) being 1 and 8 the largest (< 14 mm).
[w]Mean separation within columns by Duncan's Multiple Range Test (P = 0.05). Rows with no letters indicate no significant differences.
[v]Significant effects (P $\leq$ 0.05).

*vitro* rooting is often slower than *in vitro* (Wolfe et al., 1983). Plants obtained *in vitro* and *ex vitro* were adapted and transferred to greenhouse with a survival rate of 80% to 90%.

## *CONCLUSIONS AND GROWER BENEFITS*

The results presented in this paper provide evidence that the modified cranberry medium containing low level of zeatin (2-4 $\mu$M) and sucrose (20 g L$^{-1}$) increased the *in vitro*-shoot multiplication rate of lowbush blueberry by about 50 to 100-fold over a 12-week interval. Higher levels of 2iP (59-73.8 $\mu$M) alone or in combination with an auxin were suggested for *in vitro* culture for optimal shoot proliferation of lowbush blueberry by Nickerson (1978), Frett and Smagula (1983), and Brissette et al. (1990). Overall, the present results suggest that a threshold level of endogenous growth regulators accumulates during culture initiation which enables the explants of shoot cultures to produce shoots optimally at reduced levels of zeatin without auxin augmentation. However, shoot proliferation depended on the interaction of clone with zeatin concentration. Nodal axillary bud cultures appeared to be a suitable source

FIGURE 3. The effect of light and two subculture periods on shoot number per explant (a), shoot height (cm) (b), leaf number per shoot (c), shoot vigor (scored on a scale from 1 to 8, with the poorest shoot being 1 and 8 the best) (d), and callus growth (rated on a scale from 1 to 8, with the callus diameter > 2 mm being 1 and 8, < 14 mm) (e) of lowbush blueberry clone, 'NFB1', grown *in vitro* for 8 weeks on the modified cranberry medium supplemented with 2 µM zeatin and 30 g $L^{-1}$ sucrose. Vertical bars represent SD.

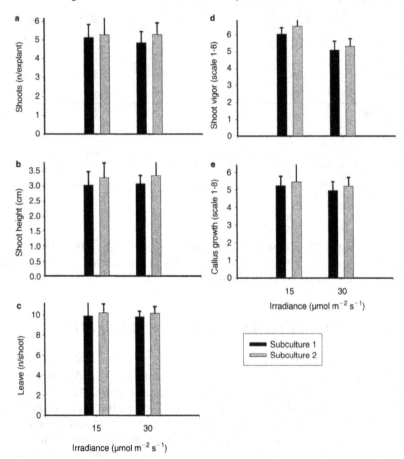

for native clones; shoot multiplication can be enhanced with repeated subculture under a mass propagation or production strategy. Such a high multiplication rate shows promise for micropropagation of selected lowbush blueberry clones. This technique produced a reasonable multiplication rate and good plant growth within 12 weeks with some selected lowbush blueberry clones in our laboratory (data not shown). The main advantage of this technique is that all the shoot tips can be used for rooting studies, whereas nodes with basal mass can be used for routine shoot multiplication. The procedure described here is of considerable practical value for the large-scale production of selected lowbush blueberry clones for increasing plant cover in commercial lowbush blueberry fields.

## LITERATURE CITED

Baraldi, R., F. Rossi, and B. Lercari 1988. *In vitro* shoot development of *Prunus* GF 655-2: Interaction between light and benzyladenine. Physiol. Plant. 74:440-443.

Brissette, L., L. Tremblay, and D. Lord. 1990. Micropropagation of lowbush blueberry from mature field-grown plants. HortScience 25:349-351.

Chandler, C.K. and A.D. Draper. 1986. Effect of zeatin and 2iP on shoot proliferation of three highbush blueberry clones *in vitro*. HortScience 21:1065-1066.

Debnath, S.C. and K.B. McRae. 2001a. An efficient *in vitro* shoot propagation of cranberry (*Vaccinium macrocarpon* Ait.) by axillary bud proliferation. In Vitro Cell. Dev. Biol. Plant 37:243-249.

Debnath S.C. and K.B. McRae. 2001b. *In vitro* culture of lingonberry (*Vaccinium vitis-idaea* L.): The influence of cytokinins and media types on propagation. Small Fruit Rev. 1:3-19.

Eccher, T. and N. Noè. 1989. Comparison between 2iP and zeatin in the micropropagation of highbush blueberry (*Vaccinium corymbosum*). Acta Hort. 241:185-190.

Frett, J.J. and J.M. Smagula. 1983. *In vitro* shoot production of lowbush blueberry. Can. J. Plant Sci. 63:467-472.

Gebhardt, K. and M. Friedrich. 1986. *In vitro* shoot regeneration of lingonberry clones. Gartenbauwissenschaft 51:170-175.

George, E.F. 1996. Plant propagation by tissue culture, part 2: In practice. Exegetics Ltd., Edington, England.

George, E.F. and P.D. Sherrington. 1984. Plant propagation by tissue culture. Exegetics Ltd., Reading, England.

Hoefs, M.E. and J.M. Shay. 1981. The effects of shade on shoot growth of *Vaccinium angustifolium* Ait. after fire pruning in southern Manitoba. Can. J. Bot. 59:166-174.

Hu, C.Y. and P.J. Wang. 1983. Meristem, shoot tip and bud culture. Pp. 177-227. In: D.A. Evans, W.R. Sharp, P.V. Ammirato, and Y. Yamada (eds.). Handbook of plant cell culture, vol.1. Macmillan, New York.

Larkin, P.J. and W.R. Scowcroft. 1981. Somaclonal variation–a novel source of variability from cell cultures for plant improvement. Theoretical Appl. Genet. 60:197-214.

Maene, L.M. and P.C. Debergh. 1983. Rooting of tissue cultured plants under *in vivo* conditions. Acta Hort. 131:201-208.

Marcotrigiano, M. and S.P. McGlew. 1991. A two-stage micropropagation system for cranberries. J. Amer. Soc. Hort. Sci. 116:911-916.

Metzger, H.B. and A.A. Ismail. 1977. Costs and returns in lowbush blueberry production in Maine, 1974 Crop. Univ. Maine Life Sci. Agr. Expt. Sta. Bul. 738.

Morrison, S., J.M. Smagula, and W. Litten. 2000. Morphology, growth, and rhizome development of *Vaccinium angustifolium* Ait. seedlings, rooted softwood cuttings, and micropropagated plantlets. HortScience 35:738-741.

Nickerson, N.L. 1978. *In vitro* shoot formation in lowbush blueberry seedling explants. HortScience 13:698.

Noè, N. and T. Eccher. 1994. Influence of irradiance on *in vitro* growth and proliferation of *Vaccinium corymbosum* (highbush blueberry) and subsequent rooting *in vivo*. Physiol. Plant. 91:273-275.

Reed, B.M. and A. Abdelnour-Esquivel. 1991. The use of zeatin to initiate *in vitro* cultures of *Vaccinium* species and cultivars. HortScience 26:1320-1322.

Serres, R.A., S. Pan, B.H. McCown, and E.J. Stang. 1994. Micropropagation of several lingonberry cultivars. Fruit Var. J. 48:7-14.

Smagula, J.M. and J. Harker. 1997. Cranberry micropropagation using a lowbush blueberry medium. Acta Hort. 446:343-347.

Smagula, J.M. and P.M. Lyrene. 1984. Blueberry. Pp. 383-401. In: P.V. Ammirato, D. A. Evans, W.R. Sharp, and Y. Yamada (eds.). Handbook of plant cell culture, Vol. 3. Macmillan, New York.

Vander Kloet, S.P. 1988. The Genus *Vaccinium* in North America. Res. Branch Agric. Can. Publ. 1828.

Wolfe, D.E., P. Eck, and C. Chin. 1983. Evaluation of seven media for micropropagation of highbush blueberry. HortScience 18:703-705.

Zimmerman, R.H. 1987. Micropropagation of woody plants: Post tissue culture aspects. Acta Hort. 227:489-499.

Zimmerman R.H. and O.C. Broome. 1980. Blueberry micropropagation. Pp. 44-47. In: Proc. Conf. on Nursery Prod. of Fruit Plants Through Tiss. Cult.-Applications and Feasibility. U.S. Dept. Agr. Sci. Educ. Admin. ARR-NE-11.

# Effects of Sunlight or Shade
# on Maturity and Optical Density
# in Blueberries

R. P. Rohrbach
C. M. Mainland
J. A. Osborne

**SUMMARY.** Sensing of maturity in blueberries (*Vaccinium darrowi* L.) using optical density measurements has long been known and used by researchers. This past year a large sample of berries (540) were visually selected with six levels of maturity; Green, White, Red, Just-Ripe, Ripe, and Over-Ripe. Subsequently, spectrographs of individual berries were obtained from about 400 nm to 900 nm. Each berry was chemically assayed for soluble solids (SS), pH, and titratable acidity (Acids). One interesting observation was that the SS/Acids ratios within each of the six carefully selected levels of maturity varied greatly and overlapped between the maturity categories. This study was initiated by one primary question. Is there an effect on the SS/Acids ratio of an individual berry if the berry is allowed to mature in either the presence or absence of sun-

R. P. Rohrbach is Professor of Biological and Agricultural Engineering, C. M. Mainland is Professor of Horticultural Science, and J. A. Osborne is Assistant Professor of Statistics, College of Agriculture and Life Sciences, NC State University.

Address correspondence to: R. P. Rohrbach, Professor, Department of Biological and Agricultural Engineering, Box 7625, Raleigh, NC 27695-7625 (E-mail: rohrbach@eos.ncsu.edu).

[Haworth co-indexing entry note]: "Effects of Sunlight or Shade on Maturity and Optical Density in Blueberries." Rohrbach, R. P., C. M. Mainland, and J. A. Osborne. Co-published simultaneously in *Small Fruits Review* (Food Products Press, an imprint of The Haworth Press, Inc.) Vol. 3, No. 3/4, 2004, pp. 409-421; and: *Proceedings of the Ninth North American Blueberry Research and Extension Workers Conference* (ed: Charles F. Forney, and Leonard J. Eaton) Food Products Press, an imprint of The Haworth Press, Inc., 2004, pp. 409-421. Single or multiple copies of this article are available for a fee from The Haworth Document Delivery Service [1-800-HAWORTH, 9:00 a.m. - 5:00 p.m. (EST). E-mail address: docdelivery@haworthpress.com].

light? A secondary issue of investigation was: If there is a sun/shade effect on a berry's SS/Acids ratio, can this effect be observed by measuring the optical density ratio at 740 and 800 nm? Results of 38 sun and 40 shade berries (Cultivars, 'NC 3103' and 'NC 3104') had mean SS/Acids ratios of 30.9 and 31.5, respectively, and were in the same ($\alpha = 0.05$) Duncan's multiple range grouping. Results of 57 'NC 3103' and 21 'NC 3104' berries had mean SS/Acids ratios of 35.6 and 19.3, respectively, and were in different ($\alpha = 0.05$) Duncan's multiple range groupings. Correlation coefficients between the SS/Acids ratios and the optical density ratio ranged from $-0.88 < r < -0.66$ for shade berries and $-0.84 < r < -0.78$ for sun berries, depending upon the optical orientation of the individual berry. *[Article copies available for a fee from The Haworth Document Delivery Service: 1-800-HAWORTH. E-mail address: <docdelivery@haworthpress.com> Website: <http://www.HaworthPress.com> © 2004 by The Haworth Press, Inc. All rights reserved.]*

**KEYWORDS.** Blueberries, maturity sensing, optical density, spectrographs, soluble solids, ripeness

## INTRODUCTION

Optical density measurements have long been known and used by researchers to nondestructively assess maturity in blueberries (Birth and Norris, 1965; McClure et al., 1975; McClure and Rohrbach, 1978). We have also known that there is variability in the optical density measurement. In the development of the M-Belt sorter, Rohrbach and McClure (1978) reported that upon repeated classification of blueberries into three ripeness categories only 66% to 73% of the berries were reclassified into the same ripeness category upon the second sorting.

In a recent effort to incorporate digital imaging of area spectrographs into optical density measurement, a large sample of berries (540) was investigated. Ninety berries were visually selected within each of six levels of maturity; Green, White, Red, Just-Ripe, Ripe, and Over-Ripe. Each berry was chemically assayed for soluble solids (SS), pH, and titratable acidity (Acids). One interesting observation from that study was that the group mean SS/Acids ratios appeared to be decreasing at each level of maturity. Figure 1 provides an indication of this association. The overlap in SS/Acids ratio between neighboring ripeness levels and the variability within ripeness levels are the motivation underlying the present study.

FIGURE 1. Blueberries were visually selected by experienced personnel to be in one of six categories. Each consecutive 90 berry group is in one of the following categories: Over Ripe, Ripe, Just Ripe, Red, White or Green, respectively.

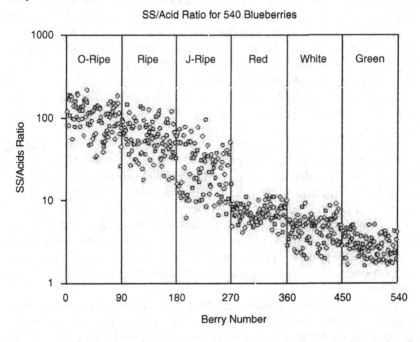

The purpose of this paper is to report on an initial investigation into why the SS/Acids ratios vary within a carefully selected maturity sample of fruit. The primary hypothesis formulated was that some berries ripening on a bush might be exposed to full sunlight while other berries may be completely shaded by leaves, and this shade may affect the SS/Acid ratio. In addition to the primary hypothesis, we also hoped to determine if the variability in optical density measurements could be explained by non-uniform distribution of SS and Acids within the berry.

## MATERIALS AND METHODS

*Experimental Materials.* For this study, two blueberry bushes, growing in a greenhouse, were selected. One bush was 'NC 3103' and the other 'NC 3104' (*Vaccinium darrowi* L.). Berry clusters were selected in pairs (on the same bush) and one of each cluster pair was shaded from the sun with a loose covering of aluminum foil while the other cluster in the pair was fully exposed to the sunlight. The berries were permitted to continue to ripen until about 10 d past the normal first harvest date. This was to insure that we could harvest a range of berry ripeness from green to over-ripe. As each berry was harvested it was identified with the following experimental variables: a sequential berry ID number, cluster pair number, treatment (Sun or Shade), and the cultivar. Each berry was subjected to a series of optical density measurements, weighed, inserted into an individual sealable polyethylene pouch, and frozen prior to subsequent measurement of acids and sugars. A total of 78 berries were used.

*Optical Density Measurements.* An ImSpector V9-090-100 imaging spectrograph with a 16 mm objective lens was mounted on a JAI CV-M300 CCD video camera. The analog video output was captured with a Data Translation DT3155 frame grabber in a PC. For example, when the imaging spectrograph is focused at a "red" target, an area spectrograph is produced. The horizontal axis of the spectrograph represents the spatial distance across the "red" target while the vertical axis is the spectral intensity. The color of sample target is "red" to the eye, but, its spectrum is not monochromatic; it contains energy in the 550 nm to 900 nm region.

Area spectrographs of individual berries in various orientations were obtained from about 400 nm to 900 nm. The optical measurement used was the ratio of the transmitted light intensity at 800 nm divided by the transmitted light intensity at 740 nm. The effective width of the spectrum at the two sampling points was 3.8 nm wide. All subsequent image analyses were conducted using MatLab software. Statistical analyses were conducted with SAS version 8.2 software.

Each berry was subjected to optical density measurements in four different positions as schematically shown in Figure 2. We have called these four positions EQ0, EQ90, EQ180, and EQ270 degrees. For purposes of describing these four positions, one can imagine a berry with the stem at the "North pole" and the calyx at the "South pole." The light source is shining into the "equator" of the berry in all four positions, hence the "EQ" designation. In the EQ0 degree berry position, the light sample is emerging from a point 135 degrees around the equator from the point at which the light entered. Similarly, the EQ180 degree berry position, the light sample is emerging from a point −135 degrees around the equator from the point at which the light entered. In the EQ90 degree berry position, the light sample is emerging from a point at

FIGURE 2. Sketch showing the camera, light source, and berry positions for the four "Equator" views.

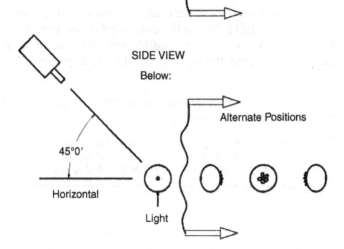

**Equator View Positions**

45 degrees "North latitude" in the stem hemisphere of the berry. And finally, in the EQ270 degree berry position, the light sample is emerging from a point at 45 degrees "South latitude" in the calyx hemisphere of the berry.

*Soluble Solids (SS) and Acid (%A) Measurement.* Berries were weighed to 0.01 g before freezing. Frozen berries were held for approximately two months at −30°C. After thawing, the berries were cut in half along the centerline axis between the stem scar and the calyx. A drop of juice was squeezed from the cut surfaces onto the measuring lens of a hand-held refractometer. The SS measurement of sweetness, read from the refractometer was calibrated as percent glucose equivalent. Glucose and fructose are the predominant sugars in blueberries. Percent acid was measured by recovering the drop of juice from the refractometer, and subsequently, homogenizing it along with both berry halves and 20 ml of distilled water. The homogenate was titrated to an end point of pH 8.1 with standardized 0.1 normal NaOH. The percent citric acid, the predominant acid in blueberry fruit, was calculated from the amount of NaOH required. A ratio of SS to percent acid was calculated. The SS/ Acid ratio is another measure of maturity that helps predict if berries will be considered to taste sweet or sour.

## RESULTS AND DISCUSSION

Intensity measurements from the equator view were made at 4 angles (EQ0, EQ90, EQ180, EQ270) and the data as a function of SS/Acids ratio are shown in Figure 3. Under the null hypothesis that sugar and acid are uniformly distributed throughout the berry, these four measure-

FIGURE 3. In this scatter plot, differences in intensity between the four optical measurement positions can be seen.

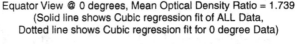

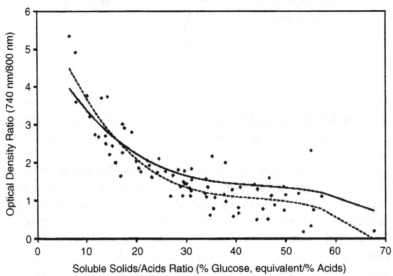

Equator View @ 0 degrees, Mean Optical Density Ratio = 1.739
(Solid line shows Cubic regression fit of ALL Data,
Dotted line shows Cubic regression fit for 0 degree Data)

Equator View @ 90 degrees, Mean Optical Density Ratio = 1.887
(Solid line shows Cubic regression fit of ALL Data,
Dotted line shows Cubic regression fit for 90 degree Data)

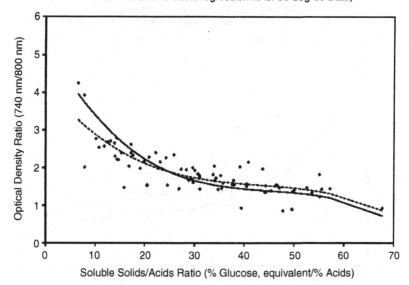

Equator View @ 180 degrees, Mean Optical Density Ratio = 1.744
(Solid line shows Cubic regression fit of ALL Data,
Dotted line shows Cubic regression fit for 180 degree Data)

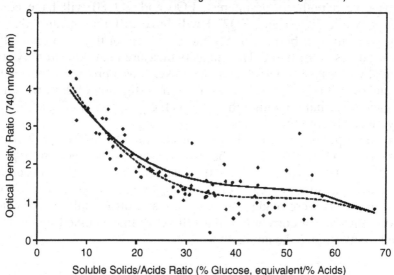

## FIGURE 3 (continued)

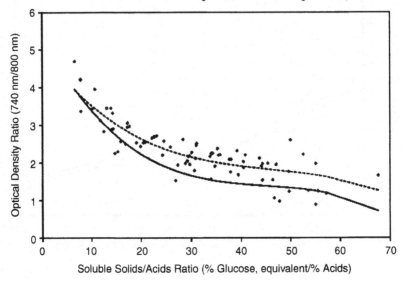

Equator View @ 270 degrees, Mean Optical Density Ratio = 2.307
(Solid line shows Cubic regression fit of ALL Data,
Dotted line shows Cubic regression fit for 270 degree Data)

ments should reflect the same mean sugar and acid contents. Under an alternative hypothesis of interest, EQ0 and EQ180 will have equal means, while EQ90 and EQ270 will have different means, due to non-uniform distributions of SS and acids within the stem and calyx hemispheres of the berry. The intensity measurements are strongly correlated with sugar and acid content and serve as a proxy for their measurement. A first thing to look at in the investigation of uniform sugar and acid distribution within the berry is the plausibility of equal intensity means for the four angles.

In the experiment, 57 berries were used from 'NC-3103' and 21 from 'NC-3104'. Of the 'NC-3103' berries, 31 were in the shade, among the 21 'NC-3104' berries, 9 were in the shade. Optical density, at four angles, was measured on each of the 78 different berries. Because four measurements were made on each berry and prior analyses indicated strong intra-berry correlation, the following linear mixed model was used to investigate the angle effect

$$y_{ijkl} = \mu + \alpha_i + \beta_j + (\alpha\beta)_{ij} + B_{l(ij)} + E_{ijkl}$$

Here, $y$ denotes the transmission intensity at angle $k$ for the berry which comes from cultivar $i$ and sun/shade $j$. $B_{l(ij)}$ is the random effect for berry, nested in the cultivar $\times$ shade treatment combination. It has variance component. $E_{ijkl}$ is the remaining experimental error, it has variance component $\sigma^2$. The model was fit using PROC MIXED. The analysis indicated differences of intensity means by angle ($p < 0.0001$), by cultivar ($p < 0.0001$) and by sun/shade ($p = 0.0021$).

The estimated variance components are $\hat{\sigma}^2_B = 0.39$ and $= 0.09$ so that the estimated intraberry correlation coefficient is 0.8. The estimated least squares means by angle appear in Table 1. Multiple comparisons among means are made using the Bonferroni adjustment. Multiple comparisons indicate that the mean intensity for EQ270 is significantly different from that for the other three angles and no other differences are significant.

To further investigate the possibility of non-uniform distribution of sugars and acids throughout the berry, we consider modeling the means of the response variables soluble solids (SS), acids (%A) and soluble solids/acids ratio (SSA) as functions of the angle intensities (EQ0-EQ270) as well as the sun/shade and cultivar treatments.

Inspection of pairwise Pearson correlation coefficients reveals moderate negative associations between soluble sugars (SS) and the intensities at the four angles, $-0.65 < r < -0.45$ with EQ180 the best single predictor. The negative correlation between SSA and intensities at the four angles is stronger ($-0.80 < r < -0.75$) than that for SS, with EQ270 as the best single predictor. Percent acid (%A) is strongly positively correlated with each intensity, with EQ0 ($r = 0.85$) the highest. The correlations computed from the residuals obtained from a linear

TABLE 1. Estimated least squares mean optical intensities, by angle, adjusted by the Bonferroni correction for multiplicitiy.

| Angle | Standard | | |
|-------|----------|-------|----------|
|       | Estimate | Error | Grouping |
| 0     | 1.9283   | 0.08985 | a |
| 90    | 1.9720   | 0.08985 | a |
| 180   | 1.9351   | 0.08985 | a |
| 270   | 2.4616   | 0.08985 | b |

[z]Two means with the same letter do not differ significantly using Bonferroni procedure.

model including sun/shade and cultivar effects did not substantially change any of these associations.

Partial and sequential F-tests for model selection indicate that different models may be appropriate to best explain variation in the different responses SS, %A and SSA. Subsets of independent variables from EQ0, EQ90, EQ 180, EQ270, and shade and cultivar effects are selected using Mallows $C_p$ criterion (Mallows, 1973) in conjunction with $r^2$. The estimated models selected by this criterion, along with $r^2$ and root mean squared error, for each of the three responses, are shown in Table 2. The Type I (sequential) sums of squares for each of these models appear in Table 3. Under the hypothesis of uniformity of soluble solids distribu-

TABLE 2. Models effects for soluble solids/acids ratio (SSA), soluble solids (SS), and percent acids (% A) selected using Mallow's $C_p$ criterion [(Number of effects + 1) > $C_p$].

| Dependent Variable: SS # of effects | $C_p$ | R-Square | Variables in Model |
|---|---|---|---|
| 5 | 5.8217 | 0.5043 | EQ90 EQ180 EQ27i0 cv3 shadel |
| 5 | 5.9637 | 0.5033 | EQ0 EQ90 EQ180 EQ270 cv3 |
| 6 | 7.000 | 0.5100 | EQ0 EQ90 EQ180 EQ270 cv3 shadel |

| Dependent Variable: %A # of effects | $C_p$ | R-Square | Variables in Model |
|---|---|---|---|
| 3 | 1.8719 | 0.7753 | EQ0 EQ90 shadel |
| 4 | 3.2384 | 0.7773 | EQ0 EQ90 EQ180 shadel |
| 4 | 3.4023 | 0.7768 | EQ0 EQ90 EQ270 shadel |
| 4 | 3.5269 | 0.7764 | EQ0 EQ90 cv3 shadel |
| 5 | 5.0673 | 0.7778 | EQ0 EQ90 EQ180 cv3 shadel |
| 5 | 5.1299 | 0.7776 | EQ0 EQ90 Eq180 EQ270 shadel |
| 5 | 5.2265 | 0.7773 | EQ0 EQ90 EQ270 cv3 shadel |
| 6 | 7.000 | 0.7780 | EQ0 EQ90 Eq180 EQ270 cv3 shadel |

| Dependent Variable: SSA # of effects | $C_p$ | R-Square | Variables in Model |
|---|---|---|---|
| 4 | 3.6142 | 0.7661 | EQ0 EQ270 cv3 shadel |
| 5 | 5.0695 | 0.7678 | EQ0 EQ90 EQ270 cv3 shadel |
| 5 | 5.5700 | 0.7662 | EQ0 EQ180 EQ270 cv3 shadel |
| 6 | i7.000 | 0.7681 | EQ0 EQ90 EQ180 EQ270 cv3 shadel |

Notes:  cv3 = 1 for 'NC-3103' or 0 for 'NC-3104'.
        shadel = 1 for shade or 0 for sun.

TABLE 3. General linear models for soluble solids/acids ratio (SSA), soluble solids (SS), and percent acids (%A).

| | | The LM Procedure | | | |
|---|---|---|---|---|---|
| **Dependent Variable: SS** | | | | | |
| Source | DF | Type I SS | Mean Square | F Value | Pr > F |
| EQ90 | 1 | 45.06335598 | 45.06335598 | 28.78 | < .0001 |
| EQ180 | 1 | 52.33184476 | 52.33184476 | 33.43 | < .0001 |
| EQ270 | 1 | 9.58845081 | 9.58845081 | 6.12 | 0.0156 |
| Error | 74 | 115.8492972 | 1.5655310 | | |
| **Dependent Variable: percent A** | | | | | |
| Source | DF | Type I SS | Mean Square | F Value | Pr > F |
| shadel | 1 | 0.11447579 | 0.11447579 | 6.62 | 0.0121 |
| EQ0 | 1 | 4.19962729 | 4.19962729 | 243.04 | < .0001 |
| EQ90 | 1 | 0.09759839 | 0.09759839 | 5.65 | 0.0201 |
| Error | 74 | 1.27869340 | 0.01727964 | | |
| Corrected Total | 77 | 5.69039487 | | | |
| **Dependent Variable: SSA** | | | | | |
| Source | DF | Type I SS | Mean Square | F Value | Pr > F |
| cv3 | 1 | 4104.762175 | 4104.762175 | 82.89 | < .0001 |
| shadel | 1 | 14.130728 | 14.130728 | 0.29 | 0.5948 |
| EQ0 | 1 | 7468.146443 | 7468.146443 | 150.81 | < .0001 |
| EQ270 | 1 | 250.393437 | 250.393437 | 5.06 | 0.0276 |
| Error | 73 | 3615.06692 | | | |
| Corrected Total | 77 | 15452.49970 | | | |

tion, any single optical path (angle) should be sufficient to explain variation in soluble solids in berries. One would not expect to see more than one angle having a significant sequential F-ratio.

Summary of observations from $C_p$ statistic and multiple regression models:

- After accounting for the effects of all other important explanatory variables, including EQ90 and EQ180, the mean SS has a significant, negative association with EQ270.
- The mean %A is decreasing in EQ0 and EQ270 and neither is by itself sufficient to explain variation. Even after accounting for the intensities at these angles, there is still evidence that SSA varies by shade.

- The mean SSA is decreasing in EQ0 and EQ270 and neither is by itself sufficient to explain variation. Even after accounting for the intensities at these angles, there is still evidence that SSA varies by cultivar.

## CONCLUSIONS

*Effects of Sun/Shade on Soluble Solids and Acids:* Results of 38 sun and 40 shade berries (Cultivars, 'NC 3103' and 'NC 3104') had mean SS/Acids ratios of 30.9 and 31.5, respectively, and were in the same ($\alpha$= 0.05) Duncan's multiple range grouping. There is no evidence of a sun/shade effect on the optical density measurements.

*Effects of Orientation on Optical Density Measurements:* The pathway of transmitted light used to measure the optical density does affect the optical density measurement. Therefore, in the same berry, different optical density measurements result from different orientations of the berry with respect to the optical transmission pathway. As expected, two similar orientation pathways (EQ0 and EQ180) resulted in pair-wise comparisons that do not indicate a significant difference.

The orientation, EQ270, is the only significant ($p < 0.001$) factor in the general model for soluble solids. EQ270 is the orientation where the optical pathway emerges from the calyx hemisphere of the berry, suggesting that the soluble solids may be concentrated in the calyx hemisphere of the berry. This conclusion is further supported by the fact that the Pearson correlation coefficient between soluble solids and EQ270 is B0.45 and is substantially lower than the other three ranging between B0.60 and B0.65. But, in the case of acids distribution within the berry, we can find no conclusive evidence suggesting that the acids may be concentrated in any particular portion of the berry.

## GROWER BENEFITS

Sorting machine-harvested blueberries for color defects (green-red-blue) is a commercial reality. However, the ability to sort blueberries according to their physiological maturity (ripeness) has only been demonstrated in the laboratory. This work is intended to provide insight into the underlying fundamentals of how optical density measurements can be used to non-destructively measure berry ripeness. The development of high-speed commercial sorting systems for blueberry maturity should be possible once the parameters affecting optical density measurements are fully understood.

## LITERATURE CITED

Birth, G.S. and K.H. Norris. 1965. The difference meter for measuring interior quality of foods and pigments in biological tissue. Technical Bulletin No. 1341. USDA, ARS, Washington, DC.

Mallows, C.P., 1973. Some comments on $C_p$. Technometrics 15: 661-675.

McClure, W.F., R.P. Rohrbach, L.J. Kushman, and W.E. Ballinger. 1975. Design of a high-speed fiber optic blueberry sorter. Trans of the ASAE 18: 487-490.

McClure, W.F. and R.P. Rohrbach. 1978. Asynchronous sensing for sorting small fruit, Agri. Eng. 59:13-14.

Rohrbach, R.P. and W.F. McClure. 1978. A production capacity conveyor for small fruit sorting: The M-belt. Trans. ASAE, 21:1092-1095.

# Quality Curves for Highbush Blueberries as a Function of the Storage Temperature

Maria Cecilia N. Nunes
Jean-Pierre Emond
Jeffrey K. Brecht

**SUMMARY.** Blueberries *Vaccinium corymbosum* cv. Patriot were harvested twice at the full ripe stage and held for 14 d at 0, 5, 10, 15, or 20°C. The objectives of this work were to (1) obtain quality curves for blueberries stored at different temperatures; (2) identify, for each temperature, which quality factor(s) limits blueberry marketability; and (3) compare the quality curves and shelf-life of blueberries based on quality evaluations with those predicted by respiration rates reported in the literature. Blueberry weight loss, instrumental color (L\*a\*b\*), visual color, firmness, shriveling, decay, taste, and aroma were evaluated every

Maria Cecilia N. Nunes is Senior Researcher of The Air Cargo Transportation Research Group, University Laval and an Affiliate Assistant Professor, Horticultural Sciences Department, University of Florida.

Jean-Pierre Emond is Professor, University Laval, Head of the Air Cargo Transportation Research Group and Affiliate Professor in the Agricultural and Biological Engineering Department, University of Florida.

Jeffrey K. Brecht is Professor, Horticultural Sciences Department, University of Florida.

Thanks to ETR Group, Sweden (http://www.envirotainer.com/) for funding this research and Nadine Béland for technical support.

[Haworth co-indexing entry note]: "Quality Curves for Highbush Blueberries as a Function of the Storage Temperature." Nunes, Maria Cecilia N., Jean-Pierre Emond, and Jeffrey K. Brecht. Co-published simultaneously in *Small Fruits Review* (Food Products Press, an imprint of The Haworth Press, Inc.) Vol. 3, No. 3/4, 2004, pp. 423-438; and: *Proceedings of the Ninth North American Blueberry Research and Extension Workers Conference* (ed: Charles F. Forney, and Leonard J. Eaton) Food Products Press, an imprint of The Haworth Press, Inc., 2004, pp. 423-438. Single or multiple copies of this article are available for a fee from The Haworth Document Delivery Service [1-800-HAWORTH, 9:00 a.m. - 5:00 p.m. (EST). E-mail address: docdelivery@haworthpress.com].

http://www.haworthpress.com/web/SFR
Digital Object Identifier: 10.1300/J301v03n03_18

2 d for a 14-day storage period. Development of an unpleasant taste and aroma were the primary limiting factors for blueberries stored at 0 and 5°C. Darkening of the color and development of unpleasant aroma were the primary limiting factors for berries stored at 10 or 15°C, and darkening of the color was the primary limiting factor for those fruit stored at 20°C. Overall, fruit stored at 0 or 5°C had shorter shelf life compared with predicted values based on the $Q_{10}$ for respiration, while those stored at higher temperatures had longer shelf life. Prediction of blueberry shelf life calculated from data reported in the literature is not precise unless the characteristics of the fruit and environmental factors involved are well known. The quality curves obtained from quality evaluations for each temperature showed that a single quality attribute cannot be used to express loss of quality of blueberries over the normal physiological range of temperatures. *[Article copies available for a fee from The Haworth Document Delivery Service: 1-800-HAWORTH. E-mail address: <docdelivery@ haworthpress.com> Website: <http://www.HaworthPress.com> © 2004 by The Haworth Press, Inc. All rights reserved.]*

**KEYWORDS.** Blueberry, *Vaccinium corymbosum*, storage, weight loss, color, aroma, taste, respiration, quality

## *INTRODUCTION*

Blueberries are exported to Western Europe via airfreight, arriving at the market within approximately 48 h after harvest (Miller and Smittle, 1987). However, the optimum fruit temperature is seldom maintained during storage or transportation, resulting in rapid quality deterioration of the fruit compared with fruit maintained at an optimum temperature of 0 to 2°C (Ballinger et al., 1978; Cappellini et al., 1983; Hardenburg et al., 1986; Miller and Smittle, 1987). Hudson et al. (1981) reported that within North America, blueberries are seldom precooled after harvest and are shipped to local markets with little or no refrigeration, while for distant markets they are usually shipped by truck at 5 to 7°C. At retail, they are often displayed without any refrigeration. Miller and Smittle (1987) reported that weight loss of blueberries during simulated airfreight was similar to that after 7 d at 3°C. Furthermore, decay of blueberries greatly increases when the fruit are stored at temperatures above 1.1°C; when held at 22.2°C they deteriorated very rapidly, reaching the minimum acceptable quality level of 15% to 20% decay within 1 to 5 d (Ballinger et al., 1978; Cappellini et al., 1982). In fact, the physiological behavior of blueberries, like other fruits and vegetables, is very much dependent on postharvest handling temperatures. Visual quality, flavor, and nutritional value of every fruit or vegetable are greatly dependent on their temperature history. Increasing the storage temperature results in increased incidence of shriveled and decayed blueberries, as well as increased fruit breakdown (Sanford et al., 1991). Therefore, besides the quality of the fruit at harvest, the use of an optimum temperature during handling and storage of fresh blueberries is a major factor that determines the quality of the fresh product.

Although some studies refer to the quality changes in blueberries during storage, no information was found regarding precise quality curves for blueberries stored at different temperatures or regarding which quality factor(s) are the most important to determine the limits of marketability. The objectives of this work were to (1) obtain quality curves for blueberries stored at different temperatures; (2) identify, for each temperature, which quality factor(s) limits blueberry marketability; and (3) compare the quality curves and shelf-life of blueberries based on quality evaluations with those predicted by respiration rates reported in the literature.

## MATERIAL AND METHODS

*Plant material and storage conditions.* Blueberries *Vaccinium corymbosum* cv. Patriot, a northern early season highbush variety, were obtained from a commercial field near Quebec City, Canada. A total of two harvests (experiments) were conducted during the 2001 summer season. Blueberries were harvested on 30 and 31 July. The weather conditions at the time of harvest (11:00 a.m.) were 25.1 and 26.0°C and no rainfall for the first and second harvests, respectively. Commercially harvested fruit were packed in 2-kg fiberboard baskets, removed from the field with minimal delay after harvest and transported to the laboratory in Quebec City within approximately 30 min. A total of 600 berries (300 berries for non-destructive evaluations plus 300 berries for taste evaluations) from four baskets were selected for uniformity of color and freedom from defects, weighed and carefully placed in plastic cups. Each individual cup was placed inside an open plastic bag to maintain a relative humidity level of about 95% to 100%. Six cups containing 20 berries each were placed in each of five temperature-controlled rooms at $0.5 \pm 0.5$°C, $5 \pm 0.2$°C, $10 \pm 0.4$°C, $15 \pm 0.2$°C, and $20 \pm 0.2$°C for a

14-d storage period. Three of the plastic cups containing 20 berries each per storage temperature were repeatedly evaluated over the 14 d of the experiment for weight loss, instrumental color (L*a*b*), visual color, firmness, shriveling, decay severity and incidence, and aroma. The other three plastic cups of 20 berries each were used for taste assessment. Four berries per storage temperature and evaluation time were sampled for taste assessment. The same trained person assessed the fruit quality throughout storage.

*Weight loss.* Weights of three individual replicated samples of 20 berries each per storage temperature were measured using a precision scale with an accuracy of ± 0.01 g (Acculab Model LT-3200, Acculab-Sartorius Company Group, Germany). Weight loss was then calculated from the weight of the fruit measured at harvest and after every 2 d of storage.

*Instrumental color.* Surface color measurements (CIE L*, a*, b*) of each individual blueberry were taken with a reflectance colorimeter on the stem end of the berry, which is the last area to develop color (Kalt et al., 1995). Numerical values of a* and b* were converted into hue angle (H° = tan$^{-1}$b*/a*) and chroma [Chroma = (a*$^2$ + b*$^2$)$^{1/2}$ (Francis, 1980)].

*Visual color.* Color of each individual blueberry was assessed using a visual rating scale where 1 = extremely overripe or senescent (purple brownish-blue or black), no waxy bloom; 2 = overripe (very dark blue); 3 = dark blue, starting to lose waxy bloom; 4 = deep blue; 5 = bright blue color, high waxy bloom (Jackson et al., 1999; Sapers et al., 1984). The waxy bloom is the grayish waxy deposit on the skin of the berries, which is a natural protective coating. The amount of bloom on blueberry fruit depends on the cultivar, physical handling, and the degree of freshness.

*Firmness.* Berries were individually rolled between the thumb and the index finger and rated for firmness on a scale where 1 = berry rupture upon touch, soft; 2 = berry surface easily depressed upon touch, but no rupture; 3 = berry surface depressed on touch, berry is more soft than firm; 4 = slight depression upon touch; 5 = berry firm, not yielding to touch (Miller et al., 1984; Miller and Smittle, 1987; Sanford et al., 1991).

*Shrivelling.* Shrivelling of each individual blueberry was assessed using a visual rating scale where 1 = none, field-fresh, no signs of shriveling; 2 = slight, minor signs of shriveling, not objectionable; 3 = moderate, shriveling evident, becoming objectionable; 4 = severe shriveling, definitely objectionable; 5 = extreme wilted and dry, not acceptable.

*Decay incidence and severity.* The incidence of decay was recorded after 12 or 14 d by counting the number of soft, leaky, or moldy berries. Decay severity of each individual blueberry was assessed every 2 d using a modified visual rating scale from Horsfall and Barratt (1945) where 1 = 0%, no decay; 2 = 1%-25% decay, probable decay (brownish/grayish sunken minor spots); 3 = 26%-50% decay, slight to moderate decay (spots with decay and some mycelium growth); 4 = 51%-75% decay, moderate to severe decay; 5 = 76%-100% decay, severe to extreme decay (the fruit is either partially or completely rotten).

*Aroma.* Aroma of three replicated samples of 20 blueberries each was assessed using a scenting rating scale where 1 = unacceptable, unpleasant odor (off-odor, fermented, product is deteriorated); 2 = poor, moderate off-odors; 3 = acceptable; 4 = good; 5 = excellent, fresh and pleasant characteristic aroma.

*Taste.* Taste of four blueberries per temperature and storage time treatment was assessed using a tasting rating scale were 1 = very poor, unpleasant taste (off-odor, fermented, very sour, bitter); 2 = poor; 3 = acceptable; 4 = good; 5 = excellent, fresh pleasant taste.

*Limiting factor.* For each temperature, a limiting quality factor(s) was established considering the rating value of 3 as the limit for acceptable quality before the product becomes non-salable. More specifically, for each temperature the factor(s) that limited the product marketability was identified from the quality curves.

*Prediction of shelf life.* Shelf life of blueberries was predicted according to Nunes et al. (2002). For each temperature, the shelf life of the fruit was predicted based on the $Q_{10}$ calculated from the respiration rates for blueberries found in the literature (Hardenburg et al., 1986). The reference for calculating the relative shelf life at each temperature based on respiration rate was the reported maximum shelf life of 14 d at the optimum temperature of 0°C for blueberries reported in the literature (Hardenburg et al., 1986).

*Statistical analysis.* Initial analysis of variance indicated minimal differences between the two harvests. Consequently, the combined data from the two harvests (experiments) were analyzed. The Statistical Analysis System computer package (SAS Institute, Inc., 1982) was used. In order to determine the primary limiting factor(s), quality attributes for each temperature were compared using the (Least Significant Difference (LSD) at the 5% significance level.

## RESULTS AND DISCUSSION

*Weight loss.* According to the literature, the maximum weight loss before blueberries become non-salable is approximately 5% to 8% (Sanford et al., 1991). After 14 d, regardless of the storage temperature, the blueberries had lost less than 5% of their initial weight (Figure 1). Results from the present study agree with those from Jackson et al. (1999), which also reported a minimal weigh loss of about 2% in blueberries after 14 d of storage at 0°C. Although weight loss was greater in fruit from the second harvest compared to the first harvest, it never attained 5%.

*Color.* In these experiments, the L* value of the blueberries slightly decreased from approximately 31.0 at harvest to 28.0 after 14 d, regardless of the storage temperature, meaning that the fruit lost their bright color during storage, becoming darker or losing their bloom. The hue of blueberries stored at 10, 15, or 20°C slightly decreased from 299.9 at the time of harvest to approximately 295.7 after 2 d, and increased henceforward to reach levels comparable to initial values. Hue of blueberries stored at 0 or 5°C slightly increased to approximately 302.8 after 14 d. At the end of storage the berries were more purplish-dark blue than blue (higher hue) compared with freshly harvested fruit. At the time of harvest, fruit from first harvest had a higher hue value (303.6) compared to

FIGURE 1. Weight loss of 'Patriot' blueberries during storage at different temperatures.

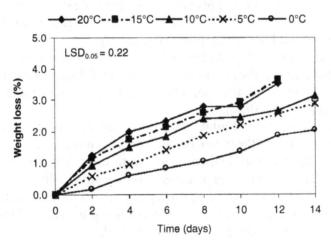

fruit from the second harvest (295.4). Chroma of blueberries decreased from 7.5 at the time of harvest to approximately 5.9 after 12 or 14 d, regardless of the storage temperature. During storage, the color of the fruit became less vivid than at the time of harvest (lower chroma values).

Blueberries attained a minimum acceptable color rating after approximately 8 d, regardless of the storage temperature (Figures 2-6). Although the berries changed color from a bright purplish-blue to a darker blue, particularly the fruit stored at 15 or 20°C, changes in the color were mainly due to loss of the waxy bloom rather than changes in the blue color of the berries. Loss of the waxy bloom might be in some way attributed to the manipulation of the berries during the quality evaluations. Results from the present study suggest that visually observed color changes during storage of blueberries at different temperatures were more evident than changes in the instrumental color measurements such as hue or chroma. However, L* value was a good indicator of loss of brightness, which was, in fact, a more important criterion for evaluating the changes in color than the blue color of the fruit. In the present study, as reported by Sapers et al. (1984), color measurements made on blueberries showed a close relationship between the L* value

FIGURE 2. Quality characteristics of 'Patriot' blueberries stored at 0°C; objectionable taste and aroma were the limiting quality factors (the dotted line corresponds to the limit of acceptability before the quality of the fruit became objectionable).

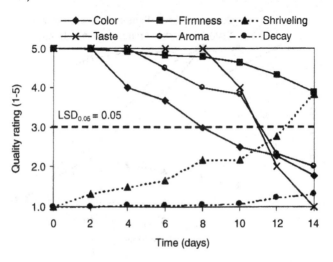

FIGURE 3. Quality characteristics of 'Patriot' blueberries stored at 5°C; objectionable taste and aroma were the limiting quality factors (the dotted line corresponds to the limit of acceptability before the quality of the fruit became objectionable).

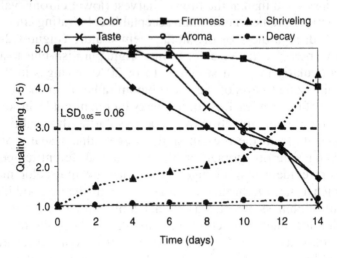

FIGURE 4. Quality characteristics of 'Patriot' blueberries stored at 10°C; darkening of the color or loss of waxy bloom and objectionable aroma were the limiting quality factors (the dotted line corresponds to the limit of acceptability before the quality of the fruit became objectionable).

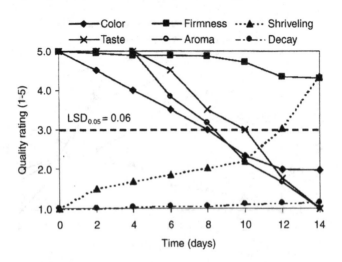

FIGURE 5. Quality characteristics of 'Patriot' blueberries stored at 15°C; darkening of the color or loss of waxy bloom and objectionable aroma were the limiting quality factors (the dotted line corresponds to the limit of acceptability before the quality of the fruit became objectionable).

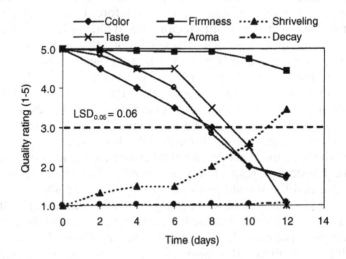

FIGURE 6. Quality characteristics of 'Patriot' blueberries stored at 20°C; darkening of the color or loss of waxy bloom was the limiting quality factor (the dotted line corresponds to the limit of acceptability before the quality of the fruit became objectionable).

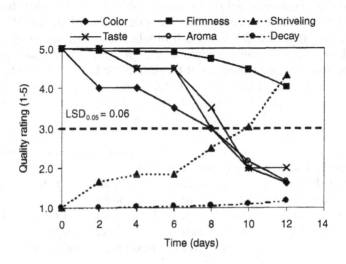

(higher L* values indicating lighter-colored fruit) and visual assessment of waxy bloom (visual color rating).

*Firmness.* Although blueberry firmness slightly decreased during storage, regardless of the storage temperature, it never reached the minimum acceptable score (Figures 2-6). Therefore, loss of firmness was not considered a critical quality factor since it did not fall below a minimum score of approximately 4. Blueberries of cultivar 'Patriot' are considered to be bigger and firmer than many other highbush blueberry cultivars (Hepler and Draper, 1976; Ødvin and Øydvin, 1999). This might explain the minimal changes in the firmness on touch of the fruit even when stored at temperatures higher than 0°C (Figures 3-6).

*Shriveling.* Shriveling of the berries increased during storage, regardless of storage temperature (Figures 2-6). Fruit stored at 15 or 20°C reached the maximum acceptable shriveling rating after approximately 11 or 10 d in storage, respectively, followed by fruit stored at 5 or 10°C, which attained the threshold value after 12 d. After about 13 d, fruit stored at 0°C attained a maximum rating of 3 for shriveling. Although, according to values found in the literature, weight loss of the fruit from this study was not considered a limiting factor, berries showed signs of shriveling after about 10 d. Shriveling is one of the visual symptoms generally attributed to loss of moisture during storage. If we consider that signs of shriveling started to be noticeable after 11 and 10 d in berries stored at 15 or 20°C, 12 d at 5 or 10°C, and after approximately 13 d at 0°C, then the corresponding weight loss at those storage times was on average 2% to 3% (Figure 1). These values might be, in our study, considered the maximum weight loss acceptable before 'Patriot' highbush blueberries became unacceptable for sale, rather than the 5% to 8% weight loss reported for low-bush blueberries (Sanford et al., 1991). Therefore, the maximum acceptable weight loss before blueberry fruit become unacceptable should not be generalized, but rather established based on each particular cultivar. In fact, Bounous et al. (1997) reported that weight loss of highbush blueberries cultivars Darrow, Coville, and Dixie stored at 1°C for 2 weeks might vary from 17.5% to 2.5% depending on the cultivar. Furthermore, Sanford et al. (1991) used wild low-bush blueberries in their work, which are smaller fruit than 'Patriot' and most likely less resistant to moisture loss.

*Decay.* Incidence of decay increased with increasing storage temperature. After 12 d at 20°C, 15% to 18% of the berries from the first and second harvests, respectively, showed minor signs of decay, while decay averaged 13% in fruit stored at 15°C. Decay ranged from 10% to 15% after 12 and 14 d at 10°C for the first and second harvests, respectively. Fruit stored at 0 or 5°C developed the smallest amount of decay compared with fruit stored at higher temperatures (5% to 7.5% decay in fruit stored at 0°C, and 7.5% to 10% decay in fruit stored at 5°C after 12 and 14 d for the two harvests, respectively). Although an increase in the incidence of decay was observed with increasing storage time and temperature, the percentage of decay was still within the limits of acceptability. Ballinger et al. (1978) considered 20% to be the level of quality deterioration above which blueberries should not be sold on the fresh market, while Cappelini et al. (1982) found an average of 15% defective fruit in consumer samples. Even though the fruit stored at 20°C showed a maximum 18% decay after 12 d, the severity of the decay was very low and remained much below the maximum acceptable score in both harvests (Figures 2-6). Therefore, development of decay was not considered a critical factor since it did not increase above a maximum rating of 1.3 even after 14 d of storage. Sanford et al. (1991) reported very low decay on wild low-bush blueberries stored at temperatures from 0 to 20°C. Generally, decay contributed to an average loss of only 1% to 2% in fruit stored at 15 or 20°C, respectively, and 0.1% to 0.4% in fruit stored at 0 or 5°C, respectively. In highbush 'Bluechip' berries stored at 21°C for 7 d, decay was reported to range from 3.6% for dry-handled berries to 63.5% when the berries were handled wet (Cline, 1997). In the present study, the surface of the berries was always maintained dry, which according to Cline (1997) might have contributed to a lower level of decay even in berries stored at 20°C (Figure 6).

*Aroma.* Loss of aroma was faster in the berries stored at 15 or 20°C than in fruit stored at 0°C (Figures 2, 5, and 6). After approximately 7 to 8 d, the aroma of the berries stored at 15 or 20°C was considered unacceptable due to off-aromas. However, fruit stored at 0°C maintained an acceptable aroma for 11 d.

*Taste.* Loss of taste followed the same pattern as loss of aroma for berries stored at 15 or 20°C (Figures 5 and 6). However, loss of taste was more pronounced (lower ratings) than loss of aroma in fruit stored at lower temperatures. In fruit stored at 0 or 5°C, the taste was no longer acceptable after 11 or 10 d of storage, respectively, while fruit stored at 10, 15, or 20°C had an unpleasant taste after 9 to 10 d in storage.

*Limiting quality factor(s) for each temperature.* Color change during storage, such as darkening of the fruit or loss of waxy bloom, was statistically the primary limiting factor for all the storage temperatures. However, it was considered that for temperatures lower than 10°C changes in the color (blueness) of the fruit were very small compared with changes in other quality factors evaluated. Changes in the hue value were also minor in fruit stored at 0 or 5°C compared with those stored at 10, 15, or 20°C. As the instrumental color measurements were limited to the stem end of the berries, differences between visual color evaluation and instrumental color were observed. In addition, the natural grayish waxy bloom of blueberries is easily removed if they are manipulated frequently. That was the case in these experiments, as every time the berries were evaluated they were taken one by one from the package and examined. For these reasons, in some cases the second most important quality factor(s) was considered as the limiting factor rather than the color change alone. Therefore, development of unpleasant taste and aroma were simultaneously the limiting factors for blueberries stored at 0 or 5°C (Figures 2 and 3). Darkening of the color and development of unpleasant aroma were the limiting factors for berries stored at 10 or 15°C (Figures 4 and 5). For blueberries stored at 20°C, darkening of the color was the primary limiting factor (Figure 6). Although weight loss and shriveling were not primary limiting factors, they may eventually contribute to loss of quality of blueberries during storage. Therefore, they should also be taken into consideration, particularly when fruit are stored or handled under inadequate conditions (high temperature and low relative humidity). Loss of firmness and development of decay were not considered major limiting quality factors at any temperature, since they remained quite constant during storage and did not attain unacceptable rating values. In fact, decay was previously reported by other authors not to be a serious cause of loss for blueberries (Jackson et al., 1999; Sanford et al., 1991). Based on the limiting factor(s) for each temperature, a final quality graph was re-plotted considering only the curves obtained for each limiting factor(s) (Figure 7). For any temperature, whenever two or more limiting factors were observed, new curves were established using the average of the limiting factors.

*Quality curves predicted from respiration rates reported in the literature.* As the temperature of a fruit increases, the rates of chemical reactions increase. The changes in the rates of the reactions due to temperature are usually characterized using a measure called the $Q_{10}$, which is the ratio of the rate of a specific reaction at one temperature $(T_1)$ versus the rate at that temperature + 10°C. Since respiration gives a very general estimate of the effect of temperature on the overall metabolic rate of the tissue, the $Q_{10}$ values often refer to respiration (Kays, 1991). Therefore, the predicted shelf life of blueberries based on $Q_{10}$ values relative to the reported maximum shelf life of 14 d at 0°C and published respiration rates (Hardenburg et al., 1986), was approximately 12 d at 5°C, 5 d at 10°C, 4 d at 15°C, and approximately 2.5 d at 20°C (Figure 8). In com-

FIGURE 7. Quality curves for 'Patriot' blueberries stored at different temperatures (the dotted line corresponds to minimum acceptable quality before the fruit became unacceptable for sale).

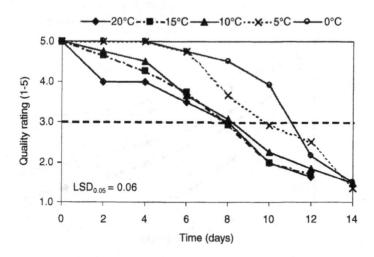

parison, the shelf life of blueberries from our experimental quality evaluations was, 11 d at 0°C, 10 d at 5°C, and 8 d at 10, 15, and 20°C (Figure 7). The predicted shelf life of blueberries based on $Q_{10}$ values for respiration rate was 3 d longer at 0°C, 2 d longer at 5°C, and 3, 4, and 5.5 d shorter at 10, 15, and 20°C when compared to the shelf life of blueberries obtained from our experimental quality evaluations.

Overall, blueberries used in the present study stored at 0 or 5°C had shorter shelf life compared with values based on the respiratory $Q_{10}$, while those stored at higher temperatures had longer shelf life. Some of the reasons for the difference between the results from this study and those from the data found in the literature, might be related to uncertain information for the published data regarding the blueberry cultivars used, the quality of the fruit at harvest or at the beginning of the data collection, the time between harvest and the beginning of data collection, the harvest season and temperatures, as well as other environmental factors during harvest, transport, or storage. In the present study, all of the above variables were well known and the quality of the blueberries at the beginning of the experiment was excellent. In fact, one of the factors that has a great influence on the shelf life of any fruit or vegetable is the quality at harvest. If the initial quality of a product is considered merely good or acceptable instead of excellent, the shelf life of the product will

FIGURE 8. Quality curves based on $Q_{10}$ values from the literature for blueberries stored at different temperatures (the dotted line corresponds to minimum acceptable quality before the fruit became unacceptable for sale).

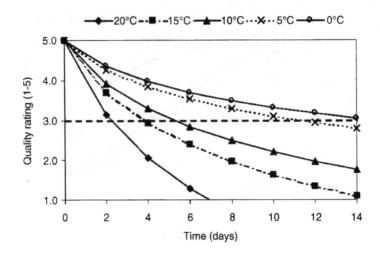

be obviously reduced. That might explain the longer shelf life of the blueberries in our study stored at 10, 15 or 20°C, compared with the data from the literature ($Q_{10}$). It is important to note that the cultivar used for this study had a particularly high firmness and probably a thicker cuticle compared to other cultivars, which might have increased the fruit resistance to water loss and tissue breakdown at temperatures higher than 0°C.

Another reason for the difference between the values for blueberry shelf life from experimental quality curves and the predicted quality curves may be that respiration rate is not always a reliable predictor of fruit quality or salability. With regard to the blueberries stored at 0 or 5°C in this study, their shorter than predicted shelf life was most likely due to the limiting quality factors at those temperatures (taste and aroma) not being closely related to overall metabolic rate as reflected by respiration measurement. When the respiration rate of a particular fruit or vegetable is measured, the only aspect considered is the $O_2$ consumed and $CO_2$ released and not the visual quality. That means that a crop may be respiring (consuming $O_2$ and releasing $CO_2$) at a low rate, but this may not relate to whether it remains acceptable in terms of sensory quality.

## CONCLUSIONS AND GROWER BENEFITS

Storage of 'Patriot' blueberries at 0 to 1°C is recommended for maximum storage life. The quality curves obtained from quality evaluations for each temperature showed that a single quality factor cannot be used to express loss of quality of blueberries over the normal physiological range of temperatures. Furthermore, if the berries are frequently manipulated, the use of a visual color rating may not be a good quality factor to estimate the shelf life of blueberries during storage. In addition, prediction of blueberry shelf life calculated from data from the literature on the respiration rates at various temperatures is not precise unless the type of cultivar and the quality of the fruit at harvest as well as environmental factors involved such handling temperatures or relative humidity are well known and the limiting quality factor is closely related to overall metabolic rate.

## LITERATURE CITED

Ballinger, W.E., E.P. Maness, and W.F. McClure. 1978. Relationship of stage of ripeness and holding temperature to decay development of blueberries. J. Amer. Soc. Hort. Sci. 103:130-134.

Bounous, G., G. Giacalone, A. Guarinoni, and C. Peano. 1997. Modified atmosphere storage of highbush blueberry. Acta Hort. 446: 197-203.

Cappellini, R.A., M.J. Ceponis, and G. Koslow. 1982. Nature and extent of losses in consumer-grade samples of blueberries in greater New York. HortScience 17: 55-56.

Cappellini, R.A., M.J. Ceponis, and C.P. Schulze, Jr. 1983. The influence of "sweating" on postharvest decay of blueberries. Plant Dis. 67: 381-382.

Cline, W.O. 1997. Postharvest infection of blueberries during handling. Acta Hort. 446: 319-324.

Francis, F.J. 1980. Color quality evaluation of horticultural crops. HortScience 15:58-59.

Hardenburg, R.E., A.E. Watada, and C.Y. Wang. 1986. The commercial storage of fruits, vegetables and florist and nursery stocks. Agr. Hdbk. No. 66. U.S. Dept. Agr., Washington, DC.

Hepler, P.R. and A.D. Draper. 1976. 'Patriot' blueberry. HortScience 11: 272.

Horsfall, J.G. and R.W. Barratt. 1945. An improved grading system for measuring plant disease (Abstract). Phytopathology. 35: 655.

Hudson, D.E., and W.H. Tietjen. 1981. Effects of cooling rate on shelf life and decay of highbush blueberries. HortScience 16: 656-657.

Jackson, E.D., K.A. Sanford, R.A. Lawrence, K.B. McRae, and R. Stark. 1999. Lowbush blueberry quality changes in response to prepacking, delays and holding temperatures. Postharv. Biol. Technol. 15: 117-126.

Kalt, W.. K.B. Mcrae, and L.C. Hamilton. 1995. Relationship between color and other maturity indices in wild lowbush blueberries. Can. J. Plant Sci. 75: 485-490.

Kays, S.J. 1991. Postharvest physiology of perishable plant products. Van Nostrand Reinhold. New York.

Miller, W.R. and D.A. Smittle. 1987. Storage quality of hand- and machine-harvested rabbiteye blueberries. J. Amer. Soc. Hort. Sci. 112: 487-490.

Miller, W.R., R.E. MacDonald, C.F. Melvin, and K.A. Munroe. 1984. Effect of package type and storage time-temperature on weight loss, firmness, and spoilage of rabbiteye blueberries. HortScience. 19: 638-640.

Nunes. M.C.N. and J.P. Emond. 2002. Storage temperature, pp. 209-228. In: J.A. Bartz and J.K. Brecht (eds.). Postharvest physiology and pathology of vegetables. Marcel Dekker, Inc., New York.

Øydvin, J. and B. Øydvin. 1999. Highbush blueberry crops in a trial in Norway, 1988-1999. Fruit Var. J. 53(3): 155-159.

Sanford, K.A., P.D. Lidster, K.B. McRae, E.D. Jackson, R.A. Lawrence, R. Stark, and R.K. Prange. 1991. Lowbush blueberry quality changes in response to mechanical damage and storage temperature. J. Amer. Soc. Hort. Sci. 116: 47-51.

Sapers, G.M., A.M. Burgher, J.G. Phillips, S.B. Jones, and E.G. Stone. 1984. Color and composition of highbush blueberry cultivars. J. Amer. Soc. Hort. Sci. 109: 105-111.

SAS Institute Inc. 1986. SAS User's Guide: Basics, Version 5 Edition, SAS Institute Inc., Cary, NC, USA.

# Index

Printed in the United States
by Baker & Taylor Publisher Services